Bio-Climatology for Built Environment

T0143284

Masanori Shukuya

Department of Restoration Ecology and Built Environment
Tokyo City University
Yokohama, Kanagawa
Japan

CRC Press

Taylor & Francis Group
Boca Raton London New York

CRC Press is an imprint of the
Taylor & Francis Group, an **informa** business

A SCIENCE PUBLISHERS BOOK

Cover illustrations provided by the author, Dr. Masanori Shukuya.

CRC Press
Taylor & Francis Group
6000 Broken Sound Parkway NW, Suite 300
Boca Raton, FL 33487-2742

First issued in paperback 2021

© 2019 by Taylor & Francis Group, LLC
CRC Press is an imprint of Taylor & Francis Group, an Informa business

No claim to original U.S. Government works

Version Date: 20181120

ISBN-13: 978-0-367-78041-8 (pbk)
ISBN-13: 978-1-4987-2729-7 (hbk)

Visit the Taylor & Francis Web site at
http://www.taylorandfrancis.com

and the CRC Press Web site at
http://www.crcpress.com

To
Yoriko and Yumi

Preface

Built environment is the most immediate space surrounding us during the whole period of our life. This is true for anyone living on the Earth. This treatise describes how the built environment functions within the flow of radiation, from the short-wavelength radiation coming from the Sun to the long-wavelength radiation going out into the Universe. This flow of radiation makes all the natural phenomena that we experience not only outdoors but also indoors. Indoor climate to be made by rational lighting, heating, cooling and ventilating systems for occupants' well-being should be consistent with how regional outdoor climate works in the flow of radiation via four paths of heat transfer: radiation, convection, conduction, and evaporation. What is compiled in this treatise is the present state-of-the-art knowledge that I have learnt through my research on the built environment and also through my teaching in the series of lectures given at the School of Environmental Studies, Tokyo City University.

Science and technology have grown to the present status since their respective births and the latter in particular has made it possible for us to have our contemporary life styles. This of course includes the built environment to be the focus in this treatise. The technology has its root in the development of a simple tool for making our life easier and convenient while avoiding the danger, health risks, and discomfort that might be caused by the harsh Nature. On the other hand, science is to reveal the true character of the Nature, both non-living and living things. In general, science has been very successful in disclosing the existence of substantial components. We may say that science has been good at atomistic approaches, while it tends to have left the holistic view behind. This seems to have caused various problems in human societies locally and globally; so-called energy issues and also environmental issues are considered to be the two of those problems.

The reach of contemporary technology is undoubtedly owing to the development of science so far developed while at the same time, the so-far developed technology has made what the science can do in more precise and detailed manners. In due course, both science and technology have developed together with each other and have reached their present state of the art.

Back in early 1970s, my university education started in the field of architecture. I was, of course, attracted by its artistic side, but not fully, and rather more attracted by its scientific side. As a consequence, I jumped into the field of research on built-environmental science and technology such as lighting, heating, cooling, and ventilating systems in relation to so-called energy issues. Almost ten years later, in early 1980s, I came across the concept of exergy, which was totally new to me;

I had no experience in learning thermodynamics at all while I was a university student. Nonetheless, or maybe because of my ignorance, I became fascinated by thermodynamics. Then my focus in the research turned to the one that aims at developing the thermodynamic understanding of lighting, heating, cooling, and ventilating systems together with the nature of human beings as building occupants. When I started my own exergy research, the use of exergy concept was not common at all in the field of building science. Therefore, I thought that it should be all right simply to apply the concept of exergy to building science for its further development. But, within a first couple of years, I came to notice that a further development of the exergy concept itself was necessary, since it had not yet been matured so that it can be fully applied to building science. I, therefore, involved myself not only in the applied research but also in the fundamental research in relation to the concept of exergy. Such circumstances then motivated me to do all my research work by coming back once to the very basics. This has also let me have a chance to rethink my way of learning and also teaching and revitalize how we should learn ourselves and teach the established concepts and also new concepts. This treatise is the integration of such work in my research and education over the last three decades or so.

The discussion in this treatise starts with the relationship between the human body and his or her immediate environmental space followed by a brief introduction of passive and active systems for indoor climate conditioning. The nature of light and heat is discussed with the focus on building envelope systems such as walls and windows, and then the associated thoughts from the viewpoints of thermodynamics and human-biology are given with some examples useful to have a better understanding of luminous and thermal characteristics of our most immediate environment. The unique feature of this treatise is, if there is, to clarify what the rational built-environment conditioning to be realized in the smooth flow from light to heat is by making connection of heat-transfer science with thermodynamics, and also of building physics with human biology. This is my little trial of holistic approach to be merged with atomistic approach very well developed in conventional science. I would like to call such holistic approach to building science as bio-climatology.

This treatise is written for those who want to develop their own connective way of thinking by learning the basics of science in relation to the most immediate environment and, in so doing, to develop their own rational approach to the bio-climatic design principle. I hope that what is written in this treatise can provide a new way of thinking in science, engineering, and design in general for their rational advancement to proceed towards a better future.

Masanori Shukuya

July 2018   Yokohama, Japan

Contents

Chapter 1

Built Environment and Human Beings

1.1 Our closest environmental space

Let us look around our surroundings. You may be in front of your desk at home, at your office, or at a library reading this book. There is a chair, on which your body is being seated, under your feet, there is a wooden floor or carpeted floor, and there is air surrounding all over your body. There are always a variety of things around you.

Whenever we discuss something associated with so-called environment, it is very important for us to specify a system to be focused on and its environmental space. In this example, you yourself are a system and all things that surround you are the environment. Define first a system and then its environment. Figure 1.1 shows the general relationship between a system and its environment. First, there is the universe, and then, you conceive a system and define it with the closed boundary surface, which is drawn with the line on a sheet of paper, as shown in Fig. 1.1. The space out of the system but within the universe is the environment. It may sound too self-evident and nothing important, but the word "environment" can imply a variety of things, space-wise and time-wise, so that it is worth mentioning what a system is and what its environment is in the very beginning of discussion.

Let us come back to the example raised above: you as the system and your environment. How long do you think you stay inside buildings as your environmental

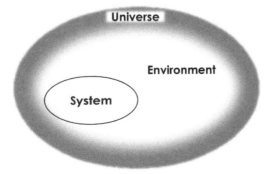

Fig. 1.1: The universe, system and environment.

space for one ordinary day? Take just thirty seconds or so and recall where you were over the last 24 hours. The answer could be different, of course, from one person to another person, but it is usually over 90%; this implies that, assuming one's life to span 90 years, we spend more than 80 years surrounded by the built environment.

Imagine your outdoor space, e.g., a space between your home and the nearest train station that you usually use or that between your working place or your classroom and a train station that you usually use. There may be some trees planted on your right or left and behind them there may be some condominium buildings, some shops or restaurants. There must be paved road under your foot, and so on. We find that most of them are man-made. Those trees along the road, of course, belong to living creatures, not to artificial objects, but most of them were planted by people involved in the road construction following the city planning of this urban area.

The pure nature that you can find is in fact the sky above our heads alone; it is especially so if you are in urban areas. Thus, we come to recognize that the closest environmental space to you as the system as one of the living systems is mostly artificial, that is, man-made. As will be discussed later in this chapter and also other chapters ahead, because of such characteristics of the built environment, I think that it is important to understand its bio-climatic aspect, that is, the relationship between the built environment and us as living systems. This is not only for those involved in building-related professionals but also for all others, since all of us as living systems spend most of the time within the built environment as discussed above.

1.2 Relationship between a system and its environmental space

"Environment" is one of the key words very often used here and there in our local and global societies. How long has it been used? In Japan, where I live, I recall that the first time I heard the word "environment", it was associated with air and water pollutions that had become apparent to almost everyone in the very last period of high economic growth rate, that is, early 1970s. The second time I not only heard but also used the word "environment" myself was in relation to so-called energy crisis that happened in late 1970s and early 1980s, and finally, as most of you are aware of and also must have used it, in relation to so-called global warming. This is after late 1980s up to the present.

In order to confirm that they are all associated with so-called energy issues, let us take a look at Fig. 1.2, the trend in fossil-fuel use on per-capita annual basis in Japan together with those in the USA and in the UK. We find here that there was a sharp increase in the rate of per-capita fossil-fuel use in Japan from 1950 to 1975. In this period, the emission of exhaust gas and contaminated water inevitably generated by various human industrial activities, all of which were for making the factory systems work by feeding on fossil fuels, was not well-controlled so that the air and water as primary environmental constituents were polluted very much. One of the dreadful problems that happened was the Minamata disease (MDA 2017).

In the UK, such environmental pollution had already emerged much earlier, from 1850 to 1900, and its problem was discussed by, e.g., Jevons (1900). This is the period in which British Empire had flourished the most.

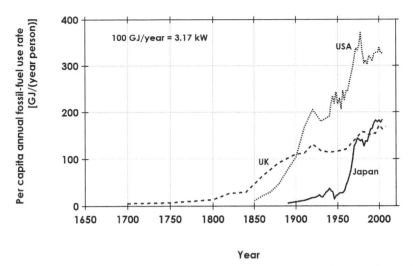

Fig. 1.2: Variation of per-capita annual fossil-fuel use rate in three countries in the period from 1700 to 2000 (made by the author referring to the data given in Tateno (2012)).

In the period of so-called energy crisis in Japan, from 1973 to 1980, the per-capita fossil-fuel use rate decreased a little. Similar trends, though the values themselves are different, can be seen in the USA and the UK. In all of the three countries, the fossil-fuel use rate increased again sooner or later till the year 2000. In Japan, this corresponds to the period in which so-called "bubble economy" flourished and then broke.

Interestingly, in all of these three countries, the overall profiles of growth look the same, although the length of time taken and the values of rate are different from each other. This suggests that the collective-learning process of human beings in a society may be the same regardless of the difference in local climate and culture.

First, the way of using fossil fuels is found, then that of exploitation follows and thereafter the rate of use increases, by which the industrialization of the society as a system grows. Sooner or later the environmental degradation becomes apparent and thereby some measure of improvement in technology emerges and it reflects the societal attitude, but again the rate of fossil-fuel use increases. Such a pattern seems to repeat, but, nonetheless, there must be a certain limit, that is, the so-called environmental capacity, within which the wastes that the systems inevitably generate and excrete in order for them to sustain themselves are processed so as to be decomposed into non-harmful environmental constituents.

1.3 What can flow through makes a system sustainable

As discussed above, looking at Fig. 1.2, a society as a system grows gradually by feeding on more and more fossil fuels, but it would not be limitless due to the environmental capacity. There are a variety of factors that determine the environmental capacity. Let us think about it, with a cooking oven as a simple system. Suppose that we burn wood to heat the oven so that we can cook something. In due course, some amount of smoke and heat necessarily disperse into the surrounding space, that is, the kitchen space as the environmental space of the oven. In order to avoid smoke and unnecessary heat that is not directly used for cooking, you may get rid of them from the kitchen by a chimney, but it does not imply that you can make them into nothing. Smoke and heat going out of the immediate environmental space simply come into a larger environmental space surrounding the kitchen environment and thereby their concentration turns out to be lower due to dispersion. Such a process applies to any working system.

Its essence is, as shown in Fig. 1.3, that a system feeds on some substance containing an ability to disperse from near or far natural resources and hence lets it disperse in order to perform its purpose, while at the same time it necessarily excretes the waste into the environmental space.

The types of fuels have changed from wood to coal, to petrol, and then to natural gas, that is, from solid to liquid or gas. This is because of ease of delivering fuels from where the resource exists to a system that feeds on them. Between petrol and natural gas, there exists uranium, which is nuclear fissile, and whose massive use was originally conceived for the purpose of massacre, that is, so-called nuclear bombs. The super-devastating capability was realized through the twice actual use in Hiroshima and Nagasaki, and quite a few number of experimental bombing performed in the cold-war period from 1947 to 1991.

Since the intrinsic danger has been gradually recognized by more and more people, the use of nuclear fission materials should have led to their abandonment, but the reality has not yet proceeded in this direction. This is because, since 1953 up to the present, uranium has been used for electricity production under the socio-political slogan of so-called "atoms for peace".

Electric-power generation with the use of chained nuclear fissile reactions inevitably produces enormous amount of heat, together with nuclear waste matters. On the one hand, the heat has to be dispersed into the environment, while on the other hand, the nuclear waste has to be kept in very carefully sealed containers because of their long lasting radioactivity, which is capable of destroying a variety of chemical bonds, with which all living systems, including us humans, sustain their life. In other

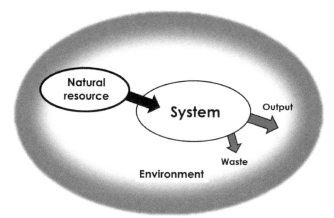

Fig. 1.3: A system sustains its state by taking in some substance from natural resource to perform its objective of making some output while at same time excreting the waste into the environment.

words, any living creatures can keep their states of dynamic equilibrium on the basis of the stability in their cellular nuclei, which contain all the genetic information, and process a variety of bio-chemical chained reactions for making life possible. This is why the nuclear waste matters have to be kept within a carefully closed system so that any radioactive waste matters do not come into the environmental space, though no leakage at all is, in reality, hardly achievable. This implies that the nuclear power plants cannot work sustainably. Therefore, any of them working for now must be abandoned sooner or later. Note that even if they are abandoned today, they have to be kept under control in order to avoid further contamination for a very long period of time: nobody knows how long it really is, maybe one-hundred, two-hundred or much more years.

On the one hand, the development of nuclear physics in the first half of twentieth century has revealed that an enormous amount of energy is hidden in the bond of nucleons: protons and neutrons, while on the other hand, the development of molecular biology has revealed that the nuclei of living cells composed of deoxyribonucleic acid (DNA) molecules contain genetic information. The word "nucleus" in the field of nuclear physics was coined by the word "nucleus" in the field of biology. What we should learn from both fields of science together is that biological nucleus can exist on the basis of the stability of atomic nucleus. And this is why, as mentioned above, the massive use of nuclear-fission materials is to be abandoned for electricity generation, let alone for super-devastating bombs.

One may regard that what has been described above is a denial of the advancement of science to be followed by that of technology, but this is not a denial at all. The advancement of science has revealed what is possible and what is not possible and the technology must be made in the direction of human well-being, not only in the present generation but also in the future generations. Ideally speaking, science and technology should advance in this direction, causing neither discrimination against nor sacrifice to both the present generation and the future generations. Such philosophy should be, I think, the concrete basis of science and technology looking into a better future.

The way towards the abandonment of nuclear fissile materials as fuel for electricity generation looks still far away due to the complicated socio-political hurdles rather than the conscientious scientific and philosophical reasons mentioned above. Nevertheless, the knowledge of what happened in Three Mile Island (1979), in Chernobyl (1986), and in Fukushima (2011), has definitely been transferred to more and more people than before as time has gone by, although it still looks slow. Therefore, together with such thought mentioned above, the abandonment must be realized sooner or later.

1.4 Nested structure of environmental spaces

We are exposed to a variety of news obtained from radio, television, newspaper and internet, thanks to the development of ICT (Information and Communications Technology). Nonetheless, this fact has not led everyone to the clear definition of "environment".

On the contrary, because of the state-of-the-art ICT, people tend to use the word "environment" in a variety of spectra: from an image of the atmospheric air, implicitly meaning its pollution, as raised in section 1.2, to another image of societal or individual behaviour influenced by other societies or other individuals, but the most contemporary word among them must be "global environment", especially over the last twenty to thirty years. This is, I think, coined by a piece of jargon, "global warming", which started appearing in various media including quite a few academic journals in late 1980s reflected by the recognition of the trends in atmospheric CO_2 concentration over the last fifty years in the twentieth century.

As already mentioned in section 1.1, whenever we say "environment", we always need to have a clear image of the "environmental space", in which a "system" to be discussed exists. Therefore, with the word "global environment", the systems to be considered can be either of all living systems on the Earth.

With this relationship between the systems and the global environment in mind, Fig. 1.4 schematically shows the relationship between a human as a system and its immediate environment together with the nested structural relationship between other larger systems and environmental spaces. The most immediate environment of human body as a system is, as described in section 1.1, the built environment. If the built environment is taken to be a system, then its immediate environment is the urban environmental space, which may instead be considered as a town or a village. If the urban environment is taken to be a system, then its environment is the regional

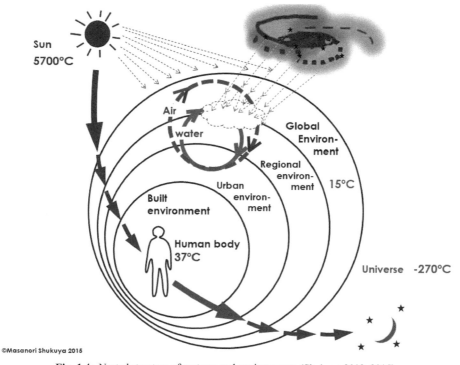

©Masanori Shukuya 2015

Fig. 1.4: Nested structure of systems and environments (Shukuya 2013, 2015).

environmental space. The largest regional environmental space is the Earth itself, that is, the global environment, since all the living creatures sustain their life within the Earth as a macrocosm. The Earth as a macrocosmic system is surrounded by the Universe. The relationship between systems and environmental spaces described above is called "nested structure".

Table 1.1 shows the relationship between systems and environmental spaces with their respective sizes. This relationship starts from the order of 10^1 m for human beings, up to the order of 10^{26} m for the whole Universal scale and also down to the order of 10^{-16} m for nucleon.

As the contemporary astrophysics developed on the basis of relativity and as quantum mechanics has revealed, the solar system itself revolves inside the Milky Way Galactic space, which is actually a tiny portion of the whole Universe that has been expanding ever since the Big Bang (Silk 1994). The Universe may be thought of

Table 1.1: Relationship between systems and their environmental space.

System	Environment	Order of the environmental size [m]
Milky-way galaxy	Universe	$\sim 10^{26}$
Solar system	Milky-way galaxy	$\sim 10^{20}$
Earth	Solar system	10^{13}
Region (forest, river, sea)	Global environment (Earth)	10^7
City (town, village)	Regional environment	10^5
Buildings	Urban environment	10^4
Human being	**Built environment**	**10^1**
Organs (brain, heart and others)	Internal environment	10^0
Tissues (a bunch of cells)	Organ (brain, heart or other)	$10^{-1} \sim 10^{-2}$
Cells	Tissue (a bunch of cells)	$10^{-3} \sim 10^{-4}$
Cellular organs (nucleus, mitochondrion)	Cell (Cytoplasm)	10^{-5}
Molecules	Giant molecule (protein, DNA and others)	$10^{-8} \sim 10^{-7}$
Atoms	Molecule	10^{-9}
Nuclei	Atom	10^{-10}
Nucleons (proton, neutron)	Nucleus	$10^{-15} \sim 10^{-14}$
Quarks	Nucleon	$\sim 10^{-16}$

as subdivided into three portions: one is our solar system, to which the Earth belongs with the order of 10^{13} m, the other is the Milky Way Galaxy, in which our solar system belongs with the order of 10^{20} m, and the rest, the whole Universal space, in which our Milky Way Galaxy exists, with the order of 10^{26} m. It is not easy to imagine, but what has been going on in the Universe seems to have been affecting the ever changing global climate, whether it is global warming or global cooling (Svensmark and Calder 2007). Such relationships between systems and environmental spaces may be conceived towards smaller scale and finally very microscopic direction, that is, starting from the size of human being with the order of 10^1 m via living cells with the order of 10^{-5} m to nucleons with the order of 10^{-16} m. Such an image, moving either toward macroscopic direction or toward microscopic direction, was first well documented and picturized by Charles and Ray Eames (1968, 1977).

Note that the size of biological nucleus, DNA, is in the order of 10^{-8} to 10^{-7} m, while that of the atomic nucleus is in the order of 10^{-15} to 10^{-14} m. As mentioned in the previous section, without the stability of nucleons with the order of 10^{-15} to 10^{-14} m, biological nucleus with the order of 10^{-8} to 10^{-7} m cannot sufficiently be stable. Without such stability, the inheritance of genetic information of living systems from older generations to newer generations could not have, and will not, proceed sustainably. It is also worth noting that Meitner and Frisch (1938), who found atomic-nuclear fission reaction, named it as "nuclear fission" mimicking biological cellular-nuclear fission found and named by Virchow (1858), eighty years before their discovery (Yamamoto 2015).

Let us take a look again at Fig. 1.4, this time with the concept of temperature in mind. The average internal temperature of the human body is approximately 37°C, at which most of the human-body cells function properly as a whole (Reece et al. 2011). As a very basic home medicine, what we do is to measure the body temperature with home medical thermometer and if the measured value goes beyond 37°C, let's say, 38°C, then we usually take a full rest and lie on the bed for a while so that we can recover from the sick condition. The value of our body temperature, 37°C, is such a reference point.

The indoor temperature ranges from 10°C to 35°C, preferably between 18°C and 30°C, depending on how the building envelopes are formed with a variety of materials. The reason why the indoor temperature has to be lower than 37°C is that we humans always emit heat at a certain rate, which is, in fact, surprisingly large, as long as we live. This is going to be discussed later in Chapter 3 and also in Chapter 10.

How about the average global temperature, which is the space-wise average air temperature near ground surface all over the weather stations on the Earth? This is considered to be roughly 15°C or so, as shown in Fig. 1.4. This nominal value of 15°C can be obtained from a rather simple calculation with a set of assumptions, in which the short-wavelength radiation coming from the Sun through the atmosphere is at the energy flow rate of 959 W/m², that is, the product of solar constant, 1370 W/m², and overall solar absorptance of the Earth, 0.7; the annual precipitation rate, 1000 mm/m²; and the long-wavelength radiation going out from the ground surface via atmosphere into the Universal space at −270°C is exactly the same as the inflow rate of 959 W/m² (Tsuchida 1992, 2006, Shukuya and Komuro 1996, Shukuya 2013).

The increasing trend of atmospheric carbon-dioxide concentration in the second half of 20th century looks consistent with the increasing trend of global environmental temperature and also with the increasing rate of fossil-fuel use, coal and petrol in particular, as was shown in Fig. 1.2. Therefore, many people have come to believe that global warming is caused by human activities such as burning of fossil fuels and a consensus was made by those who have believed that it must be true. In this sense, we may call the period of the last decade of 20th century and the first two decades of 21st century the era of so-called global warming.

But we need to be cautious because this is not yet a scientifically proven knowledge equivalent to the established pieces of knowledge such as the existence of atomic particles that was proven by the early 20th century physicists. According to the direct measurement and also the rational estimation of near-ground surface air temperature all over the world, the global environmental temperature seems to have increased since 1800s up to the end of 20th century which is probably due to the recovery from a low value of global temperature in the period of so-called Little Ice Age, which was, approximately, from 1400 to 1800, after the Medieval Warm period, which was, approximately, from 800 to 1400 (e.g., Akasofu 2008, Tsuchida 1992, 2006). We should not be preoccupied too much by such a consensus so far reached by many people that the carbon-dioxide emission due to the use of fossil fuel alone is the single possible cause of global warming. We should be able to think about how the nature works with an open-minded attitude (Akasofu 2008, Hoffman and Simmons 2008).

While the belief that the global warming is caused by carbon-dioxide emission due to human activities has spread very much over the last thirty years or so, a rather new insight has also been emerging with respect to the global climate change: that is, the global cloud formation is influenced very much by the ionization of lower atmospheric air due to the galactic cosmic-ray shower, whose occurrence is affected by the intensity of solar cosmic rays, that is, solar wind, the intensity of which correlates very well with the number of black spots on the Sun (Calder 1997, Eddy 2009, Svensmark and Calder 2007). More will be described later in the last chapter.

Anyway, it is important, I think, to develop a conscientious scientific thought while at the same time developing a rational environmental conditioning technology that requires less and less fossil-fuel input but much smarter use of immediate natural resources.

1.5 Three classes of relationship between environment and information

One other thing worth keeping in mind other than what has been described so far is the general relationship between "environment" and "information" nested in three classes as described below.

To do so, let us first make it clear that both the built environment and the urban environment are just in between the two portions of the whole universe: one existing inside, the human body being one of the living systems made possible by the nature, and the other existing outside, that is, the Earth and the Universe, which are of course the portions of the whole nature. As I pointed out in section 1.1, we humans spend more than 80 years of time inside buildings assuming that our life spans for 90 years. The reason why we live inside buildings for such long years during our life is simply due to the fact that the outdoor environmental condition is usually too harsh to live. It may be too cold in winter, too hot in summer and so on.

Such a word like "cold" or "hot", for instance, exists in order to express a thermal state of our immediate environmental condition that reflects the respective thermal state of human body. This is one of the three classified relationships between "environment" and "information": the relationship between built environment and sensory information as shown in the middle of Fig. 1.5.

We humans are all equipped with nervous system consisting of a huge number of nerve cells, all of which are more or less interconnected with each other, within our body. Such interconnection grows throughout our life. The centre of this complex system is our brain and the sensory portals such as eyes, ears, tongue, nose, all of which are connected through a lot of nerve fibre cells with the brain, and are open to our closest environmental space. There are also two other kinds of sensory portals: one spreads over our skin layers for sensing fine touch, pressure, pain, hot, cold and muscle movement; the other exists in internal milieu, that is, extra-cellular environment for sensing viscera and their associated pain. The whole of this nervous system is self-organized from the very beginning of one single fertilized egg cell to the very end of one's life. We discuss this issue more thoroughly in relation to the built environment in Chapter 3.

Such self-organization takes place as a whole within a living system, whether it is human being, monkey, bird, or earth worm. The important fact is that its basic process can only be realized within each living cell according to a series of codes embedded in the chemical structure of deoxyribonucleic acid (DNA), that is, respective complete set of genetic information, such as the human genome, monkey genome, worm genome, and so on.

An adult human body is considered to consist approximately sixty trillion cells. The interesting thing is that all those cells are not totally identical; there are a variety of cell types such as nerve cells, muscular cells, and others. Why such different cells can develop respectively from one single fertilized egg cell as the starting point of

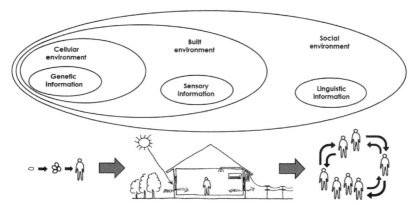

Fig. 1.5: Three classified environments together with the respective corresponding information.

each of our own life to each of our adult bodies sustaining its structure and function as a whole as human being until the end of its life is due to the fact that those cells utilize portions of the genetic information recorded within DNA nucleus from one stage to another depending on the respective states of cellular environment. This is one other level of relationship between "environment" and "information": cellular environment and genetic information, as shown in the left of Fig. 1.5.

The reason that various sensory information, whether the content of information given is positive or negative for human being to live, can be obtained through our body proper from the built environment is that our body proper and brain emerge, develop, and sustain themselves on the basis of the relationship between genetic information and cellular environment.

The third class of relationship between "environment" and "information" is the relationship between social environment and linguistic information. I am writing this book, of course, in English, which is, in fact, not my mother tongue—it is Japanese. I learned very basic English in high school, but the reason that my skill has advanced to the present level is due to the fact that I have had many occasions to use English for communicating with respect to our research activities with my international colleagues over the last thirty years or so. In other words, my whole nerve system has developed so that the neural network can be responsive to the social environment where English is being used. My brain and body proper has come up with my own neural patterns by a number of the participating nerve cells that were stimulated by English language. The same applies to my Japanese, but it is much deeper, because my basic skill of Japanese was not the one learnt first in school, but in the relationship with my mother, father, brother, and close friends in my very early childhood. Anyway, in the social environment where Japanese is being used, my brain and body proper has come up with the neural patterns by a number of the participating nerve cells that were stimulated by Japanese language.

Whether it is English, Japanese, or any other language, all of them are common as linguistic information. There is no doubt that the development of languages has made us humans survive and has allowed us to be able to develop our contemporary societies. This is the third class, which is realized on the basis of the former two classes, of the relationship between "environment" and "information": social environment and linguistic information as shown in the right of Fig. 1.5. Whichever language—English, Japanese, German, Danish, Chinese, or Korean—they all have their own respective pronunciation and characters, but their common function is to let people communicate with each other.

In any language, a whole book consists of chapters, each of which consists of sections, then paragraphs, sentences, words, and finally characters. Such structure looks very similar to that of genetic information of various living systems. For instance, a human body consists of several organs, each of which consists of various tissues such as muscles, nerves and so on, and then cells, proteins, amino acids, whose corresponding information is coded with four main chemical bases: adenine, thymine, guanine, and cytosine. We may say that four characters called adenine (A), thymine (T), guanine (G), and cytosine (C) are used in genetic language, while twenty-six alphabetical characters are used in English, fifty-one characters in Japanese and so on. We may regard that the biological structure and function performed by a variety of animals and plants are realized by the same genetic language using four characters. In other words, there are so many stories in the form of biological systems, all written in the same language coded with four characters of biological system language.

The purpose of bio-climatology is to seek the rational way of constructing and running the built environment for human well-being. In so doing, it is, I think, invaluable to take these three classes of relationships between "environment" and "information" into consideration. On the firm basis of the biological nucleus that can be stable on the stability of atomic nuclei as described in section 1.4, the living structure and function of human body can emerge properly with the genetic information within the healthy cellular environment. Then the sensory information emerges properly within the built environment. On the basis of these two classes of the relationships between "environment" and "information"—the first, biological cellular environment and genetic information, and the second, built environment and sensory information, the linguistic information including a variety of human culture that develops within the social environment can eventually emerge properly.

Bio-climatology for built environment shall hopefully be one of the sets of linguistic information contributing to the sustainability of our social environment. This is to be developed on the firm basis of the aforementioned two classes of "environment" and "information" that are embedded in our body proper and brain.

References

Akasofu S. 2008. Rational argument on so-called global warming issue—in order to be free from misleading propaganda. Seibundo-Shinkosha Publisher.

Calder N. 1997. The manic sun. Pilkington press. ISBN 1-899044-11-6.

Eddy J. A. 2009. The sun, the earth, and near-earth space—a guide to the sun-earth system. NASA NP-2009-1-066 GSFC. ISBN: 978-0-16-08308-8.

Hoffman V. L. and Simmons A. 2008. The resilient earth—Science, global warming and the future of humanity. 1st ed. Booksurge Publishing (http://www.theresilientearth.com).

Jevons W. S. 1906. The Coal Question-the 3rd edition (the 1st edition in 1865). Reprints of economic classics. Augustus M. Kelley Publisher NY.

Minamata Disease Archives (MDA). 2001. http://www.nimd.go.jp (retrieved 21st August 2017).

Reece J. B., Urry L. A., Cain M. L., Wasserman S. A., Minorsky P. V. and Jackson R. B. 2011. Campbell biology 9th ed. Pearson Education Inc. p. 80.

Shukuya M. and Komuro D. 1996. Exergy-entropy process of passive solar heating and global environmental systems. Solar Energy 58(1-3): 25–32.

Shukuya M. 2013. Exergy—theory and application in the built environment. Springer-Verlag, London.

Silk J. 1994. A short history of the universe. Scientific American Library.

Svensmark H. and Calder N. 2007. The chilling stars—a cosmic view of climate change. Icon Books Ltd. UK.

Tateno J. 2012. Threat of severe accident—how can we free from nuclear-power paradigm? Toyo Publisher, 200–203.

Tsuchida A. 1992. Another thought on thermal science, Asakura-shoten publishers (in Japanese).

Tsuchida A. 2006. The global-warming doctrine could be wrong, Hotaru-shuppan publishers (in Japanese).

Yamamoto Y. 2015. Atoms, nuclei and atomic power—what I wanted to have the students know. Iwanami-shoten Publisher (in Japanese).

Chapter 2

Passive and Active Systems for Conditioning the Built Environment

2.1 Light, heat, air, moisture, and sound: five built-environmental elements

As described in Chapter 1, we spend more than 90% of a day indoors. This time, as a start of the present chapter, let us make a rough sketch on how such indoor environmental space is conditioned.

Where are you now? And what time of day are you reading the present page of this book? You may be in a library room, or in your living room, or in your study at home. It may be early or late in the morning, or maybe in the evening.

There must be a certain amount of light being incident on this page that you are looking at; it is necessarily true, since, otherwise, you cannot read what is described here. Let us observe and think about the source(s) of light. It may be daylight coming from a nearby window, or maybe artificial light coming from a nearby lamp.

Daylight is a portion of short-wavelength radiation, which originates from the Sun and reaches the terrestrial surface taking about 8.3 minutes, where the nuclear fusion reaction of hydrogen atoms takes place in a sustainable manner due to rigorous, nearly explosive, yet seemingly controlled reaction occurring under the enormous gravitational force, which is 280 times stronger than that on the Earth. The age of the Sun is considered to be 4.6 billion years so far and is expected to last about 4.8 billion years (Silk 1994, Fleisch et al. 2013).

Artificial light is a portion of short-wavelength radiation, which originates from a lamp functioning by being connected, via the plug, the socket, and the electricity grid, either with a large-scaled electric power plant or with a rather small-scaled electric power plant: the former is operated with the combustion of a variety of fossil fuels such as coal, petrol, or natural gas, while the latter is operated with solar radiation, wind or biomass.

Fossil fuels are the resources made by nature, taking hundreds of million years under the atmosphere and the sea. A variety of microbial, plant, and animal systems which once lived during those days died and then all of their dead bodies sank, turned into the sediment, and thereby changed their forms to what we call fossil fuels. This is why they are mostly found under the ground. We may regard them to be originating from solar radiation and brought from the far past with the then minerals and water as the components to the present. The wind blowing outdoors and solar radiation reaching the ground surfaces yesterday and today are the immediate natural resources and their use is so-called renewable; the word "renewable" here means that such wind and solar radiation are available from the very immediate past and they are very much surely available tomorrow.

Indoor space is a system that surrounds your body as your closest environment. It is formed by building envelope components such as walls, windows, ceiling, and floor, with which the boundary surface of the indoor space system is determined. All of the internal surfaces receive a certain amount of short-wavelength radiation, either daylight from windows during daytime or artificial light from the light bulbs during nighttime. In addition to such short-wavelength radiation that stimulates our sensory portals inside our eyeballs, we should not forget that we are always, either during daytime or nighttime, exposed to another kind of radiation, which relates very much to our perception of warmth and coolness, that is, long-wavelength radiation that stimulates the sensory portals embeded all over our skin layers.

The long-wavelength radiation is always emitted and absorbed by the surfaces of building components, while at the same time, it is also emitted and absorbed by our skin and clothing surfaces. The absorption of short-wavelength radiation, whether it is solar radiation or artificial light, brings about, in general, an increase in temperature of building envelope components and hence it results more or less in the emission of long-wavelength radiation depending on the thermal characteristics of building materials. How much of long-wavelength radiation is emitted depends on the surface temperature.

What fills inside the space as a system defined by the boundary surface of building walls, windows, ceiling, and floor is moist air that we breathe in and out all the time. The air except water vapour is made of nitrogen, oxygen, argon, carbon dioxide, and other miscellaneous molecules, all in gaseous state. The two major constituents, nitrogen and oxygen, account for more than 99% of all the constituents and are in the ratio of four to one. Adding the third, argon, to the two major constituents, they account for 99.9%. Note that because of nitrogen being the most abundant and oxygen being the second abundant by the ratio of one-fourth, most of the chemical reactions involving combustion can proceed silently as all of the biological systems including us human beings can remain stable in their respective dynamic equilibrium state. The concentration of carbon dioxide is the fourth highest, but its value is in fact very small as can be expected from the proportion of the major constituents mentioned above, the order of 400 to 1600 ppm, that is, 0.04 to 0.16%.

The water vapour is, of course, another important constituent in addition to the four constituents described above. Its concentration ranges from 1 to 4% of the whole of the constituents depending on locations and seasons. Although the amount of water vapour is such a small fraction compared with those of nitrogen and oxygen, a change in the range from 1 to 4% affects our sensation and perception very much, whether we feel it dry or wet.

Both carbon dioxide and water molecules themselves are not harmful, but the levels of their concentration are very much associated with the air quality for maintaining the human well-being. A higher carbon-dioxide gas concentration usually indicates more occupants having resided within the room air and thereby the contaminants and the odour there must be more due to a variety of indoor activities of those occupants. Therefore, the level of contamination of the room air, indirectly though, is usually judged by the carbon-dioxide gas concentration within the room space.

The air quality has to be kept good enough for us human beings to breathe. Since we cannot stop breathing as long as we live, the amount of air existing indoors is limited, and the pollution of room air is inevitable, we always need to keep exhausting an amount of room air while at the same time take in the same amount of fresh outdoor air. This is the major purpose of ventilation, which may be made by opening and closing windows by occupants themselves or by mechanical fans, either for air removal or intake, which are driven by being connected with an electric-power plant.

The air we exhale is necessarily more humid than the one we inhale because approximately 70% of the human body is filled with liquid water. For this reason, the moisture concentration of room air increases unless the windows are kept open. Therefore, moisture is also another indoor environmental element. We need ventilation also for controlling the moisture level inside the room space.

Fresh air brought into the room may be too cold and dry in winter or too hot and wet in summer. If it is too cold and dry, it needs to be warmed up and humidified; if it is too hot and wet, it may have better be cooled down and dehumidified. Such mechanical heating and cooling devices which have been developed, taking several decades in the 2nd half of the twentieth century, are the so-called air conditioning systems to which many people have become accustomed.

Our perception of warmth and coolness is also influenced by conditioned air, either warmed-up, humidified, cooled-down, or dehumidified, whose convection takes place here and there inside the room space, in addition to long-wavelength radiation traveling between building interior surfaces, and also between our body surfaces and those building interior surfaces.

The sound travels in the air, the bulk of which moves at the velocity of about 340 m/s. The air velocity in naturally ventilated rooms ranges from 0.1 to 1.0 m/s. The sound travels much faster than the air itself that moves inside ordinary room space. Therefore, we are always exposed to the sound, which is of course one of the important elements forming the indoor environmental conditions, though the acoustic feature in the built environment is not going to be discussed further in this treatise.

Figure 2.1 schematically shows what we have so far raised and discussed as the environmental elements. In what follows, we focus on light, heat, air, and moisture in relation to human well-being.

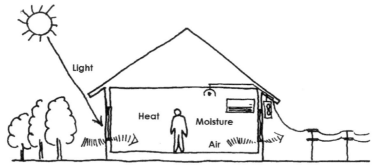

Fig. 2.1: Light, heat, air, moisture, and sound; how these five environmental elements behave in our surrounding space influence the well-being of people. In this treatise, four of them: light; heat; air; and moisture, are the major focus.

2.2 Diversity of passive systems

Technology for conditioning the built environment, which is the closest environmental space to us all, must have started to emerge some one-hundred-thousand years ago, almost together with the emergence of *Homo-sapiens*.

We define a system conditioning the built-environmental space to be safe, healthy, and comfortable as "built-environmental system". This is the system that controls light, heat, air, and moisture indoors in order to fulfil their required level of human well-being. Building elements such as walls, windows, roofs, and floors as a whole are called "building envelope systems". Ancient people must have tried making various openings with or without shutters and others on the building envelopes in addition to making use of fire as light and heat sources so that the indoor illuminance, temperature, humidity, air current, and air quality come as closer to the desired level of comfort as possible by applying a variety of the then most advanced ideas, namely the then available building technology (Banham 1984).

Building envelope systems having openings with or without shutters to condition the indoor environment within a range of comfort are called "passive systems". We call the whole of designing, constructing, and managing the passive systems "passive technology".

The significant characteristic of the passive systems is that the given forms are different from one region to another region. Whenever we discuss such characteristics of a passive system, it is necessary for us to take a look at two different aspects of the system: one is the structure, "*Katachi*" in Japanese, the given form as just mentioned above and the other is the function, "*Kata*" in Japanese, and how it performs with the given form. The structure (*Katachi*) is associated with space and the function (*Kata*) with time. We can photograph the structure (*Katachi*), but not the function (*Kata*), since the function emerges from the flow of time. We "see" the structure (*Katachi*), namely the form. On the other hand, we "read" the function (*Kata*), namely a series of changes. The structure is something to see and the function is something to read.

Let us take a look at Fig. 2.2. Suppose that there is a house standing, as shown on the left, and also a collection of all building materials used in this house, as shown on the right. What do you think is the difference between this house standing on

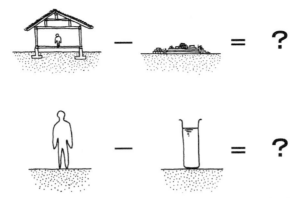

Fig. 2.2: Structure (*Katachi*) and function (*Kata*) that we should see and read. What is the difference between a house standing and the collection of all building materials dismantled? Neither heat to be released when burnt nor weight is different. The same applies to a human body.

the left and all the materials dismantled as shown on the right? Let us also think about the same question about ourselves, the human body. Neither the house nor the human body is a mere collection of matter, but has a respective structure (*Katchi*), and function (*Kata*).

Climatic patterns vary with regions. Yokohama where I live, for example, is one of the regions where it becomes hot and humid in summer while on the other hand, cold, though mild, and dry in winter. Yokohama has such annual climatic pattern. Wherever on Earth, whether it is Yokohama, Singapore, San-Francisco, or Copenhagen, each of these cities and their regions has its own climatic pattern.

Vernacular buildings in respective regions developed taking long years from generation to generation, having their own forms, namely the structure (*Katachi*), reflecting their climatic characteristics (Kimura 1993). All of such forms are closely related to the respective function, namely "*Kata*", of the buildings themselves and also their occupants' behavioural patterns, again "*Kata*" as the occupants' living styles, which will be discussed further in some chapters that follow. The structure, "*Katachi*" and function, "*Kata*", as a whole, reflected by the climatic patterns is called "architectural culture".

Let us compare a couple of examples taking a look at some buildings shown in Figs. 2.3 and 2.4. The former is a wooden-framed house with straw mat rooms with corridor space and thatched roof; the latter is a masonry house with white plaster finish. One thing that we notice immediately looking at these two sets of photographs in Figs. 2.3 and 2.4 is that there is a clear difference in their form (*Katachi*).

At first glance, where on Earth did you think each of them is located? You might have imagined that the former stands somewhere in a hot and humid region: maybe in Vietnam, Thailand, or Japan, and the latter somewhere in a sunny and dry region: maybe in Portugal, Spain, or Greece. The answer is that the former is in Yokohama, Japan, and the latter in Nisyros Island, Greece. It may be a bit hard to tell exactly where it is from the respective three choices, but if you are asked to answer which of Figs. 2.3 and 2.4 is in hot and humid region, you are very less likely to choose a wrong one. The same must be true if you are to answer a question which is in a hot

a)   b)

Fig. 2.3: Vernacular wooden-framed houses in Yokohama, Japan: (a) living room with straw mat surrounded by open corridor space effective in natural ventilation, whose roof overhangs for avoiding rain and solar radiation; (b) thick thatched roof effective in protecting the indoor space from heavy rain and intense solar radiation because of its insulating characteristic.

a)   b)

Fig. 2.4: Masonry buildings with very thick external walls painted white in Nisyros island, Greece: (a) the buildings stand collectively to form a town; (b) white colour on the exterior surfaces of the walls is effective in reflecting excess solar radiation, while the small windows allow daylighting and natural ventilation and a large heat capacity of the walls helps make the indoor temperature stable.

and dry region. We all know that such houses as shown in Fig. 2.3 must be in a hot and humid region and those in Fig. 2.4 must be in a hot and dry region. It implies that you know that there should be some difference in architectural forms reflected by respective regional characteristics of local climate.

Since the architectural culture in respective regions mentioned above was anonymously developed taking long years during the era of no media of contemporary information technology such as computers and internet, the fact that a similar form can be seen in different locations on the Earth but with similar climate to each other and also that a different form can be seen in different locations with different climate are owing, almost perfectly, to the reflection of climate represented by the outdoor temperature, humidity, solar radiation, wind, precipitation and others.

Extending this discussion on our cognitive ability of such difference and similarity and also taking the viewpoint of human neurobiology is also of much interest because these forms, in association with climate, seem to have shown the evidence that the human brain, the centre of all sensory organs, functions to read the climatic characteristics of the given region where they live and comes up with the form of buildings which best fits the local climate. We may say that the human brain together with all of our peripheral nerve systems have such universal characteristics to come up with the diversity of architectural forms dependent on the local outdoor environment.

The structure (*Katachi*) and the function (*Kata*) of passive systems seem to reflect clearly the characteristics of local climatic conditions. We can conclude that the essential feature of passive systems is such "diversity".

2.3 Metamorphosis of fire into active systems

The most dramatic event that happened over the course of human evolution is that the ancient humans started to walk by foot. The fact that the front foots, namely both hands, became free brought about the development of nervous system including brain reaching the present level of human nervous system, as will be discussed a bit more thoroughly in the next chapter. In due course, as a result of human evolutionary process, the use of fire together with various tools, the communication by language and its associated development of phonetic symbols, mathematical symbols and others emerged (more these topics in Chapter 9). Such holistic development of the human nervous system has realized the contemporary urban civilization.

Structure (*Katachi*) and function (*Kata*), which can be seen in the contemporary urban civilization realised by the state-of-the-art advanced technology, is the front edge of metamorphosis which originated from the use of fire with wood available nearby in the early stage of human history via the agricultural and industrial revolutions.

The use of fire has actually metamorphosed into the use of electricity for delivering "work" at a certain rate from the site of supply to the site of demand. You may doubt why the fire relates to the rate of work and electricity, but once you understand how the electric power is produced, you will agree with this statement. At an electric-power plant, whether it is coal-fired, nuclear-fission based, or conventional liquefied-natural-gas fired, the liquid water is heated and turned into water vapour with very high temperature and very high pressure, while on the other hand, the water vapour is cooled and condensed into liquid water again, usually by the sea or river water available nearby, or occasionally by the atmospheric air using a huge cooling tower. The flow of water vapour at high temperature and high pressure to liquid water at lower temperature and lower pressure enables a turbine (a wheel) to rotate and thereby generate work.

In essence, "work" is produced in the flow of "heat" from hot to cold. The so-called electricity in our everyday life should be recognized as the rate of work delivered from the supply site where the "work" is produced to the demand site where the "work" is used for various purposes. The whole mechanism of this production of "work" which, in fact, should be called "exergy", will be discussed more in detail in Chapter 7.

All of the contemporary mechanical lighting, heating, cooling, and ventilating system components such as lamps, fans, pumps, and heat pumps work with the electricity supplied from the power plants. A fan consisting of a wheel and a motor moves a volume of air from one place to another; a pump consisting also of a wheel and a motor moves a volume of water from one place to another. Their function is realized by the "work" delivered through the electricity grids. We call such built-environmental control systems that rely on the "work" delivered by the electricity in order to perform their purposes "active systems" and their associated building technology "active technology". Active systems and technology include those making direct use of fossil fuels, namely the fossil-fuel combustion to raise the water or air temperature for conditioning the built environment.

Figure 2.5 shows the difference in two types of ships: a yacht and a motor boat. The yacht travels on water by making use of the wind blowing in the immediate environment with the sails, while on the other hand, the motor boat does the same by feeding gasoline into the motor to turn the blades under the water. Suppose that there is wind blowing against the direction you are going to sail. In the case of yacht, it can move forward by taking zig-zag route, but in the case of motor boat, it can move forward for a shorter distance than the maximum distance that the motor boat could sail with no wind against.

The sailors on the yacht and those on the motor boat regard the wind in the vicinity differently. The former celebrates the existence of the wind unless it is stormy, while on the other hand, the latter hates or may well think that no wind against is better. Yacht sailing is a typical "passive" technology and motor boat sailing a typical "active" technology.

In Japanese, the "work" is often called "*Dou-ryoku*", which is the combination of "*dou*" to imply the movement and "*ryoku*" to imply the force; the essence of work is to move something from one place to another. A lump of matter, namely a solid

Fig. 2.5: A yacht functioning as a typical "passive" system and a motor boat functioning as a typical "active" system. Wind blowing nearby can be used by the yacht to move forward, but for the motor boat, if it blows against, it only acts as the resistance.

having a certain shape, consists of the atomic and molecular particles, each of which has its own position relative to all others so that the matter as a whole has its shape. This characteristic of solid is uniquely different from liquid and gas, both of which have no specific shape themselves so that we need a container made of solid in order to move an amount of liquid or gas collectively. Therefore, the work is performed by the collective movement of solid particles.

The fact that the shape is not destroyed in the movement of solid matter from one place to another implies that all of the atomic and molecular particles in that solid matter move parallel to each other as schematically shown in Fig. 2.6. In this case, the cause of "work" is gravitation and it makes the blades turn around. In reality, there is more or less friction so that a portion of the "work" is inevitably destroyed and turns into "heat". This implies that the parallel movement turns in part into the random motion of the particles. This process results in a temperature increase within the blades and their surroundings. What this really means can be articulated with the concept of "exergy" as its definition will be given and quantified, and some fundamental topics will also be discussed in Chapter 7.

The metamorphosis of the use of fire that started in ancient time has reached, as described above, the present state of electricity use for various purposes and in due course, we humans have become capable enough to control indoor illuminance, temperature, and humidity at almost any level that we desire. Such characteristics of active technology have realized low ceiling heights, deep room space, large glass window areas and others leading to so-called economic efficiency of buildings together with apparent, but a kind of superficial, beauty; we can see a lot of such examples, especially in high-rise buildings built in urban areas worldwide. Their basic forms are identical despite the fact that they are built under a variety of climatic conditions (Ikeda 1998).

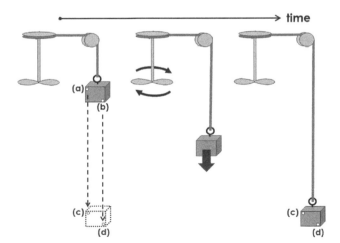

Fig. 2.6: A fall of solid matter from a higher place to a lower place. This is, from a microscopic view point, nothing other than that all atomic molecular particles composing of this matter move in parallel so that they keep their relative positions remain unchanged. The movement of a molecule from position (a) to (c) is exactly parallel to that from position (b) to (d). This movement caused by gravitation realizes the work that makes the blades turn around.

Figure 2.7 shows two such examples of high rise buildings: one built in Osaka, Japan and the other in Malmö, Sweden. If you are told that the former is in Shanghai, China and the latter in San Francisco, the USA, you may not doubt those locations at all unless you know the actual locations beforehand. This is because this kind of form can be seen either in New York, Yokohama, Shanghai, or other cities. I have used one of the photographs, twin-tower buildings, in my lecture and asked the students and the audience where these buildings are located, giving three choices of the city names, intentionally excluding the place where they really stand; for instance, Nagoya, Kuala Lumpur, and Shanghai. Nobody has ever claimed that my question is wrong and, moreover, almost all of them have tried to answer this question seriously instead. This fact proves that everybody knows, unconsciously though, such a form is possible in almost any city on Earth.

The active technology has realized such "universality" and helped grow together with "architectural civilization (or urbanism)", which is in good contrast to "architectural culture" developed together with passive technology whose significant characteristic is "diversity".

Fig. 2.7: High rise buildings realized by active technology. Such forms, which can be seen anywhere in urban areas worldwide, have become possible due to the development of active technology for lighting, heating, cooling, and ventilating the built environment.

2.4 Comparison of passive and active system components

Table 2.1 summarizes the components of passive and active technologies and their associated physical characteristics to be considered for each of lighting, heating, cooling, and ventilating systems. The components of "passive" technology are basically building envelope components themselves. Physical characteristics to be considered are how much of light, heat, air, and moisture come in or go out through each of the building envelope components. Knowing their qualitative and quantitative characteristics, one can come up with a whole building envelope system

Table 2.1: Comparison of passive and active system components together with the associated primary physical characteristics to be considered.

Objective	Passive Technology		Active Technology	
	Components	Physical characteristics	Components	Physical characteristics
Lighting	• Windows • Shadings • Glass panes • Ceiling • Internal wall surfaces	• Solar optical properties of window materials • Reflectivity of ceiling and internal wall surfaces	• Lamps • Luminaires • Ceiling • Internal wall surfaces	• Luminous efficacy • Optical properties of Luminaires • Reflectivity of ceiling and internal wall surfaces
Heating	• Windows • Walls • Floor • Ceiling	• Solar optical properties of window materials • Conductivity • Radiative and convective transfer • Thermal mass • Air tightness • Permeability	• Heat exchangers • Fans and pumps • Heat pumps • Boilers	• Convective and radiative transfer • Conductivity • Radiation • Pressure • Friction • Efficiency
Cooling	• Windows • Shadings • Walls • Floor • Ceiling	• Solar optical properties of window materials • Conductivity • Radiative and convective transfer • Thermal mass • Permeability	• Heat exchangers • Fans and pumps • Heat pumps	• Convection • Radiation • Pressure • Friction • Efficiency
Ventilation	• Windows • Doors • Vents	• Wind • Buoyancy • Pressure • Water vapour • Friction	• Fans • Ducts • Shutters	• Pressure • Water vapour • Friction • Efficiency

that realizes a certain built environment where people can reside at their certain state of well-being.

Lighting making use of daylight is called "daylighting" as a typical passive technology and that making use of torches, candles, electric lamps is called "artificial lighting" as a typical active technology. In contemporary societies, the major light sources of artificial lighting is electric light, which is realized by the lamps feeding on electric power delivered from the site of producing "work". Note that the "work" inevitably turns into "heat" in the course of lighting.

Daylight, which originates from the Sun, incident on the window openings of building façade surfaces is partly absorbed, transmitted and reflected by building envelope components, and then both transmitted and reflected daylight are sooner or later absorbed inevitably somewhere indoors or outdoors, and eventually turns into "heat". In both daylighting and electric lighting, more light with less heat is desirable.

Heating may be performed by making some amount of such heat generated as a result of the absorption of daylight transmitted through windows or it may be made with an artificial heat source obtained from burning biomass or fossil fuels. The determinant factors of "passive heating" are thermal characteristics of building envelope materials such as thermal conductivity and thermal mass, as will be discussed in detail in Chapter 6. The level of air tightness also affects the effectiveness of passive heating. The effectiveness of "active heating" is determined by how much of heat has to be delivered with a whole collective system of boilers or heat pumps together with heat exchangers, pipes, pumps, ducts, and fans. How the active-heating system configurations are to be formed is determined by how much of heat needs to be delivered into the built environment, how the heat should be distributed, and how the passive heating system is formed prior to the active heating system, which should be designed, constructed and operated in harmony with the former, passive heating system.

Thinking about cooling starts with how much of heat we can minimize inside the built environmental space. For this purpose, the reduction of solar heat gain from windows and also internal heat generation caused by electric lighting and a variety of electric and electronic appliances used in the room space is important in addition to thermal insulation of building envelope components. With the rational thermal insulation strategy, the whole of thermal mass of building envelope systems should be able to contribute to storing the coolness, which may be harvested from the immediate outdoor environment such as nocturnal sky radiation, nocturnal outdoor air, or the coolness to be found under the nearby ground.

The cultivation of the coolness to be found in our vicinity may be made firstly by passive cooling measures to be realized by a harmonious combination of thermal insulation and thermal mass in order to maximize the effectiveness of radiative and convective heat transfer within the room space. Natural ventilation as a typical passive technology should perform well for this purpose with rational control, in particular of radiative heat transfer. The effectiveness of active cooling is determined also by how much of heat has to be removed at how much rate from the built environment by a collective system of heat pumps, heat exchangers, pipes, pumps, ducts, and fans. The active cooling system configurations should be formed so as to perform in harmony with passive cooling systems.

The hurdles in making passive cooling measure truly effective are usually higher than those in making passive heating measures, since what is required in the case of cooling is not the supply of heat, but the removal of heat. Since this intrinsic characteristic in cooling differs from heating, the ventilation strategy should be carefully considered, whether it is made as a passive or active technology measure. In all of active system components, the friction is the major cause of heat generation which necessarily emerges in the course of flow and circulation of water and air. System components such as pumps, fans and heat pumps must be designed so that the unnecessary friction, that is, the unnecessary heat generation is minimized (more on this issue in Chapter 11).

The rapid growth of contemporary global human society over the last half century has been sustainable so far, dependent very much on the combustion of fossil fuels such as coal, petrol, and natural gas, and also on nuclear fissile materials, while at the same time we have come to recognize that we now face so-called energy and environmental issues and also that we need to find their solutions as mentioned in Chapter 1. With such recognition in mind, passive technology, which our ancestors developed taking long years, started to be reviewed mainly by those involved in the research on bio-climatic architecture; this movement started in early 1980s, right after the first and second oil crisis. In this course of review, some of the passive technologies were applied to the typical contemporary buildings with a full use of active systems.

In the very beginning when such trials emerged, from early 1980s to early 1990s, there was sometimes heated discussion, though not constructive, such that you were almost forced to choose either passive or active technology, but not both. One extreme was that passive technology should be fully revived in future buildings and we should get rid of the advancement of contemporary science and technology as much as possible. The other extreme was that we should make the built environment as closed as possible, or in other words, as disconnected from outdoors as possible, by a full use of active systems in order to realize a stable and constant indoor climate so that we can be free from any fluctuations caused by solar radiation, outdoor temperature, outdoor humidity, and others.

The reason that such heated disputes emerged was due to a too-easy application of the passive technology performing not sufficiently, which was based upon the old way of thinking that realized the conventional type of passive and active technology. This is in fact the problem of "human-mind", I think, to be examined whether we can review and renew our philosophy and re-establish the rational basis of science and technology looking into the future and whether we can re-create new types of passive and active technologies to be in harmony with each other.

In order to promote a healthier development of building technology for controlling the built environment, it is necessary for us to seek such active systems that can enhance the merits of passive systems or that can revitalize the forgotten passive systems to meet the requirements of indoor environmental quality.

2.5 "Work" production through "heat" flow

Passive systems function by making use of a part of natural flow of light, heat and air through them and active systems function thoroughly by utilizing "work" produced artificially. Understanding how the "work" is produced artificially or naturally helps us have a clearer and dynamic image of active systems and also that of passive systems. Therefore, in what follows, let us draw a rough sketch of how the "work" is produced and hence draw a clearer image of how the global environmental system including biological systems work. This view becomes, I believe, the foundation of bio-climatology for built environment.

2.5.1 Closed and open systems

To begin with, let us briefly review the fundamental characteristics of a "system". Any system that we defined in the beginning of Chapter 1 can be sorted into two groups: closed systems and open systems. Closed systems are the kind of systems whose boundary surfaces allow the "work" and "heat" to flow in and out, but not the "matter" to flow in and out. On the other hand, open systems allow either of them to flow in and out. Whether a certain system focussed on is closed or open depends solely on how the boundary surface is assumed. Table 2.2 summarizes these fundamental characteristics (more on "closed" and "open" will be discussed in Chapter 8).

Table 2.2: Inflow and outflow through closed and open systems.

	Closed	Open
Light and Heat	○	○
Work	○	○
Matter (air, water, moisture)	×	○

Let us make here a quick overview of a variety of open systems. Suppose that there is a man whose immediate environment is the room space. The boundary surface between him and his environment is all over the skin surface including its extended imaginary surface covering the openings such as mouth, nostrils, sweat glands, and anus.

As long as he is alive, either at resting posture or at light work, he breathes in the surrounding air rich in oxygen and breathes out the air rich in carbon dioxide at the rate of about 8.5 Litres every one minute. Even if the body is under thermally neutral or cool conditions so that he perceives no sweating, there is always natural dispersion of water, from liquid state to vapour state, through the skin surface from the inner skin layer towards the room space. Therefore, the human body is definitely an open system.

The human body consists of about sixty-trillion living cells, each of which has its respective unique structure and function and acts as an open system by transferring a variety of essential matters for its sustenance from the immediate cellar environment, while at the same time, transferring a variety of waste matters out of the cell body from the immediate cellar environment. The human body as a whole is therefore an open system as described above.

The room space, which can also be regarded as a system with the assumption of outdoor space as its environment, receives carbon dioxide and water vapour from the human body, while on the other hand, it gives off an amount of air either by natural ventilation or by mechanical ventilation into the outdoor environment.

Since the human living cells, the whole of human body, and the room space are all open systems as described above, a building consisting of a variety of room space as the basic unit of built environment is also an open system as shown in Fig. 2.8a, which represents the inner most part of the nested structure of environmental space.

The urban environment which surrounds a variety of buildings is also an open system and the regional environment consisting of the urban environment at its centre together with the nearby villages in rural plain, forests, hills, mountains, rivers, coasts, and the sea can also be regarded as an open system, since these urban and regional environment as open systems work by letting the atmospheric air and water flow in and out as natural phenomena and also by bringing in and out a variety of matters for the activities of living creatures including us humans in addition to "light" and "heat" delivered in and out, respectively, by solar radiation and long-wavelength radiation as represented in Fig. 2.8b.

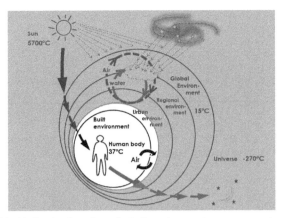

(a) Air flowing in and out the built environment as an open system

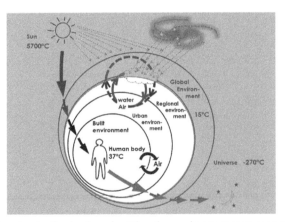

(b) Air and water flowing in and out the urban environment as an open system

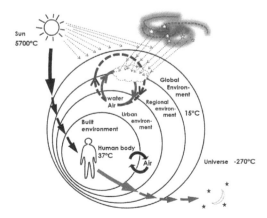

(c) No matter except cosmic rays flowing in and out the global environment

Fig. 2.8: Open and closed systems within the global environmental system.

The global environmental system absorbs solar radiation as "light", while at the same time emitting long-wavelength radiation as "heat". The Earth may, in one sense, be regarded as a "closed" system as shown in Fig. 2.8c, if we assume that it gives off no matter out from itself and also brings in no matter at all from the Universe, but in the other sense, the Earth should be regarded as an "open" system, since there is the absorption of quite a few solar and galactic cosmic rays by the atmospheric layer, which influences the long-term global climate change in spite of their total mass being negligibly small (Svensmark and Calder 2007).

2.5.2 A model heat engine producing "work"

Suppose that we have a kettle filled with water over a gas oven as shown in Fig. 2.9. The lid of the kettle is firmly closed and the opening area of the spout is quite small. Let us turn on the switch of gas oven and have fire underneath the bottom of the kettle. The water temperature gradually increases and sooner or later it starts boiling and thereby the steam comes out vigorously from the spout.

This experiment is assumed to be made under an ordinary atmospheric condition with the average pressure of 1013 hPa so that the pressure exerted on the water surface is exactly the atmospheric pressure of 1013 hPa (the meaning of atmospheric pressure and its relation to the characteristics of air and moisture is discussed in Chapter 8). The boiling temperature at which the water changes its phase from liquid to vapour under the atmospheric pressure of 1013 hPa is 100°C. The surrounding space of the boiling water contains some amount of water vapour, but its corresponding pressure is usually from 1 to 4% at the highest, that is, from 10 to 40 hPa. There is a large difference in water vapour pressure between the space inside the kettle and the surrounding space so that the vigorous flow of water vapour emerges.

As can be seen in Fig. 2.10, if a small windmill is placed in the vigorous flow of water vapour near the spout, then it starts turning around. This is since, from the microscopic viewpoint, the molecules consisting of the windmill are hit vigorously by the water molecules, whose momenta are large because of high temperature at 100°C, and thereby the windmill turns around as a collective movement.

Regarding the kettle as a system, it is exactly an open system, since the molecules of water vapour come out crossing the system boundary from the spout and the heat is transferred from the flame of burning gas around 1200°C to the boiling water at 100°C. Such a condition of water boiling in the kettle and the windmill rotating may continue for a while, but sooner or later the rotation of the windmill must stop. Why stop? There are two possible reasons: one is that the gas is used up so that the water temperature decreases and thereby the steam is not available anymore; the other is that all of liquid water turns into water vapour and the kettle is dried out, although quite a lot of fuel gas is still available.

Fig. 2.9: A kettle filled with an amount of liquid water placed on a gas oven.

Fig. 2.10: Burning gas warms up the water inside the kettle gradually and sooner or later, the water starts boiling. The lid is sealed very firmly so that the steam comes out vigorously from the spout and thereby lets a windmill rotate.

Here, let us assume the latter condition and go to the next step of discussion. If we want to continue rotating the windmill that is to secure the sustainable function of this model heat engine, then we need to keep supplying an amount of liquid water just equal to that of water vapour coming out from the spout.

Consider pouring that amount of fresh water. In order to keep the windmill rotating, we need to supply more gas to heat up the water in the kettle, since pouring a small amount of water into the volume of boiling water is very likely to decrease the water temperature and water-vapour pressure. Then the vigour of the steam may disappear all of a sudden so as to make the rotation of the wheel unstable. As an analogy, you may imagine what happens in the hot water in a pot being used for boiling noodles. When the surface of boiling water filled with lots of tiny bubbles rises up and starts almost flooding, pouring a small amount of fresh water lets those bubbles diminish very easily.

In order to avoid such instability of water-vapour pressure, it is necessary to feed on more gas and increase the intensity of flame in addition to supplying fresh water. This is nothing other than wasting both gas and water so that we should think about some other way to make a better use of water together with reducing the amount of gas to be used.

Let us first consider reusing the water vapour. To do so, we must prepare an enclosure to confine the water vapour in a finite volume of space as can be seen in Fig. 2.11. The difference in what you can see in Figs. 2.10 and 2.11 is the volume of space into which the water vapour disperses. In which case do you think the windmill turns longer than the other? With enclosure or without enclosure? A little bit of thought lets us recognize that the former case lets the windmill keep rotating much longer. This is because the volume for the water vapour to disperse in the former case is almost infinite, while on the other hand, that in the latter case is very limited. If we enlarge the volume of enclosure, the period of time, during which the windmill keeps rotating, becomes longer, but it is not the right direction to make an improvement, since the longest is just the same as what happened in the case without enclosure as shown in Fig. 2.10.

As an alternative of enlarging the enclosure space, let us apply an idea of dipping a portion of the enclosure in an amount of cold water, whose temperature is low enough to make the water vapour condensed as shown in Fig. 2.12. This is in fact equivalent to enlarge the volume of the enclosure, since the volume of a

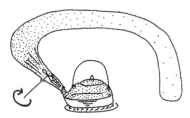

Fig. 2.11: An enclosed space is equipped to collect the water vapour flowing out after rotating the windmill. The windmill remains rotating shorter with the enclosed space than without the enclosed space.

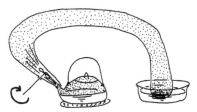

Fig. 2.12: A portion of the enclosed space cooled by chilled water contained in a washbasin. This realizes the condensation of water vapour inside the enclosed space so that more room becomes available for further condensation of water vapour into liquid water.

certain mass of water vapour shrinks into 1/1700 or so if it changes the phase from vapour to liquid under the ordinary atmospheric condition. Therefore, the rotation of the windmill in the case of Fig. 2.12 becomes much longer than in the case of Fig. 2.11. If the amount of cold water contained by a washbasin is small, then the water temperature rises easily, but here we assume that the cold water in the washbasin is always replaced with fresh cold water to be able to keep condensing the water vapour within the enclosure. In fact, this is of vital importance in parallel to the sustained availability of fuel gas.

We become happier with the case in Fig. 2.12 than with the case in Fig. 2.11, since the rotation of the windmill lasts longer, but we next come to notice that the rotation of the windmill stops within the same period of time as in the case of Fig. 2.10.

This is because once all the water turning into water vapour inside the kettle is captured at the farthest end of the enclosure dipped in the cold water in the washbasin, the kettle is dried out. This is exactly the same conclusion as the one we learnt from the process discussed in the case of Fig. 2.10. Therefore, we need to furnish something further. Let us think about a way to return the water once used for the rotation of the windmill into the kettle. To do so, we need to install a pipe connecting the space inside the kettle and the far end of the enclosure.

While the water in the kettle is being boiled, the water vapour pressure there is much higher than the water vapour pressure at the far end of enclosure, 1013 hPa versus the order of 10 hPa, so that if the pipe alone is installed, then the boiling water must start flowing towards the far end of enclosure in addition to the spout. In order to avoid this flow and let the liquid water flow towards the kettle, we need to install a pump whose outlet pressure is high enough to send the water into the kettle as shown in Fig. 2.13. The pump should be operated by supplying a portion of work produced as the rotation of the windmill.

Starting with the one way flow of water from liquid via vapour to liquid again as shown in Fig. 2.12, we now come up with the circulation of water to be sustainable as shown in Fig. 2.13.

If the amount of work required by the pump for circulating the water is comparable to the amount of work produced by the windmill, then the whole system shown in Fig. 2.13 may look fun as a kind of toy, but is not useful at all, since no amount of work is available to meet the demand outside this model heat engine. Therefore, it is very important to make the pump very efficient to have a relatively large amount of work left. All of the actual power plants in use are designed so that this essential requirement is fulfilled.

Let us summarize what we have so far discussed. In order to keep the model heat engine functioning, we need to prepare the four essential conditions summarized in Table 2.3.

The first is to prepare "hot" source, that is, the heat source. In a thermo-chemical heat engine, "chemical" exergy that is contained as the molecular structure of fuel material is consumed and its portion turns into "hot", that is, very high-temperature condition, from which "heat" is transferred to a fluid that functions as an agent to circulate.

The second is to prepare an enclosure to confine the fluid inside a closed finite volume of space to make its efficient use for expansion and contraction in the course of circulation.

The third is to prepare heat sink, which may be called "cold" source in contrast to "hot" source for heat source. In the thermo-chemical heat engine, the heat sink is for discharging an amount of heat in order to condense the water vapour into liquid water. This implies that we are never able to keep converting all of the "heat" into "work". This is of critical importance, as will be discussed later again in Chapter 7.

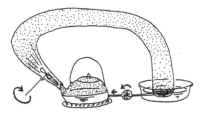

Fig. 2.13: A model heat engine consisting of a kettle as the boiler, an enclosed space for keeping the water as the agent for producing work, a washbasin as the condenser, and a circulating pump. The presence of a pump is essential for the heat engine to function in addition to the two reservoirs of heat: the gas oven and the water in the washbasin.

Table 2.3: Four requirements for a heat engine to function sustainably.

1.	To heat a fluid by 'hot' source;
2.	To confine the fluid in a closed space;
3.	To cool the fluid by 'cold' source;
4.	To let the fluid circulate by a pump.

The fourth is to prepare a pump in order to let the fluid circulate, since no circulation is made possible without pumps.

The consideration of hot source, cold source, heat flow, and the circulation of fluid with an agent of pumping is essential for understanding not only man-made heat engines but also natural systems including atmospheric air and water circulation, various living creatures, plants and animals including human body and a variety of passive and active systems for the built environment.

Focusing on the kettle, it can be regarded to be an open system, since water vapour comes out from the spout and is dispersed while at the same time liquid water comes in from the pipe connected with the far end of enclosure dipped in the washbasin. Focusing on the circulating water, there is inflow of heat from the fire of gas oven through the bottom plate of the kettle and also outflow of heat into the cold water contained by the washbasin. In due course, there is outflow of work into the windmill. Therefore, the circulating water is a closed system; this is what the second statement in Table 2.3 means. Within the closed system, the characteristic of water changing its phase from liquid to vapour and vice versa inside the kettle, the enclosure, and the pipe is essential as the agent to make the "work" extracted from the flow of "heat".

How about the windmill together with an axle as a system? There is the input of work to turn the blades of the windmill while at the same time the output of work by the rotation of the axle. If the system boundary is assumed to be the blade and axle surfaces, then there is neither input nor output of matter. Therefore, the windmill together with the axle is a closed system.

The whole of the system as shown in Fig. 2.13 is an open system since the combustion of gas requires the constant input of air rich in oxygen to the oven and also the sustenance of condensing water vapour into liquid water requires the constant supply of cold water into the washbasin. The sustainability of such an open system as the model heat engine is secured by the huge capacity of gas, air rich in oxygen, atmospheric space to discard exhaust gas, and aquatic space such as the sea or a big river for exhaust heat to be kept discarded.

The contemporary urban societies must have not been realized as they are, at present, without the electricity distribution together with large-sized power plants in the surrounding proximity and in remote places. The principle of producing work sustainably is exactly what we discussed above, whether it is fossil-fuel fired or nuclear-fuel based power plant. Their difference that we should be careful about is the characteristics of the waste matter inevitably generated as a by-product.

At the fossil-fuel fired power plants, the waste matters are mainly carbon-dioxide gas and water vapour together with some nitrogen-oxide and sulphur-oxide gases, both of which are well eliminated by the contemporary technology for de-nitrification and desulfurization. Exhaust gas containing mainly carbon dioxide and water vapour after de-nitrification and desulfurization is dispersed from the chimney of the fossil-fuel fired power plants.

In the case of nuclear power plants, there is to be almost no exhaust-gas emission, but instead the more the electricity is produced, the more the nuclear waste, which is harmful to any living creature due to its long-lasting ability of emitting alpha or beta particles and gamma radiation.

Because of such longevity and the overwhelming risk of harm, the nuclear waste has to be kept in the containers from one generation to the next generation, and then to the following generations for so many years to come. Therefore, the nuclear power plants have to be a perfect closed system, which is, in reality, not possible to build.

All of the living systems as the constituents of nested structure within the global environmental system as a closed system with an assumption that the total mass of the penetrating cosmic rays into the atmosphere is negligible, as was shown in Fig. 2.8c; they are all functioning as open systems relating to each other. Comparing this fundamental characteristic of every living system including human being with that of nuclear power plants described above, we have to admit that nuclear power plants are not the right-sized active technology that we humans can rely on. This has become much clearer to more and more people by the experiences of nuclear disasters in Three Mile Island in 1979, in Chernobyl in 1986, and in Fukushima in 2011, from which we should learn seriously for the sake of better future, and transfer what we have learnt to the future generations to come.

2.5.3 *Dynamic equilibrium of a drinking bird*

In order to confirm the knowledge we have so far established with the discussion described above and also in order to develop a dynamic image of the global environmental system, let us investigate how a toy called "drinking bird" or "peace bird", as shown in Fig. 2.14, can keep bowing and how it can sustain in an empty aquarium filled with ordinary moist air as a model of the global environmental system.

The body of a drinking bird is made of glass, which forms a closed system that contains an amount of fluid matter, dichloromethane, in the states of both liquid and vapour. The external surface of the glass body from the upper one-fourth of the long neck to the head is covered by the felted fabric. Inside the head and the neck is hollow space filled with the vapour of the fluid. The lower three-fourth of the long neck continues to go into the hollow round-shaped glass body filled with the fluid as liquid and its lower end is dipped in it.

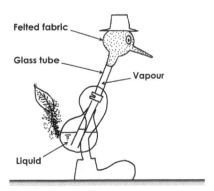

Fig. 2.14: A drinking bird made of a hollow glass tube and ball filled with liquid dichloromethane in the lower part and its vapour in the upper part. The long glass-tube neck goes into the middle of the round hollow body and its lower end is dipped in the liquid dichloromethane.

The weight of the drinking bird, which is mostly owing to the liquid held in the round-shaped body, allows it to take the upright position. The long glass tube, whose lowest part is dipped in the liquid, divides the space above the liquid surface into two portions: one is surrounded by the round-shaped glass and the glass tube, and the other is inside the glass tube, which is connected with the glass head covered by felted fabric. Both are saturated with the fluid vapour.

There is a support made of metal in the lower neck, just above the round-shaped body, and it is connected with the legs. The metal support and the legs are connected with each other by hinge joint so that the whole body of the drinking bird can swing as if it looks bowing.

Using this "drinking bird" toy, we can make an experiment to produce "work" from "heat". To do so, we need to make the felted-fabric surface of the drinking bird head get wet with liquid water and leave it for a while. The evaporation of water decreases the internal surface temperature of the glass head and lowers the saturated pressure of the fluid vapour inside the glass head. This results in the liquid surface moving upwards, since the saturated vapour pressure within the round-shaped body turns out to be higher than that within the head.

An analogy may be helpful. Imagine what happens to orange juice in a glass when you pull an amount of air inside the straw. The reason why you can drink the orange juice is that the air pressure inside the straw is lowered by sucking the air inside the straw and the atmospheric pressure exerted on the juice surface outside the straw becomes higher than that inside the straw.

The rise of liquid surface inside the glass-tube neck towards the head is necessarily accompanied by the fall of liquid surface outside the glass-tube neck. The difference in the height of liquid surface between inside and outside the neck gradually becomes larger and sooner or later, just within several seconds or so, depending on the relative humidity of the surrounding space, the liquid comes up into the drinking-bird head. This makes the gravitational centre of the drinking bird shift towards the head from the lower round-shaped body. This results in the drinking bird turning its head downwards and then eventually takes the posture of bowing deeply as shown in the lower drawing of Fig. 2.15.

When the drinking bird is taking a deeply bowing position, the two spaces filled with the fluid vapour merge into one space and their vapour pressures turn out to be the same. Then the liquid once going up inside the head flows down and thereby the drinking bird returns to its position again upright as shown in the upper drawing of Fig. 2.15. This concludes one cycle.

As we have already learnt in 2.5.2, it is necessary to have "hot" and "cold" sources. Let us consider the one-to-one correspondence of "hot" and "cold" sources, respectively, in the case of the drinking bird. What we first come to notice is that a flow of heat emerges from the fluid vapour inside the head to the external surface of the head due to the decrease in surface temperature caused by the evaporation of water. This outgoing heat flow allows the fluid vapour to contract and thereby pull up the liquid surface of the fluid at the lower neck. Therefore, the external surface of the head is the "cold" source realized by the evaporation of water.

Then, where is the "hot" source? As already described above, the rise of liquid surface of the fluid inside the glass tube results in the fall of that outside the glass

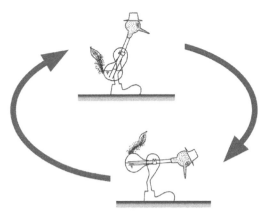

Fig. 2.15: Cyclic bowing made by the drinking bird. The upward movement of the liquid surface of the fluid inside the body due to the decrease in the internal vapour pressure for the evaporative cooling effect of liquid water spread over the external surface of the head shifts the centre of gravitation of the drinking bird. This results in the drinking bird bowing.

tube. This means that the space above the lowered surface of liquid is expanded and thereby the temperature of the fluid vapour is lowered slightly. Then, there emerges the other flow of heat from the surrounding air to the round-shaped glass body. In fact, the "hot" source is exactly the surrounding air of the drinking bird.

Bowing action of the drinking bird is provided by nothing other than the work that is produced in the two flows of "heat": one from the surrounding air into the round-shaped glass body, and the other from the inside of the hollow head filled with the fluid vapour to the external surface of the head, where the liquid water is evaporated. The drinking bird is surely a heat engine.

Is the whole of this drinking bird a closed system or an open system? Making the external surface of the head of drinking bird get wet by liquid water is to supply an amount of matter, while on the other hand, the dispersion of water vapour originating from the evaporation of liquid water is to dispose of an amount of matter. Therefore, the whole of the drinking bird is an example of open systems.

Let us next place this drinking bird, whose head is wet by liquid water, inside an aquarium filled with no water but ordinary air as shown in Fig. 2.16. Together with the drinking bird, we place a cup filled with liquid water in front of the beak of the drinking bird in order to let the drinking bird keep bowing repeatedly by getting its beak wet. We cover the top of the aquarium with a transparent plastic lid.

In Fig. 2.16, you can see a white bag on the lid, but for the first step we cover the top of the aquarium only with the transparent plastic lid and see what will happen for a period of ten minutes or so.

During the first thirty-second or one-minute period, it may look that nothing is happening except the rise and fall of the surfaces of tinted liquid dichloromethane, but sooner or later the action of bowing starts and continues. The frequency of bowing action increases gradually and it reaches a certain seemingly constant value within three to four minutes. Then, it gradually decreases and the drinking bird stops bowing within seven or eight minutes after it started bowing. Why has it stopped?

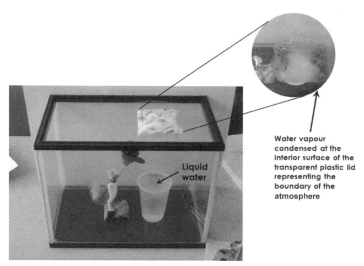

Fig. 2.16: The simplest model of global environmental system. Bowing action of a drinking bird represents a variety of atmospheric, aquatic, and biological activities on the Earth. The transparent plastic lid is for making the model a closed system. A white plastic bag on the lid is a coolant representing the Universe. The Sun may be represented by a task lamp, which is not shown in this photograph.

This is because the whole of the aquarium, in which there is the drinking bird and a glass of water, is a closed system due to the presence of the transparent plastic lid so that the immediate environmental space for the drinking bird reaches, sooner or later, the state of saturation with respect to vapour, that is, 100% of relative humidity.

No evaporation of liquid water makes no temperature difference and provides the drinking bird with no work. This terminates the action of bowing. Such a state is called static equilibrium. In other words, non-equilibrium conditions bring about actions of bowing. If we remove the plastic lid and wait for a while, then the bowing action of the drinking bird starts again. This is because the relative humidity of the air surrounding the head of the drinking bird decreases again since the space inside the aquarium is ventilated with an amount of air outside.

Let us think about a way of decreasing the relative humidity of the surrounding air space of the drinking bird without removing the plastic lid. As was already shown in Fig. 2.16, placing a coolant, whose temperature is much lower than the dew-point temperature of the air inside the aquarium, brings the condensation of water vapour and thereby the relative humidity inside the aquarium is lowered. After a while, the drinking bird starts bowing again and the rate of bowing gradually increases and soon becomes stable. Then, as can be seen at the supplementary picture shown in Fig. 2.16, we find that an amount of water condensed at the interior surface of the transparent plastic lid. More on the moisture will be discussed in Chapter 8.

The condensed water originates either from the wetted drinking bird head or from the liquid water in the cup. This implies that cooling a portion of air inside the aquarium makes a room for an amount of liquid water that exists over the drinking-bird head to transpire. We thus confirm that it is essential to sustain the state of non-equilibrium to keep producing an amount of work.

As you may have already noticed looking at Fig. 2.16, if the position of the coolant is just above the cup, then the condensed water may fall down into the cup. This completes one cycle of water circulation. The droplets of liquid water falling down from the lower surface of the transparent plastic lid may be regarded as the rain in the model global environmental system. How raindrops fall in real atmosphere will be later discussed in Chapter 9.

If, in addition, a desk-top lamp is placed near the aquarium to illuminate and heat the drinking bird and let the liquid water evaporate easier than the case explained above, then the rate of bowing of the drinking bird increases. This confirms that a larger temperature difference makes more work available.

A desk-top lamp represents the Sun, the whole of the aquarium with the transparent lid the Earth, the drinking bird the whole of biological activities on the Earth, the liquid water in a cup the Sea, and the coolant the Universe. This is the simplest model of the global environmental system (Shukuya 2003, 2013).

The experiment using the drinking bird described above together with the thought experiment described in 2.5.2 is to help us hold a clearer and holistic image of the global environmental system functioning as the chain of a variety of heat engine. The global environmental system having the nested structure shown in Fig. 1.4 and Fig. 2.8c sustains the whole of its function as a typical dynamic equilibrium, under which all of the open systems form their visible structure (*Katachi*) and perform their respective function (*Kata*). We humans and the built environmental space are no exception.

References

Banham R. 1984. The architecture of the well-tempered environment. 2nd ed. (1st ed. in 1969). The University of Chicago Press.

Ikeda T. 1998. Standing on the mother earth: for the children and the architects two-hundred years later. Bio-City (in Japanese).

Kimura K. 1993. Thermal environment in vernacular houses. pp. 1–46. *In*: Kimura K. (ed.). Buidling Physics II. Maruzen (in Japanese).

Shukuya M. 2003. The simplest model of global environmental system as a teaching material built-environmental education. Proceedings of annual meeting of Architecural Institute of Japan, pp. 669–670 (in Japanese).

Shukuya M. 2013. Exergy—theory and application in the built environment. Springer-Verlag, London.

Svensmark H. and Calder N. 2007. The Chilling Stars—a cosmic view of climate change. Icon Books Ltd. UK.

Chapter 3

Basics of Human Biology

3.1 Emergence of human body in the course of phylogeny

As mentioned earlier in Chapter 1, the human body consists of about sixty trillion cells, each of which works as an open system, as was discussed in Chapter 2. But, in fact, such cells of human body are invisible to our own naked eyes. What we can see are such collective body systems that consist of respective huge number of cells: the head with hair, eyes, ears, nose and mouth, the trunk, the arms and the legs.

The name "cell" was first conceived by Hooke, a British scientist, who watched a variety of things including the bark of cork trees, which was tiny room-space like appearance with his then using microscope (1674), the magnification of which was two to three hundred times, whereas contemporary light microscopes allow up to 2000 times and electron microscopes up to 10 million ($= 10^7$) times. Although Hooke gave the name of "cell" to what he watched, this does not necessarily imply that he immediately recognized the concept of "biological cell", which we learn in biology textbooks today. What he watched in the bark of cork trees were, in fact, not the living cells but the membranes of dead cells. Therefore, he does not seem to have noticed that what he called cell was the fundamental unit of living creatures. Following Hooke's microscopic observation, one of the scientists who is famous for having watched a variety of bacteria, red blood cells, and sperms, all in the state of being alive, was van Leeuwenhoek, a Dutch scientist, later in 1674. He also did not necessarily recognize that the living things are all made up of cells.

In 1838, more than 160 years later since the days of Hooke and van Leeuwenhoek, Schleiden (1838), who watched a number of living cells within a variety of plant bodies, came to have a thought that any plant body may be made of cells. Schleiden's thought influenced Schwan, who was watching a variety of animal bodies from a microscopic viewpoint, and Schwan (1839) came to have a similar thought with respect to animal bodies that may also have been made of cells.

These thoughts were in fact the beginning of recognition that "cells" are the fundamental units of living creatures in contemporary implication (Miyaji 1999). Twenty years after the thoughts of Schwan and Schleiden about what the living things are made of, Virchow (1859) first stated that any cell of living creatures came from a cell of living creatures. This implied that every cell is born from a previous cell except the very first one, with which the evolution must have started approximately four billion years ago. The human-body cells are no exception in this evolutional course of living creatures. Ten years after Virchow, Miescher (1869) found the nuclei of cells consisting of phosphate-rich chemical compounds. This was followed by a number of experimental research work for identifying fundamental compounds such as adenine and others, and finally in 1953, the structure of DNA (Deoxyribonucleic acid) was successfully modelled by Wilkins, Franklin, Crick, and Watson.

Each of our life starts as one single fertilized egg cell inside the mother's uterus and is grown up to the state of adult human body as you are now as a collective system consisting of about sixty-trillion cells, as shown in Fig. 3.1, and then continues growing, maybe not in the sense of physically but mentally until the end of his or her life. In order to have a holistic view of such human body system, let us take a very brief look at the evolutionary process from aggregate system consisting of eukaryotic cells. This is because the evolution of a variety of living creatures over the last four billion years, the phylogeny, is made up of a morphogenetic series of individual living organisms, ontogeny, and, although each of ontogenetic process does not necessarily delineate the exact phylogenetic process, it represents the phylogeny very well (Miki 1992, 1997, Dan 1987, 1997).

To begin with, let us first make a rough estimate of the average size of a hypothetical human-body cell. Suppose that there is a person, whose body weighs 73 kg; in fact, this is my body weight. It is equivalent to 64 to 73×10 m^3, assuming that the density is 900 to 1000 kg/m^3. Since the human body can be regarded as consisting of about sixty trillion cells, let us assume that each of them forms identical cube shape and they occupy all together the volume of 64 to 73×10^{-3} m^3. This results in the side length of a tiny cube being in the order of 10 μm (= 10^{-5} m).

The size of prokaryotes, single-celled creatures, which are considered to be the smallest among a variety of living creatures and probably the ones having been created by the living nature on Earth about four billion years ago, is in the order of 1 to 5 μm. The average size of human-body cells estimated above is, therefore, much

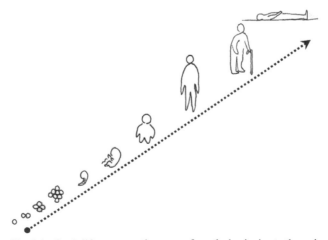

Fig. 3.1: One's life, ontogenetic process from the beginning to the end.

larger than that of prokaryotic cells. The reason of human-body cells being much larger is that human-body cells are much more complex because of being equipped with a variety of internal organelles and hence are larger than the prokaryotic cells having almost no internal organelles.

There are also single-celled creatures called eukaryotes, which are more complex for having various internal organelles. Their size is from 10 to 100 μm, larger than prokaryotes. The size of human-body cells are, in general, equivalent to the small-sized eukaryotes as far as their size is concerned. The human-body cells are called eukaryotic since they have various internal organelles as eukaryotes do. The difference between eukaryotic and prokaryotic cells is just like a difference in size between a small hut equipped with all the fundamental needs such as a kitchen to a toilet and a bed, and a contemporary complex building, which consists of offices, shops, restaurants, dwellings and others. The human being consisting of about six trillion eukaryotic cells as a collective system are intrinsically complex. If each of the human-body cell corresponds to either a residential building or an office building, then the whole of human body may correspond to a whole region or a nation.

While the life of prokaryotes seems to have started about four billion years ago, that of the eukaryotes started about two billion years ago. It took almost two billion years from prokaryotes to eukaryotes, in other words, from being simple to becoming complex, while their essential characteristic as single-celled creatures remained unchanged. Figure 3.2 shows the gradual emergence of living creatures from prokaryotes via eukaryotes to a variety of fauna and flora, from marine plants and animals to land plants and animals together with the variation of atmospheric oxygen and carbon dioxide concentrations.

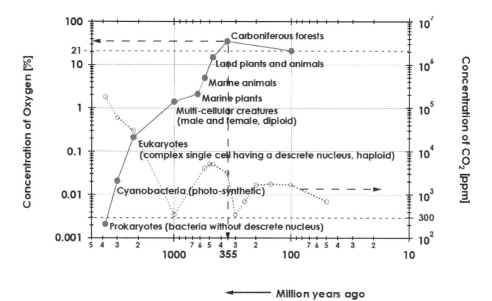

Fig. 3.2: Development of living creatures over the course of four billion years. This graph was made referring to the data given in Graedel and Crutzen (1995) and Lloyd (2012).

When the first prokaryotes emerged under the sea almost four billion years ago, there was no atmospheric oxygen above the sea surface, probably in the order of 0.002% (Lloyd 2012), so that there was no ozone layer and hence the surrounding space close to the sea surface must have been filled with a lot of ionizing radiation such as X rays, ultra-violet radiation and cosmic rays coming from the Sun and other numerous remote stars in the Universe. This implies that the open space above the sea surface was uninhabitable for any living creature.

In the course of evolution from prokaryotes to eukaryotes, a very important epoch emerged in between, that is, the emergence of cyanobacteria, which perform photo-synthesis that has been inherited by a variety of contemporary plants, and also the emergence of aerobic bacteria, a descendant of which is considered to be mitochondrion, one of the internal organelles, vitally important to both plant- and animal-body cells including human-body cells.

By the time that cyanobacteria flourished, because of their excretion of oxygen molecules into the atmosphere, the marine and then atmospheric concentration of oxygen had become much higher, 0.02%, ten times higher than the very beginning of the global atmosphere, but still very low in comparison with the present atmospheric concentration of oxygen, 21%. Oxygen molecules are chemically very reactive in general, and therefore toxic, so that the larger cells are advantageous in sustaining their life because of the surface to volume ratio being smaller. This implies that the cells were secured by minimizing the penetration of oxygen molecules into the membrane surfaces for the avoidance of oxygenic toxicity, but its drawback was that it became hard for those cells to absorb food from the environmental space and also discard waste matters into it. One behaviour of those cells for a kind of trade off was that they became larger and allowed aerobic bacteria to merge into them and to become one of the organelles, mitochondrion. Such unique behaviour in living creatures is considered as a very basic process of "symbiosis".

In the transitional course of environmental conditions from the oxygen concentration being low to high, larger cells had gradually become dominant. Some of them started to live collectively as the next stage and thereby formed multi-cellular creatures. After all, living creatures grew from simple single-celled, via complex single-celled with internal organelles including discrete nuclei, further to complex multi-cellular creatures consisting of a new type of cells, diploid.

The reason why diploid cells emerged is that the amount of genetic information necessary for life had become huge (Dan 1997, 2008). That was about one billion years ago. The essential characteristic of diploid cells distinct from haploid is that each diploid cell has two sets of genetic information inherited from both parents, though each haploid has one set. This means that diploid is more complex than haploid and also living creatures having diploid cells had become capable of diversification.

The development of diploid cells implies the emergence of sex: that is, female and male. The human body cells are all diploid except ovum and sperm. The metamorphosis from haploid to diploid diversified the possible types of plants and animals leading to the emergence of various marine plants and animals and then the development of land plants and animals followed until the carboniferous forests covered probably almost all the continents of the then Earth, almost 355 million years ago.

Due to the abundance of various plants then alive and their excretion of oxygen as a waste matter, the concentration of oxygen had once reached 35%, much higher than the present value of concentration, 21%, to which it reached about one hundred million years ago as can be seen in Fig. 3.2. Before the carboniferous forests flourished, the carbon-dioxide concentration was much higher, at the level of 3500 to 4000 ppm, than the present level of 350 to 400 ppm. The decrease in atmospheric carbon-dioxide concentration must have been caused by the flourishing of carboniferous plants which absorbed carbon dioxide gas as their primary material for photosynthesis.

Figure 3.3 shows a rough sketch of the evolutional course of animals from a single-celled to complex multi-cellular creatures, one of which is exactly us, human beings. Shown together are the trends in global temperature, which indicates the level of global temperature as high, low or medium, over a period from 500 million years ago to five hundred years ago; this was made referring to a variety of works such as Svensmark (2012) and Lloyd (2012). The medium value of global temperature is around 15°C, which corresponds to the present value. The high value is considered to be from 21 to 25°C and the low value to be 8 to 10°C.

It can be seen that bipeds and early humans emerged while the average global temperature was at a low level. *Homo sapiens* seems to have emerged under the sluggish fluctuation of global temperature between low and medium.

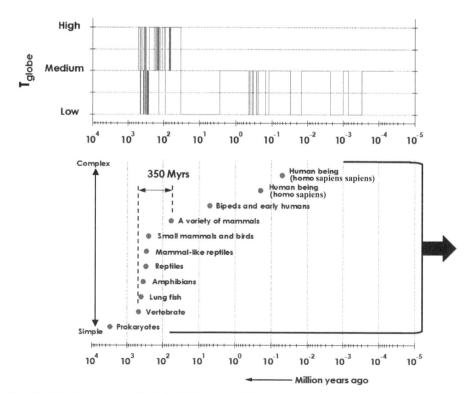

Fig. 3.3: Development of animals including human beings, from simple to complex, together with the trends of global warming and cooling for the period from 5×10^2 to 5×10^{-4} million years ago.

Until a variety of mammals emerged about 60 million years ago, there was quite a huge variation in average global temperature between low and high over the period of more than 350 million years, from 500 to 60 million years ago as shown in Fig. 3.4, which is a magnification of the corresponding portion of Fig. 3.3. During this period, it is now known that there were four large changes from warm, so-called "hothouse" condition to cold, so-called "icehouse" condition (Shaviv and Veizer 2003). The fluctuation of average global temperature must have influenced the extinction of old plants and animals and also the emergence of new plants and animals; in the case of animals in particular, the daily and seasonal fluctuations of global temperature is considered to have caused the development of a thermo-regulatory system within some groups of animals, which are called homeotherms, in contrast to the rest of animals, poikilotherms, whose body temperature swings more or less in accordance with their surrounding temperature.

Figure 3.5 shows a rough sketch of the phylogenetic course, in which animal bodies have developed their thermoregulatory system embedded within their body. In the very beginning, the body temperature of simple living creatures was the same as environmental temperature, since their body was so small that its temperature became easily equal to environmental temperature. The environmental temperature of amphibians and reptiles had become more sharply fluctuated than that of fish because the heat capacity of atmospheric air is small compared to that of sea water, only one four-thousandth for one cubic-metre of their volumes (See Table 6.1 in Chapter 6).

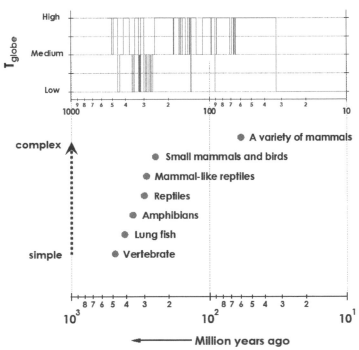

Fig. 3.4: Development of animals from simple to complex together with the trends of global warming and cooling over the period of 350 million years (from 500 to 60 million years ago).

The body-core temperature of homeotherms is known to fluctuate slightly with the one-day cycle; this is probably a trace of the poikilothermic body-core temperature fluctuation together with the environmental temperature fluctuation during ancient times. The same is true in the body-core temperature of present human beings as will be shown later in Fig. 3.11. Homeothermic animals other than human beings are exposed to harsher variation of environmental temperature than us humans. One significant difference of human beings from other homeothermic animals is that we humans have become capable of controlling indoor space temperature and humidity at any constant value, if specified, by applying active-technology measures described in Chapter 2, which necessitate a lot of fossil-fuel use. Constant indoor temperature and humidity are extrinsic and could cause trouble in our body-core temperature, since the thermoregulatory system inherent in our body controls the body-core temperature referring to outdoor temperature fluctuation.

Figure 3.6 shows the results of an experiment performed for clarifying the relationship between the activity level of human-body cells and temperature (Reece et al. 2011). The values shown on the vertical axis are the counting rate of visible light emitted from DNA molecules extracted from the examined cells incubated in petri dishes for a certain period of time with either of the environmental temperature levels from 10 to 50°C. Tritium (^{3}H), which is radioactive, was used in the agar medium so that it is considered that the higher the counting rate was, the more active was the cell division.

The highest activity level of the cells occurred at the condition of 35°C, which approximately corresponds to the actual human-body core temperature, 37°C. This indicates that the human-body thermoregulatory system has emerged through the course of phylogeny so that it keeps thermal homeostasis within the target value of body-core temperature at 37°C.

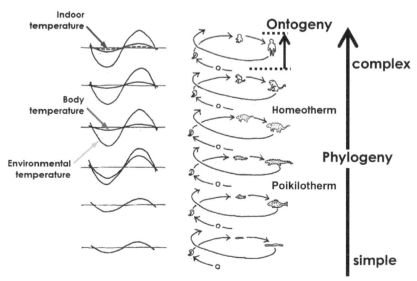

Fig. 3.5: Development of thermoregulatory system embedded within the brain of homeothermic animals.

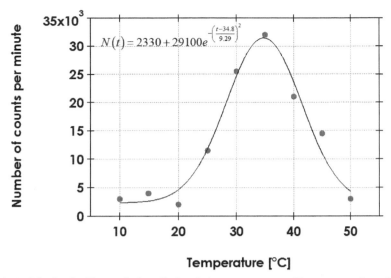

Fig. 3.6: Activity level of human body cells in relation to temperature. The count number of visible light emitted from DNA molecules soaked in scintillation fluid indicates how much active the examined human-body cells are. The data used here are quoted from Reece et al. (2011).

3.2 Development of human nervous system

The purpose of designing passive and active systems for built-environmental conditioning is to avoid human discomfort and enhance the well-being to be in harmony with what the living nature has realized, as introduced in the previous section. For the pursuit of such systems, it is necessary to know how the whole of human body, the body proper and brain, works in response to the ever-changing given environment. The human discomfort or comfort emerges within a cyclic process from sensation to behaviour, which is considered to be a basal part of adaptation that has allowed a variety of plants and animals including human beings to emerge in the course of phylogeny described in the previous section.

Here in this section, let us outline the essence of the human-body cyclic process keeping its associated morphological and thermo-physical characteristics in mind. Since the whole of human sensation, perception, cognition and behaviour is the function of nervous system, our overview is going to be in particular the general characteristics of human nervous system referring to a variety of sources (Miki 1992, Matsumoto 1992, Damasio 2000, 2012).

3.2.1 Cyclic process from sensation via perception to behaviour

As can be seen in the diagram shown in Fig. 3.7, a human being is always exposed to his or her physical environment, whose condition always varies from time to time. Among a variety of environmental factors affecting the human sensations, if the change in some of them, e.g., light and heat, is large enough to sense, or if it is beyond a certain threshold of sensation, then he or she senses it, perceives, and then

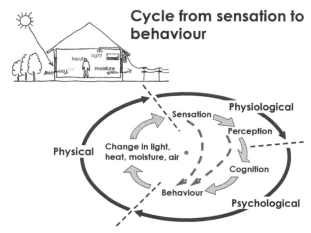

Fig. 3.7: Cyclic process from sensations via perception and cognition to behaviour.

becomes conscious. Perception and consciousness are, for example, whether it is bright or dark, warm or cold, cool or hot, and so on.

At the final stage of one cycle of the process, a certain specific behaviour upon necessity emerges in order to change the state of physical environment, which hopefully renews the sensation; for example, opening or closing a window, switching on or off a mechanical cooling unit, switching on or off light bulbs and others. These behaviours are all done by the function of skeletal muscles.

In some cases, such as while we are asleep, that is, while we are not conscious, if we sense heat because of the surrounding temperature being a little too high, then we would remove our blankets covering our bodies unconsciously and then a bit later, we may come to know the condition, that is, we may become conscious about ourselves having been feeling a little cold. We all know through our own experiences that such a phenomenon exists. Reflex is also a typical unconscious behaviour.

In such a series of manners as described above, physical, physiological, psychological, and behavioural phenomena as chain reactions take place all together through our body proper and brain relentlessly as long as we live.

3.2.2 Formation of central and peripheral nervous systems

The nervous system that performs the aforementioned cyclic process emerges in the developmental course of a single human body, that is, in the ontogenetic course of one's body including the brain.

As already shown in Fig. 3.1, a human body starts growing from the state of one single fertilized egg to the state of an adult having approximately sixty-trillion cells, and sustain the adult state, though aging inevitably, until the end of his or her life. In the early stage of development, starting from fertilization, the embryo in mother's uterus keeps changing the form unceasingly and turns into a collective multi-cellular system consisting of three portions: endoderm, mesoderm, and ectoderm (Wolpert 1999). Their relative positions to each other are as shown in the upper left drawing of Fig. 3.8, though actual shapes are much more complicated. As already mentioned, the development of human body looks as if it reflects the evolutionary process from a single-celled simple living creature emerged about four billion years ago to the *homo-sapiens* as one of multi-cellular complex living creatures as mentioned in the previous section.

The adult human body can be regarded as consisting of two major portions: the internal organs and the whole of skeleton and muscular system enveloped by skin layer, within which various sensory portals of nervous system are embeded. The central parts of internal organs such as bronchus, lungs, esophegus, stomach, and intestines, all of which are for the inflow of food and the outflow of faeces, come from the endoderm. Some other parts of internal organs such as the whole of blood vessels, heart, kidneys, bladder, ovary and testicle come from the mesoderm. They are mainly for the circulation of matter in order to deliver nutrients on the one hand, while on the other hand to dispose of waste matter. The whole of skeleton and skeletal muscles are also developed from the mesoderm. From the ectoderm, skin layer together with the sensory portals of the nervous system, epidermis and dermis, emerges.

The unique metamorphic formation of endodermic cells is to mould the inlet and outlet: that is, mouth and anus. Along with these two openings being formed, the ectoderm metamorphoses eventually into a structure called neural tube, which grows into the whole of nervous system with the skin tissues as schematically shown in Fig. 3.9. The anterior of the neural tube expands and thereby becomes brain and cranial nerve. Brain consists originally of three parts: forebrain, midbrain, and hindbrain. The rest of the neural tube, the posterior side, becomes spinal cord and spinal nerve.

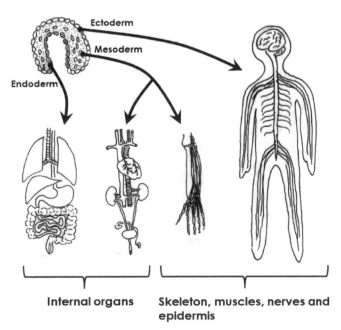

Fig. 3.8: Human-body portions developed from the early stage of embryo consisting of three major portions: endoderm, mesoderm, and ectoderm.

Human nervous system

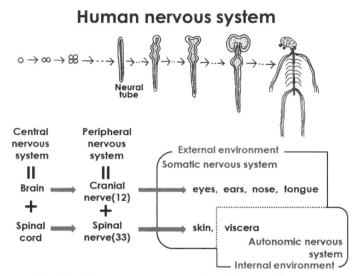

Fig. 3.9: Human nervous system and its relation to environment.

Cranial nerve as twelve pairs of neurons stretches out from the brain to sensory portals such as eyes, ears, nose, and tongue, whose respective information is visual, auditory, olfactory and gustatory. From the spinal cord, there are thirty-three pairs of neurons, spinal nerve, stretching out all over to the skin from the top of our head to the soles of our feet. They are responsive to fine touch, pain, pressure, warmth, coldness and kinesthesia, that is, somatosensory information.

Figure 3.10 depicts some of typical sensory portals in the skin layer, epidermis and dermis (Matthews 2001). Most of the portals are located below the epidermis but those associated with pain and cold are inside the epidermis. A group of the sensory portals called mechanoreceptors are associated with fine touch and pressure. Their axons are covered with myelin sheaths. Those axons with sensory portals associated with pain, nociceptors, are either covered with smaller myelin sheaths or without sheath. The same is true for thermoreceptors. Myelin sheaths are known to make the axons transfer electric signals faster than the axons without myelin sheaths.

The axon with "cold" receptors are covered with smaller myelin sheaths and that with "warm" receptors are not covered with any sheath. The depth and density of the "cold" and "warm" receptors are very different. The "cold" receptors are located 0.1 mm deep from the skin surface and their density is 6 to 23 receptors/cm^2, while on the other hand, the "warm" receptors are 0.5 to 0.7 mm deep from the skin surface and their density is much lower at 0–3 receptors/cm^2. Such difference is probably owing to the fact that major human evolution took place in cold era as described in 3.1. Therefore, the human body reacts differently against two types of thermal environmental conditions. This is considered to be the basis of different thermal sensations and resulting thermal perceptions such as "coldness" and "warmth".

In addition to these kinds of information so far described-fine touch; pressure; pain; warmth and coldness, respectively-there is a collection of information from the sensory portals embedded within the dermis and skeletal muscles to let the

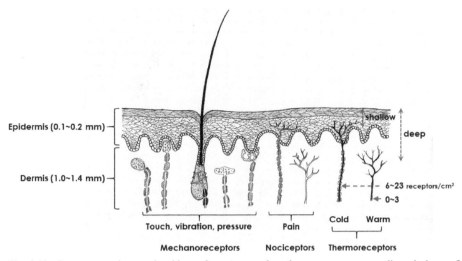

Fig. 3.10: Sensory portals near the skin surface. Axons of mechanoreceptors are myelinated, those of nociceptors and thermoreceptors are either covered with smaller myelinated sheath or with no sheath (Matthews 2001).

skeletal muscles make a variety of coordinated movement which enables the human body to perform various expression as speaking, writing, walking and so on, that is, kinesthesia. Similar to this collective motion, but as opposite bodily expression, is the balance or equilibrium, which is for the human body to take a certain posture such as standing still for a period of time. For the equilibrium, a collection of sensory information received by ears in addition to that received by skin layer is responsible.

The whole of brain and spinal cord is called "central" nervous system and forty-five pairs, the sum of twelve and thirty-three, of neurons stretching out from the central nervous system is called "peripheral" nervous system. In addition to most of the neurons of peripheral nervous system stretching out beneath the epidermis in order to detect the external environmental information, some of the peripheral nerves branch onto the internal organs in order to detect and control the functions of heart, stomach, small intestine, and others. The former is called "somatic" nervous system and the latter "visceral" nervous system. They may also be called voluntary nervous system and involuntary or autonomic nervous system, respectively.

Table 3.1 summarizes the nervous system in their association with the location of sensory portals and a variety of sensory information. So-called "five senses" are in fact to be regarded as "thirteen senses", since the sensory information can be classified into thirteen categories as listed in Table 3.1 (Nakamura 1979).

Looking again at Fig. 3.9, we may regard that the nervous system, whose centre is brain, is surrounded by two sub-environmental spaces: one is the external environment, mostly the built environment, since the human beings spend more than 90% of their life time there, and the other the internal environment, which is the whole of internal organs including the whole of skin layer, that is body proper. The interaction between the brain, the body proper, and the built environment through the whole of nervous system is, therefore, of primal importance.

Table 3.1: Nerves, sensory portals and information.

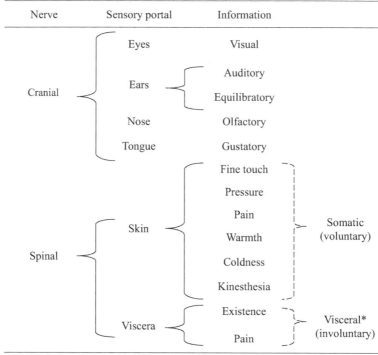

Nerve	Sensory portal	Information

* Visceral is also called autonomic.

3.2.3 *Nested structure of nervous system*

From the viewpoint of comparative morphology and physiology, the brain may be regarded as a kind of nested structure consisting of three sub-structures as shown in Fig. 3.11.

The first sub-structure is the deepest portion of the brain, brain stem, which consists of medulla oblongata and pons, that is, located at the upper edge of spinal cord, and cerebellum; this portion is equivalent to the brain developed in the course of evolution from fish via amphibians to reptiles, in which the primary constituent of the environmental space changed from water to air and thereby animals developed their whole-body system so that they control not only the quantity but also the quality of both blood and lymph. The medulla oblongata and pons are responsible for breathing, heartbeat, water and food ingestion, digestion, blood-pressure control, coughing, sneezing, swallowing and vomiting. The cerebellum is the centre for the smooth well-coordinated body movement in swimming, bicycling and so on, and also for the muscular movement in the throat and face in relation to speaking. These portions are considered to be very old since they correspond to the nerve types which emerged in the course of evolution from vertebrates, amphibians to reptiles.

The second sub-structure consists of thalamus, hypothalamus, fornix, amygdala, hippocampus, and others. This is equivalent to the brain developed in the evolutional course from reptiles to lower mammals. It is called all together limbic system. The

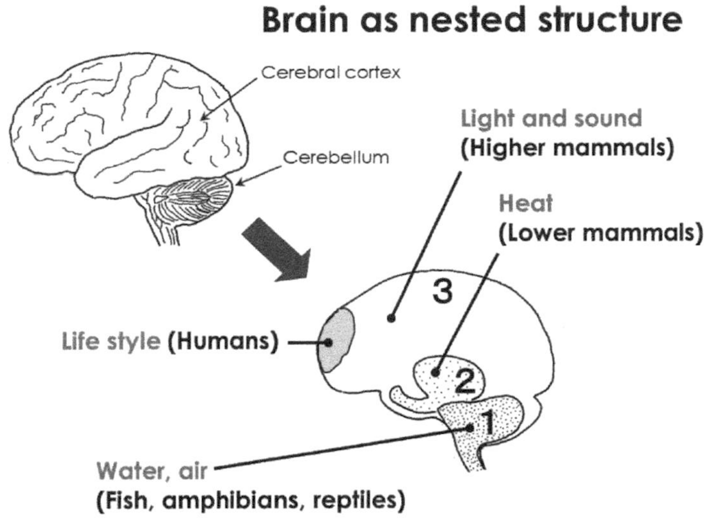

Brain as nested structure

Fig. 3.11: Brain as nested structure and its relation to environmental factors and lifestyle.

The thermoregulatory system is embedded within the hypothalamus, whose purpose is to keep the thermal homeostasis making use of the sensory information of warmth and coldness; this must have emerged with the development of small mammals and birds almost 250 million years ago, up until which there seems to have been quite a large fluctuation of environmental temperature as shown in Fig. 3.4. The emergence of thermoregulatory system may be considered to be the origin of space heating and cooling systems.

The third sub-structure of the brain, cerebral cortex, exists as if it laps up the first and the second sub-structures. All pieces of external information, somatosensory, gustatory, olfactory, auditory, vestibular, and visual, are supplied to the respective associated portions of the third sub-structure, via the second sub-structure such as hypothalamus, amygdala and others. While the external information are sent to the brain, the visceral information is also sent to the brain as well.

The whole of cerebral cortex corresponds to the brain developed at the evolutional stage of higher mammals and the prefrontal cortex in particular, which is located in the very front of the cerebral cortex, is the most highly developed part in the human brain. Rational consciousness and also human behaviour are very much associated with the function of the prefrontal cortex. Nevertheless, it is important to keep in mind that the consciousness is built up on the basis of emotion, which emerges as the function of the first and the second sub-structures. Should we re-examine the "lifestyle", then it must become important to recognize it as the function of, first, emotional brain on the basis of body proper and then conscious brain (Damasio 2000, 2012).

Note that we are unconscious to the function of the first and the second sub-structures, while on the other hand, we are partly subconscious and partly conscious to that of the third sub-structure; we all know these features through our own experiences of sleeping, dreaming, being awake but not alert, being fully alert and so on.

3.3 Evaluation of discomfort and comfort

In addition to looking at the whole of nervous system focusing on its structural aspect as mentioned in the previous section, we may also look at it focussing on its functional aspect, which is the process from emotion to consciousness in order to keep the dynamic equilibrium, that is, homeostasis.

The brain evaluates the information given from the peripheral nervous system and thereby responds to generate the outgoing information attached with a certain value, that is, whether or not it is advantageous for life: advantageous are positive signs represented by pleasure, reward, or comfort; and non-advantageous, negative signs are represented by pain, punishment, or discomfort.

The outgoing information is expressed as the function of internal organs made with the movement of involuntary muscles, a change of facial expression by the movement of facial skeletal muscles, the voice made by the movement of facial and throat muscles, or the whole body actions including written words made by the movement of fingers, hands, and arms.

Whether the environmental information given is attached with negative or positive signal mentioned above is first evaluated very quickly mostly by the second sub-structure and then followed by the third sub-structure (Matsumoto 1992).

There is a kind of feedback-loop evaluation in the brain as shown in Fig. 3.12. New pieces of information obtained from the external and internal environments are always compared with the old pieces of information piled up as memory within the brain so that one's life can be sustained as long as possible.

The evaluation is made in two steps: the first is by the second sub-structure, whose characteristic of function is fast but not so precise, and the second by the third sub-structure, whose characteristic is slow but precise (Matsumoto 1992). In these two courses of evaluation, the conscious mind on the basis of emotion is built up (Damasio 2000, 2012).

Cycle of evaluation, learning, and memory

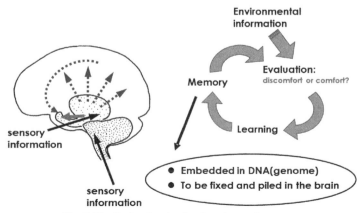

Fig. 3.12: Cycle of evaluation, learning, and memory.

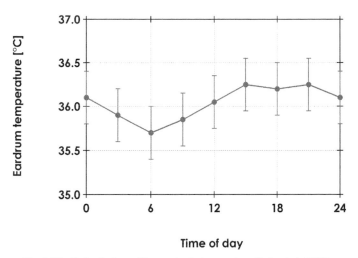

Fig. 3.13: Daily rhythm of human-body temperature (Saito et al. 2000).

The old pieces of information recorded in the body proper and brain are categorized into two kinds: one is the genetic information, the human genome, stored in the molecules of DNA, the media for the record of evolutionary history, each of which is in the core of every human-body and brain cells, and the other is the information obtained and stored as one's own particular neural-network patterns inside his or her brain, piled up since the birth to the present. The former is inherent and the latter acquired. Both are recorded in the course of long-running cyclic process of evaluating, learning, and memorizing throughout the phylogeny and one's ontogeny. The diagram shown in Fig. 3.12 schematically represents this process.

There are some pieces of environmental information inherited from the far past, of which we are not conscious. One example is the slight diurnal variation of the human-body core temperature, which usually becomes the lowest in early morning and the highest in late afternoon; such a tendency can be confirmed by measuring human-body temperature. Figure 3.13 shows, as an example, that there is a daily rhythm in human-eardrum temperature. It indicates that there is a fluctuation of eardrum temperature whose maximum occurs late in the afternoon and is about 0.5°C higher than the minimum, which occurs early in the morning. This may be regarded as one of the phylogenetic memories embedded in the whole of human body, which inherited the experience in the era of poikilotherms, cold-blooded organisms. The thermoregulatory system that we have as a part of our body system is considered to have developed in the evolutionary process from poikilotherms to homeotherms. Such a memory effect must apply also to other circadian rhythms inherent in our body system as will be described in Chapter 4.

3.4 Thermal homeostasis

The hypothalamus, a portion of the second sub-structure of our brain, is the centre of thermoregulatory system for our internal environment, whose temperature has to be 37°C, at which most of the human-body cells work properly as discussed referring to Fig. 3.6. Whenever the thermal environmental condition is likely to increase or decrease the body-core temperature more or less from the set-point reference value, a variety of reactions take place, from physiological to behavioural as summarised in Fig. 3.14.

Under a thermal environmental condition, which could sooner or later cause a slight decrease in body temperature, such behaviours as closed posture, putting on a jacket or sweater, or moving to a warmer place are taken. The passive and active heating system solutions for built environment are positioned somewhere extrapolated, since they are also a kind of human behaviour in general.

The same applies to a thermal environmental condition which could cause a slight increase in body temperature. Those behaviours as open postures, fanning with a portable fan, taking off a jacket or a sweater occur for restoring thermal comfort and the passive and active cooling system solutions are considered as the kind of behaviours extended. In contrast to these types of behaviour, all of which cause changes outside our body and to which we are more or less conscious, there are always various unconscious physiological behaviours functioning. For example, vasoconstriction and vasodilation are taking place tirelessly depending on the variation of thermal environmental condition. If cold conditions become severe, the human brain commands the human-body muscles, through the peripheral nervous system, to shiver for thermogenesis, that is, to keep the body-core temperature at 37°C. On the other hand, if the hot conditions become severe, human brain commands itself to turn off the function of brain to be conscious, that is, to be faint so that the metabolic heat emission rate can be minimum.

Physiological reactions to thermal environmental changes—vasoconstriction, vasodilation, non-shivering thermogenesis, shivering thermogenesis, or sweat secretion—start unconsciously. We may come to know the results of vasoconstriction with the feeling of coldness at your fingers afterwards, but the opposite does not

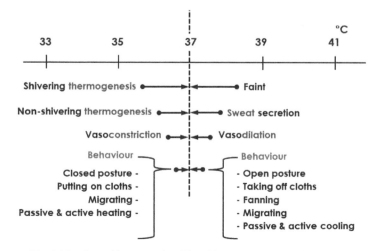

Fig. 3.14: Thermal homeostasis achieved by a variety of adaptive process.

happen. We are not able to command some blood vessels to shrink with conscious mind. Unconscious physiological behaviour is of vital importance since it influences, more or less, the state of body proper and right thereafter, the emotion to be built up. The consciousness, which we can express by words in association with these physiological reactions, emerges later. For example, an expression with words such as "I am feeling cold at my hands" is brought by our conscious mind and it appears followed by the change of body state that is required to make the capillary blood vessels embedded inside the fingers of both hands contract and hence to keep the body core temperature at 37°C. We, thereafter, become conscious of the temperature decrease in our hands.

"Cold" or "hot" that we express in words is a typical example of conscious behaviour, but there is also a subconscious behaviour such as taking off the blanket due to the sensation of hot while sleeping, or having a closed body-posture due to the sensation of cold while sleeping. Having a closed or open body-posture as behaviour also appears sub-consciously. We can easily find other persons looking hot or cold by seeing their appearances. These behavioural patterns are the results of the comparison of thermo-physiological information given at present with that recalled from the memory of thermal history embedded within the second sub-structure of the brain, the hypothalamus.

3.5 Water flow and circulation

Homeostasis is the state of equilibrium achieved by continuous struggle of human body, the brain and body proper, so as to control the internal environmental condition, namely, the temperature, pressure and chemical constituent, at respective narrow ranges of values. This is not static but dynamic equilibrium, which is realised by continuous flow of energy and circulation of matter within the human body.

The homeostasis is realized by feeding on water and nutrients while at the same time excreting urine and faeces almost regularly. Figure 3.15 demonstrates the daily rate of water that goes in and out in an average adult human body for a typical one-day period. Two kg water per day is absorbed, while at the same time, a half of it is excreted as urine, and from the other half, a half disperses by exhalation, and the other half by sweat secretion.

Water is considered to occupy approximately 70% of body mass. If the body mass of a person is 70 kg, then there is 49 kg of water within the body. It implies that the rate of water change inside the human body is about 1.2 times per month. With this value in mind, we come to realize why the availability of clean water is essential.

Blood circulating through the whole of human body is approximately one-ninth of water existing inside the human body. Figure 3.16 demonstrates the relative circulating rates of blood inside the human body. An amount of blood, which is assumed to be at the relative flow rate of 100%, is sucked into the right-hand side of heart and then pumped out towards lungs. At the lungs, the blood absorbs oxygen and desorbs carbon dioxide. The blood, which is rich in oxygen, comes back again to the left-hand side of heart and is pumped out towards various parts of the body: 25% to stomach and intestines, 20% to kidneys, 15% to brain, 5% to heart itself, 10% to other internal organs, 20% to skeletal muscles, and 5% to skin layer.

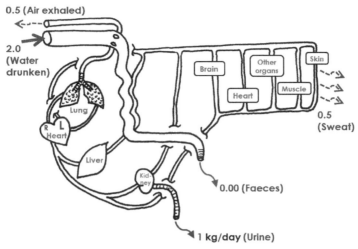

Fig. 3.15: Inflow and outflow of water through a human body.

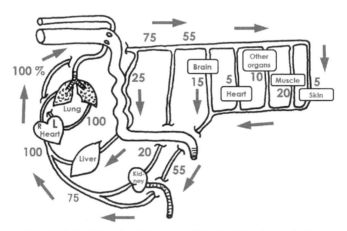

Fig. 3.16: Relative circulating rates of blood within a human body.

Assuming that one beat of heart outflows approximately 70 mL of blood and the number of beat is 65 times per minute, the blood flow rate through the heart is about 4.6 L/min. The heart as a circulating pump changes the amount of blood existing within the whole of internal space of blood vessels distributed all over the body at the rate of 0.82 times/min. It means that any substance dissolved in blood can reach the far end of our body within a minute. This is why we get easily drunk by drinking strong alcoholic beverage within a couple of minutes and also an urgent treatment is necessary if bitten by a poisonous snake.

References

Damasio A. 2000. The Feeling of What Happens—Body, Emotion and the Making of Consciousness. Vintage books London.
Damasio A. 2012. Self Comes to Mind. Vintage books New York.

Dan M. 1987. Phylogeny and ontogeny of animals. Tokyo-University Press (in Japanese). Tokyo.

Dan M. 1997. How have living creatures become complex? Iwanami-Shoten Publisher (in Japanese). Tokyo.

Dan M. 2008. Self-organizing power of cells. NHK books (in Japanese). Tokyo.

Graedel T. E. and Crutzen P. J. 1995. Atmosphere, Climate, and Change, Scientific American Library. p.63, p.66 and p.92.

Hooke R. 1674. Micrographia-Some Physiological Descriptions of Minute Bodies Made by Magnifying Glasses with Observations and Inquiries Thereupon. http://www.gutenberg.org/files/15491/15491-h/15491-h.htm (2005). ISO-8859-1.

Lloyd C. 2012. What on earth happened. Bloomsbury Publishing, London.

Matsumoto G. 1992. Brain, mind, and computer, physical society of Japan. Maruzen (in Japanese). Tokyo.

Matthews G. G. 2001. Neurobiology—Molecules, Cells, and Systems. 2nd ed. Blackwell Science. New York.

Miki S. 1992. Introduction to human morphology—Fundamental life form and metamorphosis. Ubusuna-Shoin Publishers (in Japanese).

Miki S. 1997. Human body—its phylogenetic investigation. Ubusuna-Shoin Publishers (in Japanese).

Miyaji Y. 1999. Living things and cells—Development of scientific thought on the existence of cells. Kasetsu-Sha publishers (in Japanese).

Nakamura Y. 1979. Thoughts on common sense. Iwanami-Shoten Publisher (in Japanese).

Reece J. B., Urry L. A., Cain M. L., Wasserman S. A., Minorsky P. V. and Jackson R. B. 2011. Campbell—Biology. Global Edition. Pearson. p.80, pp. 219–223.

Saito M., Matsuoka H. and Shukuya M. 2000. Study on the relationship between occupants' lifestyle in summer and their psycho-physiological response. Annual Meeting of Architectural Institute of Japan, pp. 497–500.

Shaviv N. and Veizer J. 2003. Celestial driver of Phanerozoic climate? GSA Today, pp. 4–10.

Shukuya M. 2001. Built environment formed by sustainable architecture-discussion from the viewpoint of the structure and the function of a human nerve system. Proceedings of Annual Meeting of Architectural Institute of Japan, 437–438 (in Japanese).

Shukuya M. (ed.). 2010. Theory of Exergy and Environment—Revised edition, Inoue Publishers (in Japanese). Tokyo.

Shukuya M. 2013. Exergy—Theory and Applications in the Built Environment, Springer-Verlag London.

Shukuya M. 2015. An overview of the cyclic process from sensation to adaptive behavior—interaction between body proper, brain and built environment. Annual Meeting of Architectural Institute of Japan, pp. 467–470.

Svensmark H. 2012. Evidence of nearby supernovae affecting life on Earth. Monthly Notices of the Royal Astronomical Society 423(2): 1234–1253.

Chapter 4

Solar and Lunar Effects on Built Environment

4.1 The Sun and the Earth

We live from one day to another in the twenty-four-hour cycle wherever on Earth we are. Roughly speaking, we are conscious for two thirds of one day and unconscious for the rest of the day. Being unconscious is of course different from being dead and there are always a variety of physiological phenomena taking place within each of our bodies whether we are awake or asleep. We humans usually get up at a certain time, whether it is early or late in the morning and go to sleep at a certain time, whether it is early or late in the evening. This pattern appears following the movement of the Sun over the sky. This must have been known since the early stage of human history, long before the geocentric system was conceived by ancient natural philosophers and also the heliocentric system later by modern and contemporary scientists. This is because we humans are basically born to be able to sense the light coming from the Sun with our visual sensory portals embedded in our eyes.

As we have already confirmed in Chapter 1, we humans spend more than 90% of a day indoors. Everybody knows that it is not comfortable at all spending long hours inside window-less rooms by mere experience. Daylight availability indoors is thus of primal importance. For our health and well-being, indoor environment has to be designed so that daylight is available sufficiently, no less and no more.

Therefore, it is worth knowing the basic relationships between the solar path over the sky vault depending on the time of day and month.

4.1.1 Solar position determined by the rotation and revolution of the Earth

Figure 4.1 shows the position of the Sun on the imaginary sky vault at an arbitrary time of a certain date somewhere on the northern hemisphere of the Earth. The solar position is specified by two angle values: one is solar altitude and the other solar azimuth.

The solar altitude is defined to be an angle subtended by two straight lines: one drawn horizontally from the point, where you stand on the horizontal ground surface, toward the direction of the Sun and the other drawn between the Sun and the point

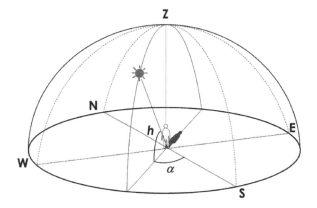

h : Solar altitude, α : Solar azimuth

Fig. 4.1: Solar position over the sky vault.

where you are. This is some value from 0 to 90°; "zero" degree is when the Sun just rises or sets and ninety degrees is when the Sun is located at the zenith.

The solar azimuth is defined to be an angle subtended by a horizontal line toward due south and that toward the direction of the Sun: it is measured to be positive in clockwise direction as you imagine looking at the horizontal surface from the zenith, that is, 90° for due west, 180° for due north, –90° for east and so on.

Anywhere on the Earth surface, the Sun looks rising from somewhere in eastern orientation and setting in somewhere in western orientation. During daytime, the Sun looks moving on the sky vault from east to west, crossing some point on the meridian, an imaginary circle passing through due north, the zenith, and due south.

Imagine that we draw solar trajectory starting from east and ending in west each day for one year on the sky vault shown in Fig. 4.1. There should be 365 different trajectories for one year and we should be able to find that those on summer days are closer to the zenith and those on winter days closer to the horizon.

With such simple consideration together with the observation of constellation with their own eyes, ancient natural philosophers must have developed their geocentric system theory. Here in our case, in order to understand the seasonal difference of solar trajectory over the sky, let us guess where the Sun locates on the sky vault at 12:00 with your watch on each of the first day of twelve months together with the knowledge of heliocentric system.

If we take the time at 12:00 to be defined when the Sun just passes the meridian, then the twelve positions of the Sun changes from one month to another, somewhere just on the meridian circle. The measurement of time according to this manner is the apparent solar time, with which the length of one day is different from one day to another throughout the year. But the time we use nowadays is not the apparent solar time, but the mean solar time as standard time, which is defined to elapse constantly anywhere on the Earth; this is Coordinated Universal Time (UTC), which is determined to be equal, anywhere, in each of the universal time zones all over the world. Consequently, the positions of the Sun at the noon of standard time are not necessarily on the meridian.

The priority of mean solar time rather than local apparent solar time may look almost self-evident, since many people travel around the world in contemporary global society and also the international communication is taking place all the time, day and night, through the internet, but the local apparent solar time remains important with respect to how the built environment should be designed for well-being of the people living there.

With the characteristic of mean solar time in mind, let us plot twelve points on Cartesian coordinates, that is, rectangular coordinates having horizontal axis for solar azimuth and vertical axis for solar altitude. Figure 4.2 exemplifies two sets of twelve points: one in Yokohama, where its standard time is +8 hours of UTC, and the other in Copenhagen, where its standard time is +1 hours of UTC. As can be seen, the lines connecting twelve plots either in Yokohama or in Copenhagen take the shape of an Arabic numeral, "eight".

This is mainly due to the fact that the orbit of the Earth is elliptic and also the rotational axis of the Earth is not perpendicular but tilted to the orbital plane as shown in Fig. 4.3. The Sun locates at one of the two centres of ellipse and the Earth rotates along the elliptic orbit taking 365.24 days, that is, one year. The position of the Earth when the distance between the Earth and the Sun is the longest is called aphelion and the shortest perihelion. The distance between the Earth and the Sun is 1.52×10^{11} m at the aphelion and 1.47×10^{11} m at the perihelion. The length between the middle point of the two elliptic centres and the perihelion or aphelion, which is called semi-major axis, is 1.495×10^{11} m; the Sun's position is 1.67% away towards the perihelion from the middle of the two elliptic centres. The eccentricity, 1.67% at present, is known to vary from 0 to 6% in the period of 100 thousand years. This rate of variation is so small that it does not matter in the discussion here in this chapter.

The dotted line between the aphelion and perihelion divides the elliptic orbital plane symmetrically but the other dotted line perpendicular to the former divides

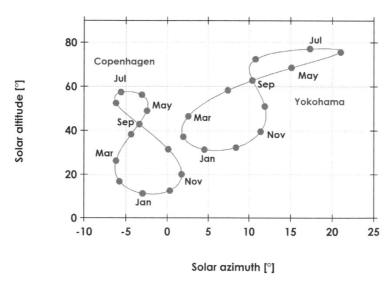

Fig. 4.2: Solar positions at the noon on every first day of twelve months according to respective standard times in Yokohama and in Copenhagen. Daylight saving time in Copenhagen is not taken into account.

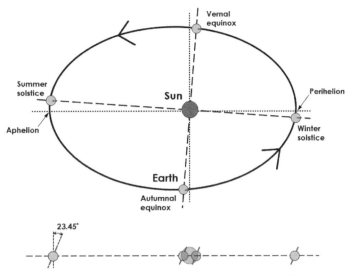

Fig. 4.3: The Earth's orbit of rotation and tilt angle of the rotational axis.

it asymmetrically; the side of aphelion is larger than that of perihelion. A dashed line, which is tilted slightly with respect to the dotted line between aphelion and perihelion, is the line that just overlaps the projection of the Earth's rotational axis onto the orbital plane. Where the Earth crosses this dashed line near the aphelion is the summer solstice, 21st or 22nd June and near the perihelion the winter solstice, 21st or 22nd December. Consequently, the Earth passes through the aphelion around 3rd of July and through the perihelion around 3rd of January. Perpendicular to the dashed line between the two solstices is the other dashed line between vernal and autumnal equinoxes, 20th or 21st March and 22nd or 23rd September, respectively. At these equinoxes, the length of day and night are exactly one half of one day, respectively.

It is known that, some 750 years ago, the winter and summer solstices were exactly at the perihelion and at the aphelion, respectively, and also that, almost five thousand years ago, the vernal and autumnal equinoxes were at aphelion and at perihelion, respectively, as demonstrated in Fig. 4.4 (NAOJ 2015).

Note that the geometrical relationship between the Sun and the Earth described above is what we can imagine with our mind's eyes, thanks to the series of work done by Copernicus (1473–1543), Kepler (1571–1630), Galilei (1564–1642), and Newton (1642–1726), but not by our own eyes alone.

As shown in Fig. 4.3, the angle between the rotational axis of the Earth and the line perpendicular to the orbital plane is 23.45° at present and it is known to vary between 22.1 and 24.5° in the period of 41 thousand years. It is also known that the rotational axis undergoes the precession with its period of 26 thousand years. These variations in addition to that of eccentricity are important in astronomical science and long-term global climate change but not much in building science.

As mentioned earlier, when the Earth is almost the farthest away from the Sun is summer and the closest to the Sun is winter, but it may sound odd to those who live

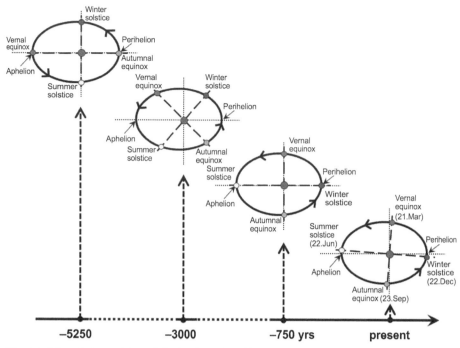

Fig. 4.4: The relative position of the Earth on its orbital plane with respect to the solstices and the equinoxes. During about a five-thousand-year period, the equinoxes and solstices shift approximately 90°.

somewhere in the northern hemisphere, because hot summer days may be considered owing to the position of the Sun being closer to the Earth. In fact, the total amount of energy delivered by solar radiation and received by the whole of Earth becomes the largest at the aphelion, but the obliquity of the Earth's rotational axis brings about a difference in the local availability of solar energy. In the Earth's northern hemisphere, solar radiation is available more on the horizontal surface in summer seasons than in winter seasons; the opposite is true in the Earth's southern hemisphere.

According to one of the three laws that were empirically discovered by Kepler, who carefully analysed the astronomical data measured and collected by Brahe (1546–1601), and also that were later confirmed theoretically by Newton, who established the law of gravitation, the rate of an area on the orbital plane swept by an imaginary hand rotating around the Sun remains constant. Namely, the speed of the Earth moving along the orbit is the fastest, 30.3 km/s, at perihelion and slowest, 29.3 km/s, at aphelion.

As already mentioned above, since the left-hand side of the orbital plane, i.e., the cumulative area from vernal equinox via summer solstice to autumnal equinox is larger than that from autumnal equinox via winter solstice to vernal equinox, the number of days corresponding to the former must be longer than those to the latter. We can confirm their difference as shown in Table 4.1; the answer is seven days. This difference in fact comes from the Earth's orbit being elliptic.

Together with this fact in mind, we come to recognize that the variation of solar azimuth angle at 12:00 of mean solar time shown in Fig. 4.2 is mainly due to the

Table 4.1: The number of days between equinoxes.

Vernal to autumnal		Autumnal to vernal	
March	10 (11)*	September	7 (8)
April	30	October	31
May	31	November	30
June	30	December	31
July	31	January	31
August	31	February	28
September	23 (22)	March	21 (20)
Total	186	Total	179

$$186 - 179 = 7$$

* The number of days in the brackets are for the equinoxes of 20th March and 22nd September.

eccentricity of the Earth's orbit. On the other hand, the tilt of the Earth's rotational axis, the obliquity, influences mainly the solar altitude changing from low values in winter to high values in summer, as was also shown in the change of solar altitude values in Fig. 4.2.

If the rotational axis was exactly perpendicular to the orbital plane and the Earth's orbit was exactly circular, then the apparent solar time could track constantly and the twelve points shown in Fig. 4.2 would merge into one point. But the reality is not as we have confirmed through our discussion so far. Thanks to the obliquity in particular, we have four seasons.

4.1.2 The sizes of the Earth and the Sun together with their distance

The sizes of the Earth and the Sun as well as their distance cannot be measured directly, but can be obtained from a rather simple series of calculation with the pieces of information available from a simple measurement, even including the ones possible for us to perform our own calculation, which assures you of having a clearer image of geometrical relationship between the Earth and the Sun described in the previous section.

(a) Earth's size

First, suppose that you are standing at some point on the Earth, A, on the Tropic of Cancer, where the latitude angle is 23.45° as shown in Fig. 4.5. The Sunlight incident upon there from the Sun at the meridian on the day of summer solstice comes from the direction of zenith, which is exactly parallel to the Earth's orbital plane shown in Fig. 4.3; in other words, the solar altitude of 90°. If there is a vertically deep well, the Sunlight can illuminate its bottom as shown in Fig. 4.5. Suppose next that you are at some other point, B, which is not so far away from point A and due north, and if there is also a vertically deep well there, you will find that its bottom is not illuminated, but a portion of the side-wall is illuminated at noon on the day of summer solstice.

With these facts, we can develop an idea of the geometrical relationship that the Earth is a huge sphere and the Sun is so remote that the Sunlight available at points

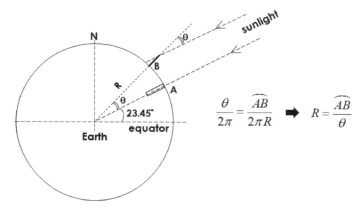

Fig. 4.5: Two points on the surface of an imaginary Earth for the estimation of the Earth's size.

A and B is parallel to each other. Then, if we measure both the distance between A and B and the shadow length of a vertical stick at B, the diameter of the Earth can be estimated.

This is what was first done by Eratosthenes, who lived some two-thousand and three-hundred years ago in Egypt. He had the figure of distance between Syene and Alexandria and also knew that at the noon of summer solstice, the bottom of a deep well in Syene was illuminated by Sunlight, but not in Alexandria. The shadow length of a vertical stick in Alexandria was one-eighth of the vertical stick; this corresponds to 7.2° subtended at the top of the stick by the bottom of the stick and the edge of the shadow.

We can confirm where these two places are located on the map; Syene is now called Aswan and its latitude is 24.05°, very close to the Tropic of Cancer, and Alexandria is located in the north of Aswan, 3° westward.

The figure of distance between Syene and Alexandria that Eratosthenes knew was 5000 stadia, which was the ancient unit of length used in Egypt, equivalent to 158 or 185 m in SI unit. Since 7.2° is one-fiftieth of 360°, the circumference of the Earth turns out to be 250000 stadia. Then, the diameter of the Earth is estimated to be 79600 stadia, which is only 1.3% smaller than the diameter of the Earth known today, 12740 km, with the assumption of one stadia to be 158 m, or 16% larger with the assumption of one stadia to be 185 m. This is, I think, a remarkable research result, even if the method applied looks crude and the accuracy of answer was not sufficient from the present view of geometrical science.

(b) Distance between the Earth and the Sun

The distance between the Earth and the Sun can be estimated in a similar way as follows. According to a series of astronomical observation having been done since the ancient time, we now know that the Earth is rotating around the Sun as described in the previous section. But our ancestor came to know the following two things first based on the geocentric-system theory: one is that the relative positions of the stars are always constant, although the stars that you can see up in the sky vault during night time vary with seasons of the year; the other is that such planets as Venus and Mars change their positions relative to other stars on the sky vault occasionally in some complicated manners.

The fact that the relative positions of stars are always constant allows us to conceive that all of the stars are much farther away from the Earth than other planets and the Sun are. Assuming such characteristics together with the idea of the Earth rotating around its centre, we find that there emerges a slight difference in the angle subtended by a target star and the Sun according to the rotation of the Earth as shown in Fig. 4.6. This is called a parallax. Since the parallax becomes the largest for a point on the equator of the Earth, the measurement should be made there, for example, somewhere close to Singapore. The difference in angle values is taken between when the point is at Sunrise, the solar altitude at 0° and when it is at the noon, the solar altitude at 90°, either on vernal or autumnal equinox. During daytime, we cannot see any starlight directly because of intense daylight, but as mentioned above, we know that the relative positions of the stars over the sky vault are constant so that those angles are obtained from a combination of measurement and calculation.

According to a sufficiently precise measurement, the parallax observed at the equator is 0.00243°, which is equal to 0.00004249 radian. We can estimate the distance between the Earth and the Sun by dividing the radius of the Earth, 6370 km, by the sine value for 0.00004249 radian based on the geometrical relationship shown in Fig. 4.6. The result turns out to be 1.5×10^{11} m, which is quite close to the semi-major axis of the Earth's orbital ellipse.

The distance between the Earth and the Sun can also be obtained from geometrical relationship between the Sun, the Venus and the Earth. The maximum angle subtended by the Sun and the Venus at a point on the ground surface of the Earth has been known to be 47° due to astronomical observation. Provided that the shortest distance between the Earth and the Venus appearing at conjunction is measured directly with laser-light application, a simple trigonometry calculation gives us the distance of the Earth and the Sun. The result becomes of course the same as the one obtained from the other calculation mentioned above.

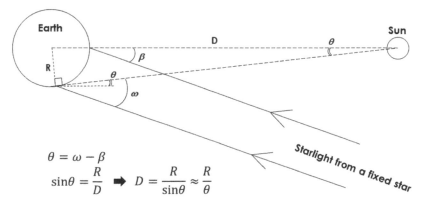

Fig. 4.6: Parallax emerged due to the rotation of the Earth with respect to the angle between the Sun and a target star being very much farther away.

(c) Sun's size

With the knowledge of the distance between the Earth and the Sun, we are now to estimate the diameter of the Sun. The angle subtended by the diameter of a solar disc that can be seen from the ground surface is 0.532°, which is almost equivalent to the angle subtended by the diameter of the hole of a fifty-yen coin (5 mm) held by your thumb and index finger at a stretch of your right or left arm, for example 540 mm, as shown in Fig. 4.7. Since we now know the distance between the Earth and the Sun as 1.5×10^{11} m, the diameter of the Sun can be easily estimated by multiplying the distance value, 1.5×10^{11} m, and the angle value in radian, 0.00928. The result is 1.39×10^9 m.

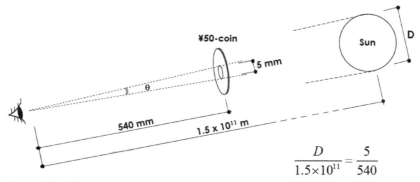

$$\frac{D}{1.5 \times 10^{11}} = \frac{5}{540}$$

Fig. 4.7: Relationship between a coin having a hole and the distance of the Sun, assuming the distance between the coin and one of your eyes to be 540 mm.

4.2 The Moon and Earth

The movement of the Sun over the sky vault is perceived by our minds starting with the primary visual sensation at our eyes and its qualitative effects on daylight availability indoors must be quite obvious, but before discussing how they can be evaluated quantitatively in the next section, there is one other invisible but important characteristic in relation to the celestial body closest to the Earth, the Moon. During night time, if the sky is free of cloud, the Moon can be seen somewhere on the sky vault unless the Moon is somewhere under the horizon. Where it is and how it shapes depends on the date and time and it affects the human biological clock.

4.2.1 Moon's revolution and Earth's rotation

Figure 4.8 illustrates a close-up view of the relative positions of the Earth and the Moon to each other for about a one-month period from mid-November to mid-December. Since the Moon is luminous by reflecting solar radiation, not by emitting visible radiation on its own, its appearance in the sky varies owing to its relative position to the Earth and the Sun. When the Moon is on the line between the Earth and the Sun, the Moon is at a totally dark phase and is called "New Moon". On the other hand, when the Earth is between the Sun and the Moon, the bright side of the Moon faces the Earth so that the Moon looks totally bright and this phase is called "Full Moon".

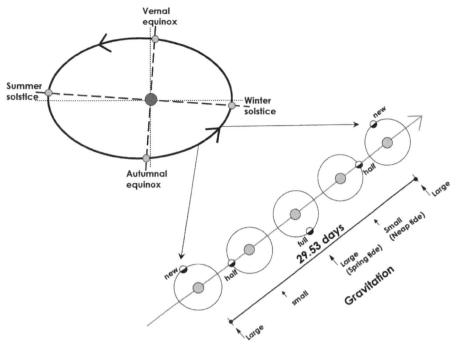

Fig. 4.8: The Earth's rotation and the Moon's revolution for the period of one lunar month.

The gravitational force exerted on the Earth is the largest when the Moon is either at the phase of New Moon or Full Moon because of three celestial bodies being in a row. The gravitational force is the smallest when it is in the exact mid of waxing and waning Moon because of the Sun and the Moon being perpendicular to each other relative to the Earth.

The change in tidal height is the most obvious effect of this gravitational force from the largest to the lowest and vice versa. Figure 4.9 shows an example of measured tidal height in Yokohama port, Japan, from mid-February to late March in the year 2016 (JMA 2016). As can be seen, the length of period from a day of spring tide, which records the largest daily variation of the tidal height, to the next day of spring tide is 14.77 days and that to the third day of spring tide is 29.53 days. The same applies to the days of neap tide. Such pattern exactly fits the number of days of the three celestial bodies being in a row from a day of New Moon to the next day of New Moon as was illustrated in Fig. 4.8.

Figure 4.10 is a close-up view of the variation of tidal height in Fig. 4.9; this is for a three-day period, from 22nd to 24th of February. As can be seen, high tide appears twice a day and low tide also twice a day. This pattern of high and low tides happens in relation to the change in gravitational force, which emerges due to the relative positions of the Earth, the Moon, and the Sun. Figure 4.11 illustrates the movement of the Earth for a one-day period; this is a close-up view of Fig. 4.8.

According to the apparent solar time described in section 4.1, one day is the period from the time when the Sun is on the meridian, due south, and appears

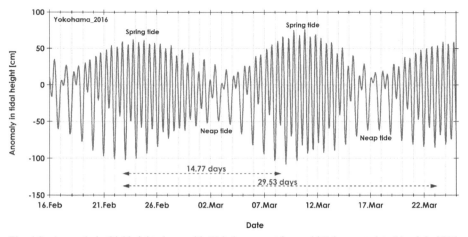

Fig. 4.9: Anomaly in tidal height observed in Yokohama port from mid-February to late March in 2016.

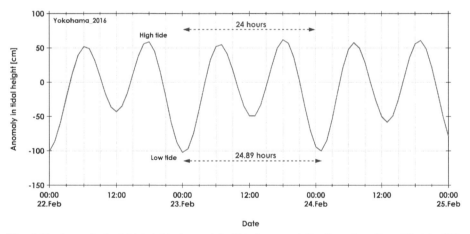

Fig. 4.10: Anomaly in tidal height observed in Yokohama port for three days from 22nd to 24th mid-February in 2016.

again next time on the meridian on the following day. It is called one solar day. As can be seen in Fig. 4.11, there is one other day called lunar day. Its length is from when point P on the Earth faces the Moon exactly on the meridian, due south, and to when it faces again exactly the Moon. It is obviously longer than one solar day because the Moon moves a bit while one solar day lasts. The Moon revolves once taking 29.53 days as was shown in Fig. 4.8. Provided that one solar day is defined as 24 hours, then one lunar day is 24.89 hours. When the line drawn between point P and the centre of the Earth exactly overlaps the line between point P and the Moon, the gravitational force exerted by the Moon at point P is the largest and when the two lines are perpendicular to each other, it becomes the smallest. This is why high and low tides appear twice, respectively, for the period of one lunar day.

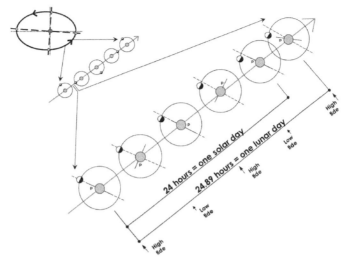

Fig. 4.11: One cycle of Earth's rotation, one solar day, and the movement of the Moon for one lunar day.

4.2.2 The variation of tidal height and biological clock

Figure 4.12 shows the variation of tidal height for the whole-year period of 2016 in Yokohama. It can be seen that the highest level of tidal height, which is spring tide, occurs in two-week cycles; the same applies to the lowest level of tidal height, neap tide. Summarizing the discussion so far, the variation of tidal height emerges certainly corresponding to the variation of gravitational force due to the change in relative positions of three celestial bodies, the Earth, the Moon and the Sun.

As mentioned in the beginning of this chapter, we humans usually live in the diurnal cycle of being awake and asleep within and it is obvious that this cycle emerges following the diurnal cycle of brightness and darkness given by solar radiation. How it emerges can be observed by measuring the natural cycle of being awake and asleep, for example, of a new born baby. A diagram shown on the left-hand side of Fig. 4.13 is one such example, observed first by Kleitmann and Engelman (Kleitmann and Engelman 1953, Winfree 1987). The pieces of horizontal bars indicate while the baby was asleep, open areas in between while awake, and the dots the time of nursing.

As can be seen, after about 150 days from the birth, the baby attains the cycle of waking up about 8 o'clock in the morning and going to sleep about 8 o'clock in the evening. But before the 150th day since the birth, there seems to have been no pattern following the diurnal cycle of daylight availability.

The right-hand side of Fig. 4.13 is another expression of the variation of tidal heights shown in Fig. 4.12. Closed circles indicate the time of high tide and open circles the time of low tide. Because of one lunar day being a little longer than one solar day as demonstrated in Fig. 4.11, there appears a pattern in which the whole of plots follows a downward slope. This pattern looks very much similar to what we can see in the pattern of the baby being asleep and awake, in particular until ninety days

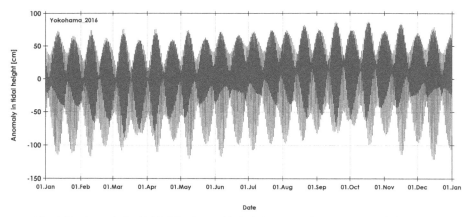

Fig. 4.12: Anomaly in tidal height observed in Yokohama port for a one-year period in 2016.

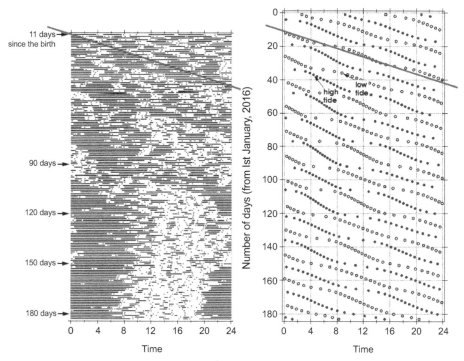

Fig. 4.13: Development of the circadian rhythm in the body of a baby and the daily variation of high and low tides for the period of a half year. The graph shown here on the left-hand side was drawn using the data illustrated in a diagram provided by Winfree (1987).

had passed since the birth. Note that two straight lines parallel to each other placed on these two charts overlap the respective patterns very well.

This illustrates that there are two kinds of biological clock embedded within our body: one following the variation of gravitational force due to the revolution of the Moon and the rotation of the Earth, and the other following the variation of daylight availability due to the rotation of the Earth. We may call the former to be implicit type and the latter to be explicit.

This proves why daylight availability through windows in buildings is important. Without windows, our circadian rhythm would have to run only with the implicit type of biological clock. We humans can adjust the biological clock in harmony with the Sun by exposing ourselves, more or less, to daylight available from windows, almost unconsciously about the presence of these two kinds of biological clock.

4.3 Sun-path diagram

With the discussion so far, it must have, I hope, become clearer how we are affected by Earth's rotation and Moon's revolution. Let us here turn our discussion again on the solar position, especially in terms of how it affects the built environment.

As we have described in 4.1.1, solar position is defined by a set of two angles: solar altitude and solar azimuth. Through the explanation of Fig. 4.3, we have come to know that the number of hours during daytime and night time become exactly the same as each other on either vernal or autumnal equinox. In turn, the length of daytime during summer season becomes longer than that during winter season. This is what people usually know by their own primordial experience, regardless of knowing the reason. But how about annual total of Sunshine duration? Are there some places on the Earth where it is longer or shorter than other places? Let us confirm this by calculating the number of Sunshine hours, which is defined to be the length of hours from the Sunrise to the Sunset every day.

Figure 4.14 shows annual variation of daily Sunshine hours in the city of Yokohama, Singapore, and Copenhagen. In Copenhagen, where the latitude is 55.68° and the highest among the three cities, the seasonal difference in Sunshine hours is quite significant; there is about ten-hour difference between summer and winter. On

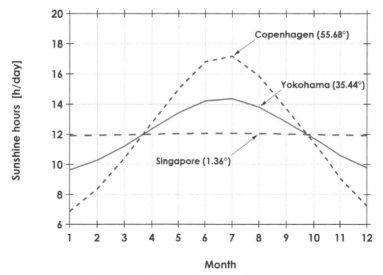

Fig. 4.14: The variation of daily Sunshine hours in three cities, Copenhagen, Yokohama and Singapore, whose latitude angles (the numbers in brackets) are different from each other.

the other hand, in Singapore, where the latitude is 1.36° and the lowest, daily Sunshine duration is almost constant at twelve hours throughout one year. In Yokohama, where the latitude is in between Copenhagen and Singapore, daily Sunshine hours varies moderately; there is about seven-hour difference between summer and winter.

What we can also know from Fig. 4.14 is that, since the areas under the curves are the same, annual total of Sunshine duration is the same as each other in these three cities. It is true everywhere on the Earth. This fact leads us to recognize that, under annual equality of total sunshine duration anywhere on the Earth, a variety of climates from one region to another emerge, depending on the locality with its unique seasonal variations of daylight availability, cloud coverage, precipitation, wind pattern and others.

4.3.1 Projected solar positions

It is important for us to know the overall movement of the Sun, which varies from one season to another, in order to select the location of a building, the orientation of its façade, or the form of shading devices such as overhang and side fins to be designed along the edges of the window openings. These features should be known by professional building designers and engineers, but they should also be of interest and useful for non-professional people as well.

For this purpose, the Sun-path diagram based on the equi-distant projection principle is the simplest to produce among other principles such as orthographic or equi-solid angular projection. It is the easiest to read and apply to planning and designing the built environment. Let us describe first what the principle of equi-distant projection is, and then demonstrate some examples of the use of equi-distant Sun path diagram.

Figure 4.15a depicts the solar position already defined in Fig. 4.1 with an imaginary hemisphere on which the Sun is projected. The equi-distant projection is to specify the Sun's projection within a horizontal circle so that the ratio of solar altitude to the right angle, 90°, is proportional to the ratio of the length of a line drawn from the circumference of the horizontal circle to the position of the Sun as the end of the line to the radius of the horizontal circle. With this principle, for example,

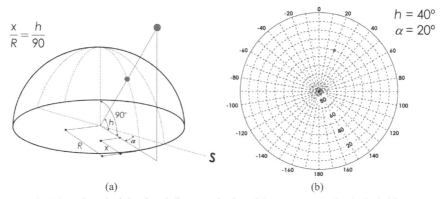

(a)   (b)

Fig. 4.15: The principle of equi-distant projection of the Sun as a small spherical object.

the solar altitude and azimuth of 40° and 20°, respectively, can be indicated as shown in Fig. 4.15b.

Since the solar position at a certain place on the Earth can be calculated with a sufficient accuracy by a computer code such as the one shown in Table 4.A.1 of appendix 4.A.1 (Shukuya 1993), it is possible to produce the Sun-path diagram for wherever on the Earth, whose latitude and longitude are given.

Figure 4.16 shows three examples of Sun-path diagram in Copenhagen, Yokohama, and Singapore. Comparison of the Sun path in one season to another reveals the basic characteristics of respective regional climate. For example, the length of daytime is very short during winter season in Copenhagen, while on the other hand, it is very long during summer season. In Singapore, there is no essential difference in the solar positions between the winter and summer solstices, which results in hot and humid climate there throughout the whole year. Comparing one diagram with another lets us know the regional difference in solar trajectory leading to the difference in their climatic characteristics.

The fact that the patterns of solar trajectory in three cities are quite different suggests that the overhangs and side fins to be designed in respective regions should have different forms reflecting the regional solar characteristics as was discussed earlier (Olgyay and Olgyay 1957, Olgyay 1963, Olgyay et al. 2015).

4.3.2 Projected vertical and horizontal bars

The idea described in previous subsection 4.3.1 in terms of the solar position can be extended to express the equi-distant projection of a vertical or horizontal bar. Suppose that there is a vertical bar standing from the horizontal flat ground surface as shown in Fig. 4.17a. This bar can be first projected on the curved surface of an imaginary hemisphere and then it can further be projected as a piece of straight line within the horizontal circle. The length of the straight line is determined so that the ratio of the line length to the radius of the circle is equal to the ratio of the altitude angle of the top end of the vertical bar to the right angle.

Imagine that we extend the vertical bar high up in the sky. Its limit is to bring the projected end of the bar to reach the centre of the horizontal circle. This corresponds to the altitude angle of the bar which is 90°.

Next, suppose that there is a horizontal bar as shown in Fig. 4.17b. Two ends of the horizontal bars are determined in the same manner as the solar position is indicated as a point on the hemisphere and then the corresponding point on the horizontal circle. The horizontal bar becomes one corresponding curve projected on the hemisphere and then the other corresponding curve within the horizontal circle.

Imagine that both ends of horizontal bar are stretched infinitely as we imagined that the top end of a vertical bar was extended towards the zenith of the sky. This results in both ends of the projected curve on the horizontal circle reaching the circumference of the horizontal circle.

Figure 4.18 illustrates the collection of infinitely long vertical bars and also of infinitely long horizontal bars; the former are indicated by dashed straight lines and latter by solid curved lines. Imagine that you are surrounded by thirty-six infinitely long vertical bars, each of which is standing apart from one to another subtending by

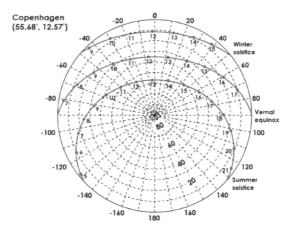

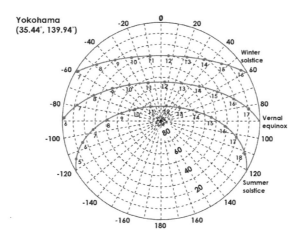

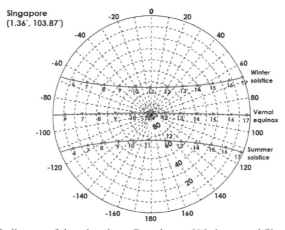

Fig. 4.16: Sun path diagram of three locations: Copenhagen, Yokohama, and Singapore. The time on summer solstice in Copenhagen is according to daylight saving time, which is one hour earlier than the time in other seasons.

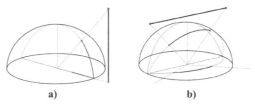

a)   b)

Fig. 4.17: Equi-distant projection of a vertical bar and a horizontal bar.

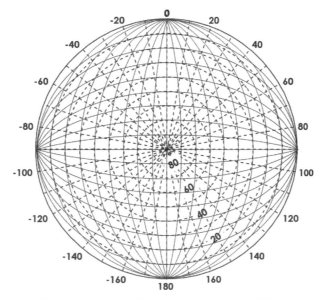

Fig. 4.18: Collection of infinitely long vertical and horizontal bars cast within a circle by the principle of equi-distant projection. Thirty six straight dashed lines concentrating towards the centre of the circles indicate the infinitely long vertical bars. There are seventeen solid curved lines corresponding to horizontal bars in the direction of azimuth angles of 0° and 180° and seventeen in the direction of azimuth angles of –90° and +90°.

10° in horizontal angle. On the equi-distant diagram, they look as radiated from the centre of the circle, or as concentrated towards the centre of the circle, as shown by dashed straight lines.

There are seventeen solid curved lines corresponding to horizontal bars in the direction of azimuth angles of 0° and 180° and seventeen in the direction of azimuth angles of –90° and +90°. The horizontal bars are parallel to one another so that each of them is positioned apart from one another by 10° in their altitude angles.

4.3.3 Shadow cast by surrounding objects

The equi-distant projection chart of vertical and horizontal bars just explained above can be used to examine which portion of the sky is shaded by an object such as an overhang or side fins attached along the edges of a window, or an adjacent building standing in front of the site of a building to be planned.

Figure 4.19 demonstrates an example of examining the shading effect of an overhang and two side fins attached to the edges of a window, whose height and width are 1.5 m and 3 m, respectively, as shown in Fig. 4.19a. We assume here that there are an overhang and two side fins, the depth of which are 0.8 m and 0.5 m, respectively. To make the projection of these overhang and the side fins, we first calculate the respective angular position of the horizontal and vertical edges of the overhang and the side fins as tabulated in Fig. 4.19a. With this information, we can determine the equi-distant projection of the lines representing the edges of the overhang and the side fins as shown in Fig. 4.19b.

Then, by overlapping, for example, four solar trajectories in Yokohama, we come to know the following: during winter season, sunlight is not obstructed at all, while on the other hand, during summer season, from summer solstice to late August, sunlight is never incident on the centre of the window sill.

Another example is to examine the shadow cast by an adjacent building at a point on the horizontal surface as illustrated in Fig. 4.20. Let us assume that the

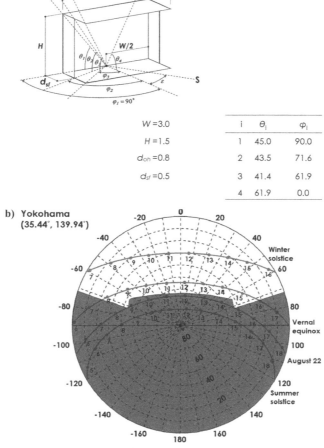

i	θ_i	φ_i
1	45.0	90.0
2	43.5	71.6
3	41.4	61.9
4	61.9	0.0

$W = 3.0$
$H = 1.5$
$d_{oh} = 0.8$
$d_{sf} = 0.5$

Fig. 4.19: Shading effect of a overhang and two side fins.

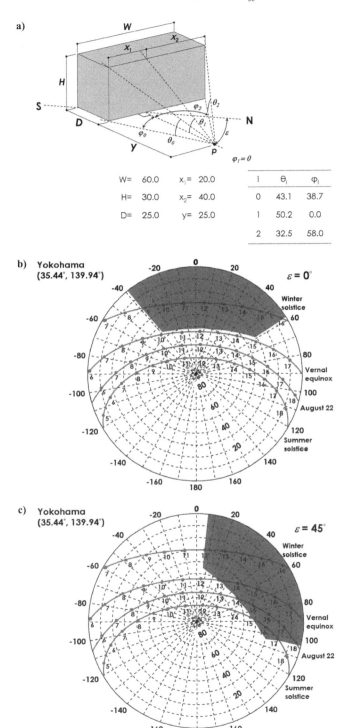

Fig. 4.20: Shading effect of an adjacent building in front of a point on the horizontal ground surface.

height and width of the adjacent building are 30 m and 60 m, respectively, and the point of concern is positioned 20 m from the left-hand corner of the adjacent building and 25 m away from the façade. In the same manner as we have done in the case of an overhang and side fins demonstrated in Fig. 4.19, we first determine the angular position of the edges of the adjacent building as tabulated in Fig. 4.20a. With this information, horizontal edge and two vertical edges of the building are drawn and then four solar trajectories are overlapped as shown in Fig. 4.20b and c: (b) is a case that the adjacent building wall surface facing the point of concern is oriented towards due north; (c) is another case that the wall surface is oriented towards north-east.

By constructing these diagrams, we come to know about when the point of concern is shaded by the adjacent building. In the case of adjacent building facing due north, sunlight on the day of winter solstice is obstructed during most of daytime, while on the other hand, in the case of adjacent building facing north-east, sunlight is available during the morning period but not during the afternoon period on the day of winter solstice. On the day of vernal equinox and 22nd August, sunlight is shaded for four to five hours during the afternoon period. With such information about the relationship between the Sun path and adjacent building, we proceed rational planning of a building on the site including the point of concern.

4.4 Overall solar control effect of overhang and side fins

Although Sun path diagrams are useful as described in the previous section, they can give little information on how much of daylight is available on the window surface or how much of excessive solar heat is protected by overhangs and side fins. One way to know such information is to compare the daily rate of energy delivered by solar radiation incident on the window surface with an overhang and side fins and that incident on the bare window.

Figure 4.21 shows one such example of a comparison in Yokohama. The width and height of a window facing south or west are assumed to be 3 m and 1.5 m, respectively, and the depth of both overhang and side fins to be 0.75 m as shown in Fig. 4.21a. The vertical axis indicates the daily rate of solar energy available from the clear sky, both direct solar radiation and diffused solar radiation, assuming the atmospheric transmittance of 0.8. The details of calculation procedure are given in section 4.A.2 and 4.A.3. The horizontal axis indicates the days of one year.

Solid lines are the cases of south-facing vertical window and dashed lines are those of west-facing vertical window. In either south or west orientation, there is a large seasonal variation in solar energy available on the window surface from the clear sky. In winter, south-facing window receives more solar radiation than west-facing window does and in summer, the opposite is true. These facts suggest that south-facing window can be an appropriate source for passive solar heating and that installing appropriate shading devices is necessary, especially for west-facing window in summer to reduce the cooling demand caused by excess solar heat.

In the case of southern orientation, having the overhang and side fins of the depth 0.75 m along the edges of window can significantly decrease the amount of solar energy from the clear sky during summer season, while on the other hand, less significantly during winter season. In the case of west-facing window, the relative

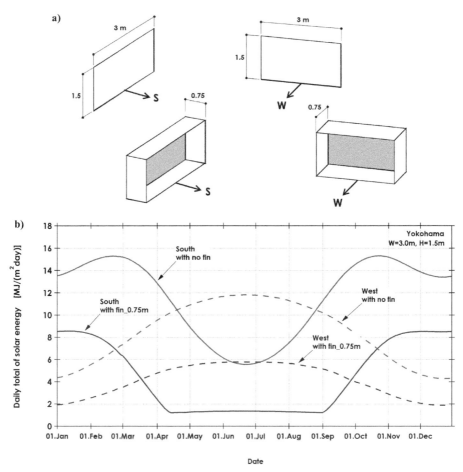

Fig. 4.21: Daily rate of total solar energy available from clear sky on south- and west-facing windows with and without overhang and side fins.

reduction of solar energy by the overhang and side fins is more or less the same throughout the year.

Thus, we confirm that overhangs and side fins are important elements and their proper design suiting to regional solar characteristics dependent on the window orientations assures the built environmental space to be created for well-being of the people residing there.

References

Japan Meteorological Agency (JMA). 2016. http://www.data.jma.go.jp/kaiyou/db/tide/suisan/suisan.php?stn=QS.

Kleitmann N. and Engelman T. G. 1953. Sleep characteristics of infants. Journal of Applied Physiology 6: 269–282.

National Astronomical Observatory of Japan (NAOJ). 2015. http://eco.mtk.nao.ac.jp/koyomi/wikiB6E1C6FCC5C0A4CEB0DCC6B0.html.

Olgyay V. and Olgyay A. 1957. Solar Control and Shading Devices. Princeton University Press, NJ.

Olgyay V. 1963. Design with climate—Bio-climatic approach to architectural regionalism. Princeton University Press, NJ.

Olgyay V., Lyndon D., Reynolds J. and Yeang K. 2015. Design with climate—Bio-climatic approach to architectural regionalism—New and expanded edition. Princeton University Press, NJ.

Shukuya M. 1993. Light and heat in the built environment: An approach by numerical simulation. Maruzen Publishers, Tokyo (in Japanese).

Winfree A. T. 1987. The Timing of Biological Clocks. Scientific American Library. W. H. Freeman Company NY.

Appendix

4.A Calculation of solar position, Sun-lit area, and solar energy

4.A.1 Solar position

Solar altitude and azimuth angles can be determined by a set of equations derived from spherical trigonometry principle. Table 4.A.1 shows a code for the calculation of solar position written with Visual Basic for Applications (VBA).

4.A.2 Sun-lit area of a window with overhang and side fins

An overhang and side fins as shown in Fig. 4.19a cast a shadow on the window surface, which is partly exposed to direct solar radiation. The ratio of the sunlit area to the whole window area can be calculated by the procedure described as a VBA code shown in Table 4.A.2.

The amount of diffuse solar radiation incident on the window surface is also decreased by the overhang and side fins since they partly protect the incidence of diffuse solar radiation. In the case of a window assumed in Fig. 4.21a, the ratio of diffuse component of solar energy incident on the window with overhang and side fins to that incident on the window without is 0.575; this value can be determined by using a VBA code shown in Table 4.A.3, which is for the calculation of form factor between two rectangular surfaces perpendicular to or parallel to each other. This code can be used to calculate the value of form factor with respect to each of the overhang and the two side fins.

4.A.3 Estimation of the rate of energy carried by solar radiation

The rate of energy carried by direct and diffuse solar radiation, the irradiance, under the clear sky conditions can be determined by a variety of empirical formulae. To produce Fig. 4.21b, Bouguer's equation for direct solar radiation and Nagata's equation for diffuse solar radiation, as shown in Table 4.A.4, were used (Shukuya 1993).

Table 4.A.1: VBA code for the calculation of solar position.

```
Sub solps(slng, xlat, xlng, im, id, jh, jm, H, Alf)
'This is a subprogram for the calculation of solar altitude and azimuth angles
'at a certain time of a day in a year
'                      17June2010->11April2013->05May2016 MS
'
' Input: slng -> longitude representing the time zone (degree)
'         xlat -> latitude (degree)
'         xlng -> longitude (degree)
'         im -> month, id -> day, jh -> hour, jm -> minute
' Output: H -> solar altitude (degree)
'             Alf -> solar azimuth (degree)
'
Dim NDY(12) As Integer
NDY(1) = 0: NDY(2) = 31: NDY(3) = 59: NDY(4) = 90: NDY(5) = 120: NDY(6) = 151
NDY(7) = 181: NDY(8) = 212: NDY(9) = 243: NDY(10) = 273: NDY(11) = 304: NDY(12) = 334
PAI = 3.14159265: Pi = 2 * PAI/360
SXL = Sin(xlat * Pi)
CXL = Cos(xlat * Pi)
MDY = NDY(im) + id
WW = 2 * PAI * MDY/365
D = Pi * (0.3622133 - 23.24763 * Cos(WW + 0.153231) - 0.3368908 * Cos(2 * WW + 0.2070988) -
                                     0.1852646 * Cos(3 * WW + 0.6201293))
CD = Cos(D)
SD = Sin(D)
E = 60 * (-0.0002786409 + 0.1227715 * Cos(WW + 1.498311) - 0.1654575 * Cos(2 * WW - 1.261546) –
                                     0.00535383 * Cos(3 * WW - 1.1571))
TS = jh + jm/60
T = (15 * (TS - 12) + xlng - slng + 0.25 * E) * Pi
SH = SXL * SD + CXL * CD * Cos(T)
If SH < 0 Then
   H = 0
   Alf = 0
Else
   CH = Sqr(1 - SH * SH)
   ca = (SH * SXL – SD)/CH/CXL
   sa = CD * Sin(T) CH
   H = Atn(SH/CH) /Pi
   A = Atn(Sqr(1 - ca * ca)/ca)/Pi
    If sa < 0 Then
       If ca > 0 Then
         Alf = -A
       Else
         Alf = -(A + 180)
       End If
    Else
       If ca > 0 Then
         Alf = A
       Else
         Alf = A + 180
       End If
    End If
End If
End Sub
```

Table 4.A.2: VBA code for the calculation of sun-lit area on the surface of a window with an overhang and side fins.

```
Sub Overhang_Sidefin(WW, HH, d_o, d_s, w_eps, H, Alf, k_AS, kkk)
'Calculation of normalized surface area of a rectangular window
'with overhang and sidefin
'           16th April, 2013 MS
' Input: WW -> Window width (m)
'        HH -> Window height (m)
'        d_o -> Depth of overhang (m)
'        d_s -> Depth of sidefin (m)
'        w_eps -> Wall azimuth angle (degree)
'        H -> solar altitude (degree)
'        Alf -> solar azimuth (degree)
' Output: k_AS -> Normalized sun-lit area (0<=k_AS<=1.0)
'         kkk -> Index of sun-lit pattern (No sunshine, A, B, C, or D)
'
PAI = 3.14159265: Pi = 2 * PAI / 360
Gamma = Alf - w_eps
If 270 < Abs(Gamma) Then
    Gamma = 360 - Abs(Gamma)
End If
cos_g = Cos(Gamma * Pi): Phi = Atn(Tan(H * Pi) / cos_g)
ABS_Gamma = Abs(Gamma * Pi)
'
If d_s <= d_o Then
        GoTo 100
Else
        GoTo 101
End If
'
'
100     'For the case of "d_s <= d_o"=======================================
'
If cos_g < 0 Then
    k_AS = 0: kkk = "No Sunshine": GoTo 11
Else
    GoTo 10
End If
'
10     Dtg = d_o * Tan(ABS_Gamma): Dtp = d_o * Tan(Phi)
'
If Dtg < WW And Dtp < HH Then
'(A)
kkk = "A"
    k_AS = ((WW - Dtg) * (HH - Dtp) + 1 / 2 * (2 * HH - (d_o + d_s) * Tan(Phi)) * (d_o - d_s) *
Tan(ABS_Gamma)) / (WW * HH)
ElseIf Dtg < WW And HH <= Dtp Then
'(B)
kkk = "B"
    k_AS = 1 / 2 * Tan(ABS_Gamma) / Tan(Phi) * (HH - d_s * Tan(Phi)) ^ 2 / (WW * HH)
    If HH < d_s * Tan(Phi) Then
            k_AS = 0
    End If
ElseIf WW <= Dtg And HH <= Dtp Then
```

Table 4.A.2 contd. ...

...Table 4.A.2 contd.

```
     '(C)
     kkk = "C"
        xc = Tan(ABS_Gamma) / Tan(Phi) * HH
        If WW < xc Then
           k_AS = 1 / 2 * (2 * HH - (WW / Tan(ABS_Gamma) + d_s) * Tan(Phi)) * (WW /
Tan(ABS_Gamma) - d_s) * Tan(ABS_Gamma) / (WW * HH)
           If WW < d_s * Tan(ABS_Gamma) Then
                k_AS = 0
           End If
        Else
           k_AS = 1 / 2 * Tan(ABS_Gamma) / Tan(Phi) * (HH - d_s * Tan(Phi)) ^ 2 / (WW * HH)
           If HH < d_s * Tan(Phi) Then
                k_AS = 0
           End If
        End If
     ElseIf WW <= Dtg And Dtp < HH Then
     '(D)
     kkk = "D"
        k_AS = 1 / 2 * (2 * HH - (WW / Tan(ABS_Gamma) + d_s) * Tan(Phi)) * (WW /
Tan(ABS_Gamma) - d_s) * Tan(ABS_Gamma) / (WW * HH)
        If WW < d_s * Tan(ABS_Gamma) Then
                k_AS = 0
        End If
     End If
11      If H = 0 Then
     k_AS = 0: kkk = "No Sunshine"
     End If
     GoTo 999
     '
     '
101        'For the case of "d_s > d_o"=========================================
     '
     If cos_g < 0 Then
       k_AS = 0: kkk = "No Sunshine": GoTo 21
     Else
        GoTo 20
     End If
     '
20      Dsg = d_s * Tan(ABS_Gamma): Dsp = d_s * Tan(Phi)
     '
     If Dsg < WW And Dsp < HH Then
     '(A)
     kkk = "A"
        k_AS = ((WW - Dsg) * (HH - Dsp) + 1 / 2 * (2 * WW - (d_o + d_s) * Tan(ABS_Gamma)) * (d_s -
d_o) * Tan(Phi)) / (WW * HH)
     ElseIf Dsg < WW And HH <= Dsp Then
     '(B)
     kkk = "B"
        k_AS = 1 / 2 * (2 * WW - (HH / Tan(Phi) + d_o) * Tan(ABS_Gamma)) * (HH / Tan(Phi) - d_o) *
Tan(Phi) / (WW * HH)
```

Table 4.A.2 contd. ...

...Table 4.A.2 contd.

```
        If HH < d_o * Tan(Phi) Then
                k_AS = 0
        End If
    ElseIf WW <= Dsg And HH <= Dsp Then
    '(C)
    kkk = "C"
        yc = Tan(Phi) / Tan(ABS_Gamma) * WW
        If yc <= HH Then
          k_AS = 1 / 2 * Tan(Phi) / Tan(ABS_Gamma) * (WW - d_o * Tan(ABS_Gamma)) ^ 2 / (WW *
HH)
            If HH < d_o * Tan(Phi) Then
                k_AS = 0
            End If
        Else
          k_AS = 1 / 2 * (2 * WW - (HH / Tan(Phi) + d_o) * Tan(ABS_Gamma)) * (HH / Tan(Phi) - d_o) *
Tan(Phi) / (WW * HH)
            If HH < d_o * Tan(Phi) Then
                k_AS = 0
            End If
        End If
    ElseIf WW <= Dsg And Dsp < HH Then
    '(D)
    kkk = "D"
        k_AS = 1 / 2 * Tan(Phi) / Tan(ABS_Gamma) * (WW - d_o * Tan(ABS_Gamma)) ^ 2 / (WW *
HH)
            If WW < d_o * Tan(ABS_Gamma) Then
                k_AS = 0
            End If
    End If
21    If H = 0 Then
    k_AS = 0: kkk = "No Sunshine"
    End If
    '
999   End Sub
```

Table 4.A.3: VBA code for the calculation of form factor between two rectangular surfaces.

```
Function SSHV(A, B, C, IC)
'    This is a program to calculate the ratio of the diffuse light
'    flux reaching a rectangular surface to that leaving the other
'    rectangular surface, namely the configuration factor.  The
'    two rectangular surfaces are either parallel or perpendicular
'    to each other.
'
'    A is the width and B is the height of the surface emitting diffuse radiation.
'    In the case of 0<IC for the surface receiving the radiation to be "parallel", C is the distance between the two
surfaces;
'    In the case of IC<=0 for "perpendicular", C is the depth.
'            October 12, 1985//revised 12 May, 1992 M. Shukuya'
'            for VBA    25th September, 2013//additional explanation 06 December, 2015
```

Table 4.A.3 contd. ...

...Table 4.A.3 contd.

```
PAI = 3.141593
x = A / C
y = B / C
RX = Sqr(1# + x * x)
RY = Sqr(1# + y * y)
RXY1 = Sqr(1# + x * x + y * y)
If x = 0 Or y = 0 Then
      SSHV = 0#
ElseIf IC <= 0 Then
      RXY = Sqr(x * x + y * y)
      SSHV = (Atn(x) / y - RY * Atn(x / RY) / y + Atn(x / y) + 0.5 / x / y * Log(RXY1 / RX / RY) + 0.5 * y / x *
Log(RXY1 * y / RXY / RY) + 0.5 * x / y * Log(RXY * RX / RXY1 / x)) / PAI
    Else
      SSHV = 2# * (RY * Atn(x / RY) / y + RX / x * Atn(y / RX) - Atn(x) / y - Atn(y) / x + Log(RX * RY /
RXY1) / x / y) / PAI
    End If
    End Function
```

Table 4.A.4: VBA code for the estimation of solar energy available from clear sky.

```
Sub Bouguer_Nagata(P, SH, IDN, ISH)
'Estimation of solar energy available on the horizontal surface
'under a clear-sky condition by Bouguer's equation for direct component
'and Nagata's equation for diffuse component.
'              15th April, 2013 MS
'
' Input: P -> Atmospheric transmittance (0<=P<1.0)
'        SH -> Sine value of solar altitude
' Output: IDN -> Direct normal component of solar energy (W/m^2)
'         ISH -> Diffuse component of solar energy (W/m^2)
'
  I0 = 1370
  IDN = I0 * P ^ (1 / SH)
  U_Nagata = (0.5 + (0.4 - 0.3 * P) * SH) * (0.66 - 0.32 * SH)
  ISH = U_Nagata * I0 * (1 - P ^ (1 / SH)) * SH
End Sub
```

Chapter 5

Visible Light and Luminous Environment

5.1 Visible light as electromagnetic wave

Light, whether it originates from the natural source, the sun and the sky, or from artificial, man-made sources such as candles or electric lamps, is utilized to make the indoor environmental space sufficiently luminous for a variety of purposes from working to resting indoors. We all know that visible light is necessary for our everyday life and our eyes are its sensory portals, though we hardly pay attention to such fact because of being so self-evident.

Here let us first pay attention to very fundamental phenomena with respect to the physical characteristics of visible light before discussing what indoor lighting is and how it works and affects human perception and resulting behaviour, which affects how the "exergy" resources for lighting is utilized.

Let us start thinking about a small experiment shown in Fig. 5.1. Prepare two pipes and hold one with your right hand and the other with your left hand. We assume that the pipe held by left hand is made of cardboard and the other by right hand of

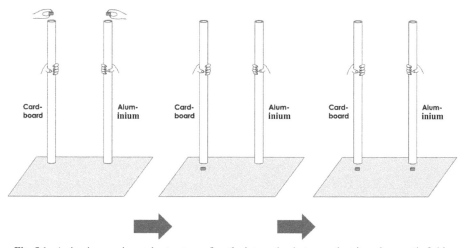

Fig. 5.1: A simple experimental setup to confirm the interaction between electric and magnetic fields.

aluminium. You ask your friend or your family member to hold two identical pieces of magnet at the same height and just above the openings of respective pipes. The diameter of these magnets is a little smaller than the diameter of the two pipes. Next you ask him or her to let the two magnets off all at once. How do they fall then?

The answer is as shown in the second and third drawings in Fig. 5.1: the magnet falling down through the cardboard pipe reaches a little earlier on the desk surface than that through the aluminium pipe does. This is not a mere imagination but surely happens (I have confirmed it many times). In fact, so much of our everyday life in contemporary societies is dependent on this fundamental phenomenon: small motors propel the blades of fans to heat pumps for cooling and refrigeration.

The fall of two magnets occurs because of attractive force existing between the Earth and each of the magnets. The space in which such attractive force emerges is called gravitational field. It exists everywhere, not only where we live near the ground surface, but also interplanetary and interstellar space. Similarly, but differently, there exists electric and magnetic fields all over the space not only near the ground surface, but also the Universal space far beyond the Moon, the Sun and the Milky-Way galaxy.

The result of the magnet falling down through the aluminium pipe a little slowly than the other falling down through the cardboard pipe is owing to the interaction of the electric and magnetic fields. The interaction between electric and magnetic field is called "electromagnetism".

Everybody knows that two magnets attract or repel each other depending on how the surfaces of two magnets face each other. The strength of attraction or repulsion depends on the distance between two magnets. Such tendency of attraction or repulsion can be schematically expressed with a bunch of lines as shown in the left-top of Fig. 5.2. The lines shown in Fig. 5.2a are called magnetic-field lines; they are defined to originate from the terminal of north and end at the terminal of south.

The strength of magnetic field is represented by the number of lines going through a unit area of imaginary plane perpendicular to the lines. Suppose a reference point and imagine you move it towards the magnet. This results in the magnetic-field strength at the reference point becoming larger.

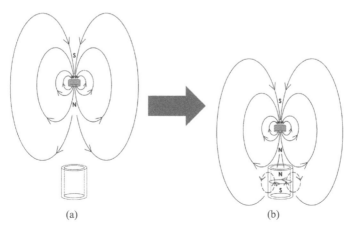

(a)   (b)

Fig. 5.2: A change of magnetic field strength within an aluminium pipe causes an electric current, which necessarily generates circular magnetic field.

The magnetic field exists all over the space, near and far, and it exists, of course, through a portion of the aluminium pipe, which exists below the magnet. Similar to the magnetic field, we may conceive that there is electric field that can also be represented by a bunch of lines flowing out uniformly from a positive electric charge and also those flowing uniformly into a negative electric charge. Since all of atoms are made of nucleus with positive charge and electrons with negative charge, there are electric fields all over the space of the atoms inside and outside.

As the magnet falls down as shown from a) to b) in Fig. 5.2, the magnetic field strength at some reference point within the aluminium pipe changes; it becomes stronger. This change of magnetic field strength induces the electric current to occur through the pipe body as shown in Fig. 5.2b, which was experimentally discovered by Faraday (1791–1867) and led him in 1830s to conceive the idea of electric and magnetic fields.

The electric current then newly induces the change of magnetic field also shown with two dashed, circular closed lines in Fig. 5.2b, which was discovered by Oersted (1777–1851) in 1820s. The induction of electric current followed by the change of magnetic-field strength was later mathematically theorized by Maxwell (1831–1879) in 1860s.

Since the magnetic field emerged in this course of circular electric current within the pipe body is necessarily repulsive to the magnetic field of the falling magnet, the magnet falling down through the aluminium pipe is required to take longer to reach the surface of the desk than the other magnet falling down through the cardboard pipe. In the cardboard, all the electron clouds formed in the surrounding tiny space of nucleus of atoms are fixed, though their centres always vibrates, so that no electric current emerges and thereby no change of electromagnetic field.

Figure 5.3 schematically demonstrates the change of electric field together with the change of magnetic field to be observed as a wave phenomenon. The reason that light is also called electromagnetic wave is that its fundamental cause is the interaction between electric and magnetic fields. This is exactly what was theorized

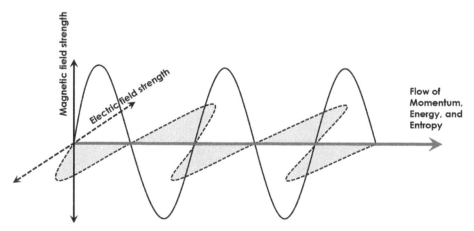

Fig. 5.3: The interacting change of electric and magnetic field strengths propagating as transverse oscillating wave, by which the momentum, energy and entropy are transferred. This is the electromagnetic radiation.

by Maxwell in 1860s. Towards the direction of wave, the momentum and energy are delivered and also as the collective waves, the entropy is delivered. Another name of electromagnetic wave, radiation, reflects its tendency of propagation in the space. The essence of energy and entropy concepts will be discussed later in Chapter 7.

There is another small experiment that we can confirm that light is really a wave phenomenon. It is to look at a bright surface through a slit. As a crude trial experiment with nothing to be prepared, you may make a slit with your right or left hand by closing two fingers and look at a bright surface, such as a window or a lamp, through the slit in between these fingers. You would find a few dark lines parallel to the fingers. They appear due to the fact that two sets of electric and magnetic field strength interfere with each other and cancel into nothing at our eyes; this is why it looks dark. This typical wave phenomenon is called "interference".

Such crude experiment can be improved to the one making use of two pencils and a couple of small pieces of hard paper, the thickness of which is about 0.25 mm, as shown in Fig. 5.4. Again, let us look at a bright light source through the slit. Of course, you would find a dark line. Knowing the distance between your face and the two pencils and also the slit width in between the two pencils, we can estimate the order of wavelength of visible light, which is the distance between adjacent two tops of magnetic or electric fields. The result comes into a range from 0.5 to 2 μm (= 500 to 2000 nm), that is, in the order of 10^{-6} m.

The velocity of electromagnetic wave propagating in vacuum space is known to be the maximum at 3×10^8 m/s and that propagating within a matter, which is denser than the vacuum is necessarily smaller. This is the cause of another wave phenomenon called "refraction". In the course of refraction, the light bends at the boundary surface of two matters so that the light travels through the two matters with the least time. How much it bends depends on the wavelength. Making use of this characteristic of refraction, we can split, for example, a beam of sunlight transmitting through a glass prism into a spectrum as demonstrated in Fig. 5.5.

What you can see on a sheet of white paper is the visible light split into a set of colourful bands, from left to right: violet, indigo, blue, green, yellow, orange and red.

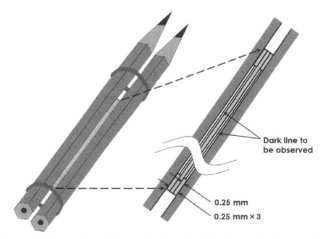

Fig. 5.4: A simple experimental setup for the estimation of wavelength of visible light.

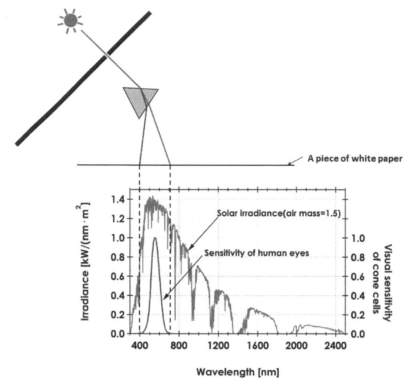

Fig. 5.5: Separation of a beam of sunlight transmitting through a prism into a spectrum. The data of solar irradiance with air mass 1.5 (solar altitude of 42°) is quoted from http://rrede.nrel.gov/solar/spectra/am1.5/.

The average visual sensitivity of human eyes differs from one colour to another: the highest is somewhere between green and yellow, the wavelength of 555 nm, and the lowest is either in violet, about 400 nm, or red, near 700 nm. It implies that the violet or red light has to be far more intense in order for the human eyes to sense the same brightness as with yellow-green light.

A graph shown at the bottom of Fig. 5.5 demonstrates an example of the distribution of spectral irradiance, the rate of energy carried by electromagnetic wave per unit length of wavelength, which is incident on a surface having the area of one squared meter. This is the case of a beam of solar radiation under a typical clear sky condition, which is incident on the horizontal surface at 42° of solar altitude. Such split of sunlight was first made by Newton in 1690s and later made by Hershel (1738–1822), an astronomer and musician, who happened to discover that there is also invisible light, "long-wavelength radiation".

By comparing the relative sensitivity of human eyes with the spectral distribution of solar irradiance, we come to know that the sensitivity of human eyes emerged so as to fit the characteristic of solar radiation in the evolutionary process. There is no doubt that eyes are important for each of us, but the emergence of eyes owned by a variety of animals as a whole in the course of evolution was also crucially important for the global ecosystem to develop as it is at present (Parker 2003).

5.2 Human eyes and visible light

Figure 5.6 shows a horizontally-cut view of the human right eye ball looking from above. The eye ball is about 24 mm in diameter and weighs about 7.2 g. Visible light comes into the eye through cornea, pupil and lens. The muscles connected with the cornea and lens react so that visible light transmitting through the cornea and lens is focussed at the layer of light sensing portals, rod and cone cells. They are distributed as shown with the curved coordinate in Fig. 5.6. Cone cells are concentrated near the fovea, while on the other hand, rod cells are scattered. The light-sensing terminals, rods and cones are embedded right above the bottom of retinal layer.

The cone and rod cells are connected with the nerve cells called horizontal cells, bipolar cells, amacrine cells and ganglion cells, all of which come together to form receptive fields. The optical nerve cells, each of which has its own receptive field, stretch their own axons out of eye balls into the brain as shown in Fig. 5.6.

There is neither rod nor cone cells in the region where the axons of optical nerve cells go out from the eye ball and stretch towards the brain. This fact can be easily confirmed with your right eye using two simple sets of circles and lines as shown in Fig. 5.7. First, you close your left eye and gaze at circle A, to which your sight line is exactly perpendicular. As shown in Fig. 5.6, this line of sight is slightly rotated inward at 3 to 5° compared with the axis of lens (Ikeda 1988). At a certain distance between your right eye and circle A, you will find that circle B disappears. This confirms where the optical nerve cells go out through the eye ball.

Next, you gaze at circle C in the same manner as you did with circle A. Then you will find that the two straight lines look crossing each other, though in reality there are four straight lines disconnected from each other as shown in Fig. 5.7. This implies that the brain processes the two lines so that they look as if they are parts of a single straight line. In other words, our eye-brain systems are capable of looking at something that we do not sense. This is the fundamental process of "perception" being different from "sensation". It is, I think, important to recognize this difference in order to understand the whole of cyclic process from sensation via perception and cognition to behaviour discussed with Fig. 3.7 in Chapter 3.

The number of optical nerve cells is known to be approximately in the order of 1.2×10^6, while on the other hand, that of rod and cone cells all together to be in the order of 1.2×10^8. This implies that the pieces of visual information sensed originally by each of cone and rod cells are reduced necessarily to one hundredth. Such process of reducing the pieces of information may well be related to the emergence of geometry as the most classical branch of mathematics in human history, although it is a kind of speculation. In Euclidean geometry, "point" is defined as one of the axioms which is something that has no area at all no matter how much you magnify it. In real world, a point that we draw on a piece of paper has a certain size of area so that, strictly speaking, it is not the point as defined in Euclidean geometry. The same applies to a "line" and a "plane" as the extension of a point. The reduction of the pieces of information from cone and rod cells to optical nerve cells must well be the basis of such axioms.

On the one hand, there are three different types of cone cells: S-cones, M-cones, and L-cones, while on the other hand, only one type of rod cells. S-cones are the most

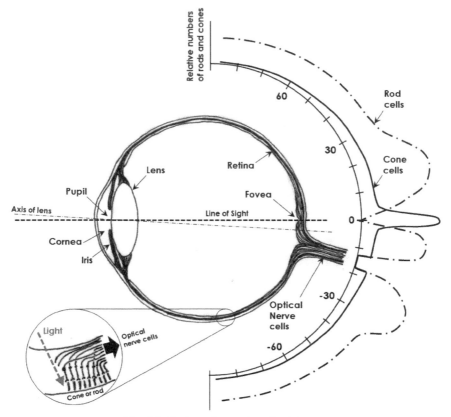

Fig. 5.6: Horizontally-cut view of a right eye.

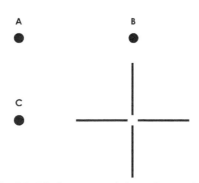

Fig. 5.7: Blind spot, no sensation and perception.

responsive to violet and blue portions of spectrum, M-cones to green and yellow, and L-cones to yellow and red (Matthews 2001).

The visual sensitivity shown in Fig. 5.5 is the overall averaged sensitivity of S-cones, M-cones and L-cones. The fact that three types of cone cells exist sounds consistent with the contemporary invention of light-emitting-diode lamps, with which three types of diode are used in order to generate white colour. This could be very much similar to what I mentioned in terms of geometry in relation to reducing the pieces of information from cone and rod cells to optical nerve cells. Creativity and innovation coming out from human brains seem to be very often realized as the mimicry of what has already happened in the nature. More examples are given in Chapters 10 and 12.

Figure 5.8 shows a comparison of the sensitivity of cone cells as a whole and that of rod cells. The peak wavelength for rod cells shifts about 50 nm shorter, from 555 nm to 507 nm. The luminous environmental condition in which rod cells mainly work is called "scotopia" and that, in which cone cells mainly work, is "photopia". Our eye-brain systems have scotopic vision under dark conditions and photopic vision under bright conditions.

In order to express the brightness or the darkness within the luminous environment, the primary quantity that we should know is illuminance, which is the rate of visible light flux incident on a surface having one squared meter. Its unit is

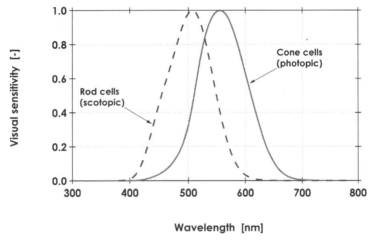

Wavelength [nm]

Fig. 5.8: Comparison of the sensitivities between cone cells for photopic and rod cells for scotopic vision.

lm/m², or lux or lx by abbreviation. "lm" is the abbreviation of "lumen" and one lumen is the visible light flux propagating through a cone of one steradian (sr) of solid angle; it is called one candela, which is equal to 1 lm/sr. The rate of visible light flux is measured as the rate of energy carried by electromagnetic radiation weighted by the visual sensitivity characterized by the photopic curve shown in Fig. 5.8 and by the maximum visible light flux emerged at 555 nm of wavelength, which is defined to be 683 lm/W.

The definition of illuminance, E, may be expressed as a mathematical formula as follows.

$$E = K_m \int_0^\infty V_\lambda \cdot I_\lambda \, d\lambda \tag{5.1}$$

where K_m is maximum luminous efficacy, which is the maximally available visible light at the wavelength of 555 nm, 683 lm/W; V_λ is visual sensitivity at the wavelength of λ; I_λ is the rate of energy carried by electromagnetic radiation at the wavelength of λ incident on the surface having one squared meter of area [W/(m²nm)]; and dλ is the infinitely small width of wavelength [nm]. The result of calculating Eq. (5.1), making use of the spectral visual sensitivity and the spectral rate of energy carried by solar irradiance under the condition of air mass 1.5 shown in Fig. 5.5, turns out to be 48780 lx or 48.8 klx.

Figure 5.9 demonstrates a wide range of illuminance, to which our eyes are usually exposed. The scotopic levels of illuminance range from 10^{-3} to 10^{0} lm/m² (= lx) and roughly correspond to such conditions as dark night outdoors with or without Moon. Under the condition of scotopia, the colours cannot be clear to our eyes and they look rather dark blue. Between scotopia and photopia, there are such conditions in which both rod and cone cells are active; it is called "mesopia" (Eloholma and Holonen 2008). This corresponds to such conditions of moderate side-walk lighting or road lighting.

The photopic levels of illuminance ranging from 10^{0} to 10^{5} lx correspond to such conditions provided by a variety of illumination available during daytime outdoors and indoors in general, and also provided by a variety of artificial lighting.

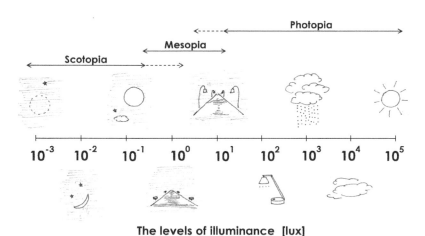

The levels of illuminance [lux]

Fig. 5.9: Scotopia, mesopia, photopia and their associated levels of illuminance.

5.3 Natural and artificial lighting

What is artificial lighting? People, whether they are non-professionals or professionals such as lighting designers, architects and engineers, usually reply that it is electric lighting. But, according to our discussion in Chapter 4 with respect to biological clock embedded within human body, artificial lighting should be redefined to be a smart combination of daylighting, which is one of the passive technologies, and electric lighting, one of the active technologies as schematically demonstrated in Fig. 5.10.

Table 5.1 overviews the characteristics of both daylighting and electric lighting together with natural lighting outdoors and traditional candle lighting.

Lighting performs in the flow of "light", which starts either at natural light sources such as the Sun and the sky or artificial light sources such as incandescent,

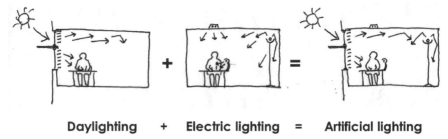

Daylighting + Electric lighting = Artificial lighting

Fig. 5.10: Artificial lighting is a smart combination of daylighting, one of the passive technologies, and electric lighting, one of the active technologies.

Table 5.1: Natural and artificial lighting.

Type of Lighting		Redirection and Scattering of Light	Light Source	Exergy Source [Fundamental Phenomena]
Natural Lighting (outdoors)		Rotation and revolution of the Earth, the atmosphere and the clouds	The Sun and the sky	The Sun [Nuclear fusion]
Artificial Lighting	Day-lighting	Windows and shadings	The Sun and the sky	The Sun and the sky [Nuclear fusion]
	Candle Lighting	Translucent glass covers, plates and louvers	Candles	Wax produced by plants and animals [Combustion]
	Electric Lighting	Louvers, fins and light shelves	Incandescent, fluorescent and LED lamps	Hydro and fossil-fuel fired (or nuclear) power plants [Gravitation and Combustion (or nuclear fission)]

fluorescent or light-emitting-diode lamps. It is important for us to have a clear image of what the flow of light is. Suppose, for example, when you wake up in the morning and open the curtain that covered a transparent glass window during night time. The sunlight comes inside and all things surrounding you in the room space such as the ceiling, wall and floor surfaces and also the furniture start to be illuminated. Suppose that you then close the curtain again. It becomes dark of course.

The same applies to the case of electric lighting. Pushing a button for turning on a lamp makes the room space bright since the light starts to flow and pushing the button again makes the room space dark since the light stops flowing.

All of these flows of light are the results of electromagnetism. The task of lighting designers, interior designers, architects and engineers is to redirect, scatter and thereby optimize the visible light with a variety of transparent, translucent or opaque reflective materials in order to create a luminous environment, within which sufficient brightness leading to luminous well-being is available with the least but necessary exergy consumption.

5.3.1 Daylight availability outdoors and indoors

Natural light source is made available by nuclear fusion, which is taking place all the time inside the Sun that has a huge gravitational force so that nuclear fusion reactions are being confined to the space within the Sun. The combination of huge gravitational force and ferocious nuclear fusion reaction keeps the state of dynamic equilibrium be formed by the Sun.

Thanks to such nuclear fusion reactions taking place in the Sun so far away from the Earth, but still sufficiently close to the Earth, and thanks also to the thickness of atmosphere to be neither too thin nor too thick, daylight available at the ground surface is to be regarded as the primary light source for artificial lighting.

Daylight consists of direct sunlight and diffuse sky light in general. Under clear sky conditions without clouds, there are both direct sunlight and diffuse sky light and under totally cloudy sky conditions, there is diffuse sky light alone. As demonstrated in Fig. 5.9, the range of daylight illuminance, which depends on the local sky conditions, the time of day, and the seasons, is very wide.

In the same manner as the illuminance was defined by Eq. (5.1), the rate of energy carried by solar radiation, irradiance I, the unit of which is W/m^2, incident on the surface of one squared meter is defined by the following equation:

$$I = \int_0^\infty I_\lambda \, d\lambda \tag{5.2}$$

where the symbols used are the same as in Eq. (5.1). The result of calculating Eq. (5.2) with respect to the spectral irradiance shown in Fig. 5.5 turns out to be 450 W/m^2.

The ratio of illuminance, E, to the irradiance, I, is called luminous efficacy, the unit of which is lm/W. In the case of the solar irradiance shown in Fig. 5.5, it becomes 108.4 lm/W (= 48780 lx/450 W/m^2). Table 5.2 summarizes the range of luminous efficacy of solar radiation together with the corresponding values of artificial light sources. For direct sunlight, the luminous efficacy is 60 to 120 lm/W and for diffuse sky light, it is 100 to 130 lm/W (Shukuya 1993).

Any kinds of visible light, whether it is emitted either by natural or by artificial sources, necessarily accompanies invisible light so that they have their own luminous efficacy values. Daylight, both direct sunlight and diffuse sky light, has remarkably higher luminous efficacy values than any of artificial light. Note that true luminous efficacy of electric lamps are those indicated in the brackets, which includes the

Table 5.2: Luminous efficacy of a variety of light sources.

Direct sunlight	60 ~ 120	lm/W
Diffuse sky light	100 ~ 130	
Incandescent bulb	10 ~ 20 (3 ~ 6)	
Fluorescent bulb	50 ~ 70 (14 ~ 20)	
Light emitting diode	60 ~ 180 (17 ~ 51)	
Candle	0.3 ~ 2	

■ The numbers in the brackets are the luminous efficacy including the rate of energy input to the power plant.

inevitable thermal energy emission at the power plants. They are much lower than those of daylight.

Candle light has the lowest luminous efficacy because it is the direct use of light emitted in the course of the combustion of wax in open air space. Note that the candle flame with a typical size emits approximately 12.6 lm (= 1 lm/sr × 4π sr) from the whole flame of a single candle into the surrounding space. With the luminous efficacy of candle being assumed to be 2 lm/W, we can estimate that a candle discharges all together visible light and thermal energy by radiation and also convection at the rate of 6.3 W, which alone is too small to provide sufficient warmth.

We can estimate daylight availability throughout one year based on the annual data set of solar irradiance on hourly basis using the luminous efficacy values of direct sunlight and diffuse sky light as the functions of solar altitude and sky condition (Shukuya 1993). Figure 5.11 is one such example generated from the annual weather data of year 2000 in Yokohama (AIJ 2005). The empirical formulae of luminous efficacy of direct sunlight and diffuse sky light used to produce the daylight availability in Fig. 5.11 is given as VBA function codes in 5.A.

The graph at the top shows the annual pattern of daylight availability on horizontal surface. The highest daylight illuminance on clear days ranges from about 60 klx in winter to 120 klx in summer, while on the other hand, that on cloudy days from about 20 klx in winter to about 35 klx in summer. There are quite a few days with low daylight illuminance in June; this is because of rainy season right before the summer season starts. The graph at the bottom demonstrates a variation of daylight illuminance for twelve days in October. Daylight availability varies from one day to another for ever changing sky conditions. The order of the highest daylight illuminance varies from 10 klx to 80 klx.

Figure 5.12 shows the occurrence of daylight illuminance with the bins of daylight illuminance at two-klx intervals and also cumulative occurrence of daylight illuminance for the whole year. The lower the daylight illuminance is, the more is the number of hours; daylight availability has such characteristic. The detail of such profile may be different from one place to another owing to the local climatic characteristics, but its essence that the number of hours for daylight illuminance gets fewer as the daylight illuminance becomes higher must be rather universal because anywhere on the Earth, the Sun rises early in the morning and sets late in the afternoon.

Within 8760 hours (= 24 h/day × 365 days), there are 4558 hours of daylight available as can be seen in Fig. 5.12. As was shown in Fig. 4.14, the number of average sunshine duration for one day is twelve hours, which is defined to be the length of hours when solar altitude is higher than zero. Therefore, annual total of sunshine hours is 4380 hours. The number of hours in which daylight is available was 178 hours more than 4380 hours in the year 2000 in Yokohama. Since those hours are either right before sunrise or right after sunset, the sky is not fully dark, that is, in the condition of mesopia. The outdoor daylight illuminance occurring for 75% of the total hours, in which daylight is more or less available, is up to 50 klx and the rest, 25%, is from 50 klx up to 120 klx. This characterizes the daylight availability in Yokohama.

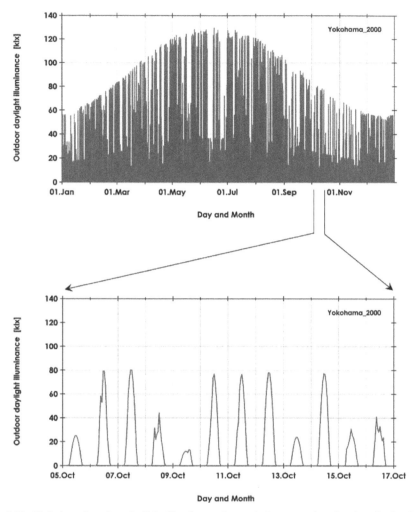

Fig. 5.11: Variation of outdoor daylight illuminance for a whole year and twelve days in October in Yokohama 2000 (AIJ 2005).

With the daylight availability outdoors as demonstrated in Fig. 5.11, one may estimate the daylight availability indoors by making indoor illuminance calculation. Figure 5.13 shows two examples of indoor horizontal illuminance calculated at 0.75 m above the floor surface in the centre of a small room having the floor area of 36 m² (6 m wide × 6 m deep × 3 m high) as shown at the top of Fig. 5.13. The window is assumed to be 3 m wide and 1.5 m high in the centre of south oriented facade of building envelope. The window is equipped with a transparent glass pane of 6 mm and an internal shading device having the daylight transmittance of 0.5. The overall average reflectance of interior surface of the room is assumed to be 0.36. The whole calculation procedure is summarised in 5.A.2 with VBA codes.

The indoor daylight illuminance appears in response to the outdoor daylight illuminance. Annual variation of the highest indoor illuminance shown in Fig. 5.13

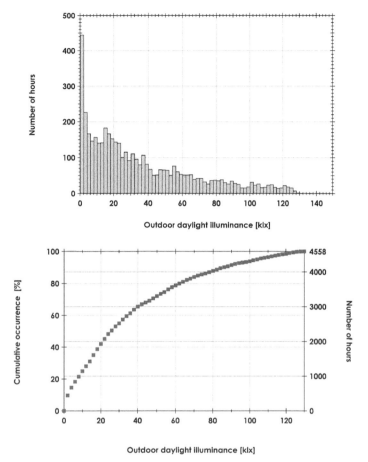

Fig. 5.12: Occurrence of outdoor daylight illuminance.

looks different from that of the highest outdoor illuminance shown in Fig. 5.11; this is because the indoor daylight illuminance is realized by the dispersive penetration of daylight through the vertical window. The overhang and side fins with the depth of 0.75 m significantly reduce the penetration of daylight and contribute to settling the indoor daylight illuminance within low but sufficient levels.

Figure 5.14 shows the cumulative occurrence of indoor daylight illuminance in the centre of the room with or without overhang and side fins. In the case of no overhang and side fins, 60% of indoor daylight illuminance is lower than 300 lx, but the rest, 40%, is between 300 and 1600 lx. With the overhang and side fins, the highest indoor illuminance is reduced to 1000 lx and for 90% of the hours, during which daylight is available, the indoor daylight illuminance stays below 400 lx.

If the efficiency of daylighting is discussed with how much of daylight is available alone, then the use of overhangs, side fins and other shading devices may look nonsense. But, it is not so, because the overhangs and side fins are very effective in reducing the unnecessary solar heat gain, as demonstrated in Chapter 4, in addition to controlling the availability of daylight so as to fill the indoor space with sufficient

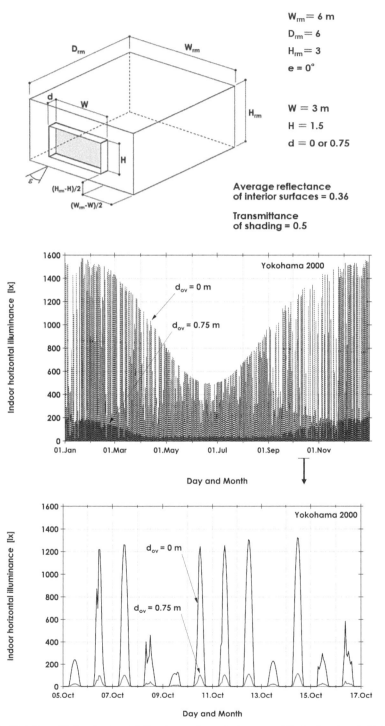

Fig. 5.13: Variation of indoor horizontal illuminance for a whole year and for twelve days in October in Yokohama 2000.

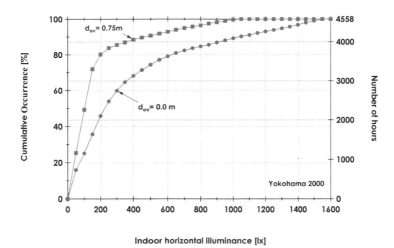

Fig. 5.14: Occurrence of indoor horizontal illuminance on hourly basis.

amount of daylight. Therefore, how much of daylight indoors is sufficient for the perceived brightness of occupants is of crucial importance. This issue is going to be discussed in the last section of this chapter.

5.3.2 Electric light and exergy consumption

In the case of electric lighting, we must not forget the fact that there is the flow of "work" at a certain rate behind the flow of light in the room space, since electric lamps perform lighting usually by being connected with the electric power plants through the electricity grids. Upon necessity, refer to how the "work" can be produced sustainably, discussed in Chapter 2.

The origin of electric lighting dates back to the year of 1879 when T. A. Edison (1847–1931), an American inventor and businessman, together with W. J. Hammer (1858–1934), who worked as Edison's chief consulting electrical engineer, succeeded in the generation of practical incandescent light with the carbonized bamboo filament durable for long hours. The basic form of a lamp, a glass ball with a metal base to be plugged into a socket, must have been determined so that it mimics the form of candle light, which had been used since very long until then.

Fifty to sixty years after the emergence of incandescent lamps, an experimental fluorescent lamp was set up by G. E. Germer (1868–1936), a German scientist, in 1927 and then practical fluorescent lamps of tube type to be sold in markets were produced first by an American team of engineers led by G. E. Inman (1895–1972) in 1938.

In our society at present, more than 130 years since the invention of incandescent lamps and more than 70 years since that of fluorescent lamps, the generation of people born before 1970s are all familiar with these types of lamps commonly used in the built environment, but younger generations born after 1990 may know little about incandescent lamps, since incandescent lamps are less used than fluorescent lamps and also one other, third type of lamps.

This third type of lamps are the ones based on the electroluminescence given by light-emitting diode (LED). A variety of LED lamps has started to be commercially available and to spread since the mid-2000s. The invention of LED dates back to when the effect of electroluminescence was observed by H. J. Round (1881–1966), a British engineer, in 1907 and later by O. Losev (1903–1942), a Russian scientist, in 1927. The LED with the intensive emission of visible light, mainly with red colour, was first made by N. Holonyak (1928 ~), an American scientist, in 1962. Later, I. Akasaki (1929 ~) and H. Amano (1960 ~), Japanese scientists, succeeded in the generation of blue LED in 1989 and S. Nakamura (1954 ~), a Japanese engineer, invented the LED that emits visible light with blue colour at high brightness enough for the purpose of indoor lighting in 1990s.

The emergence of blue LED triggered the intensive development of the third type of electric lighting as the follower of the first type with incandescence and the second type with fluorescence. This is because a variety of colours including white have become available with the combination of LEDs, which can emit the visible red, green, and blue light. The forms of LED lamps can be freer than the former two types because of the size of diodes is much smaller than the sizes required for the filament for incandescence and the electrodes and low-pressurized space for fluorescence.

Figure 5.15 shows three lamps of respective types in almost the same form. They fit into the same sockets, but require different electric power: 40 W for the incandescent lamp, 8.8 W for the fluorescent lamp and 5.5 W for the LED lamp. Figure 5.16 shows a comparison of these three lamps in terms of exergy-wise characteristics (Asada and Shukuya 1995, 1996, Yamada and Shukuya 2010, Shukuya 2013).

Assuming the rate of visible light to be 1000 lm, the required numbers of respective lamps shown in Fig. 5.15 turn out to be 1.7, 2.0 and 2.7, respectively. Figure 5.16 shows the total rate of exergy input to those lamps and the relative rates of consumption, heat and light for the illumination of 1000 lm. As can be seen, less electricity is required in the cases of fluorescent and LED lamps than in the case of incandescent lamp.

A large rate of exergy consumption within the incandescent lamps to give off 1000 lm is due to a large temperature difference between the filament made of tungsten, the melting temperature of which is the highest among a variety of metal, about 3400°C, and the environmental space of the lamp around 25°C. High temperature around 2000°C at the tungsten filament is required for the emission of visible light by incandescence, which is realized by the thermally least-conductive argon gas filled in the surrounding space of the tungsten filament inside the glass ball.

In the case of fluorescent lamp, the temperature of cathodes and anodes inside the low-pressurized folded tube has to be raised to let the cathodes emit electron particles for fluorescence, which requires much lower level of temperature than for incandescence. This makes the exergy consumption rate in the fluorescent lamps much smaller than in the incandescent lamps.

The electroluminescence taking place in LED lamps is realized by the difference in voltage between two semi-conducting materials connected to form p-n junction that lets the electron particles and the quantum-mechanical holes meet together and thereby emit visible light. This does not require high temperature so that the electricity

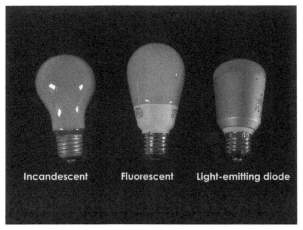

Fig. 5.15: Three types of lamp having the same basic form. Here in these examples, the electricity input required is: 40 W for the incandescent lamp, 8.8 W for the fluorescent lamp, and 5.5 W for the light-emitting diode lamp.

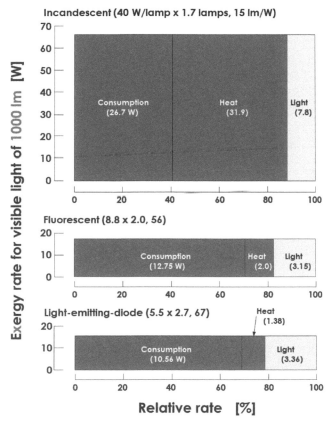

Fig. 5.16: Comparison of exergy input of incandescent, fluorescent and LED lamps for the illumination of 1000 lm and the relative rate of exergy consumption, heat and light. The numbers in the brackets within the bars indicate their actual figures.

input to LED lamps is small and also the amount of visible light is relatively large than that of heat.

The ratio of light to the total of heat and light in terms of exergy is 19.6% for incandescent, 61.2% for fluorescent, and 70.9% for LED lamps. The LED requires the least exergy consumption, 10.56 W, among the three types of lamps and also the least thermal exergy output while at the same time, the maximum visible light. This is consistent with what can be seen in Table 5.2; the luminous efficacy of LED lamps is the highest among the three types of electric lamps.

Keeping the exergy-wise characteristics of these lamps in mind, a rational design method for lighting in a variety of room spaces, which requires less exergy consumption rate with more luminous comfort than the present status of indoor lighting as one of the active systems, is to be re-developed. For such purpose, human psycho-physiological and behavioural characteristics have to be taken into consideration. What follows in the next section is the discussion on this issue.

5.4 Perceived brightness and adaptive behaviour

Figure 5.17 shows the transition of nominally required illuminance in Japanese office spaces, which has been mainly used for determining the number of electric lighting fixtures to be mounted on the ceiling, together with the transition of annual electricity use in Japanese office buildings. The median of required illuminance increased from 200, to 500, even to 1000 lx, but it decreased a bit to 750 lx in the year 2010. The annual rate of electricity use in office space has increased in a manner consistent with the increase of required illuminance.

Figure 5.18 compares the indoor required illuminance for office spaces to be made by electric lighting in various countries (Mills and Berg 1999, Yamada 2011). We can see that there is a large difference between countries, from the lowest of 100 lx to the highest of 1000 lx. It seems that high values includes both purposes of task lighting and ambient lighting and low values the purpose of ambient lighting alone.

The values shown in Figs. 5.17 and 5.18 are, anyway, the reference illuminance for the design of electric lighting in office spaces, but not the illuminance realized by electric lighting together with daylighting. In actual built-environmental space, the building occupants are exposed to ever-changing visible light as was demonstrated with respect to indoor daylight illuminance shown in Fig. 5.13. As we discussed in Chapter 3, we humans always live in the course of cyclic process of sensation, perception, cognition and behaviours. Therefore, we humans are consistently responding to a series of changes in illuminance given in the built environment both by daylighting and by electric lighting and also by our behaviours such as coming in and out the room spaces, staying here and there for a while for reading, writing, eating, resting and others (Naoi et al. 2003).

5.4.1 Variation of luminous environment and perception of brightness

Let us now move on to discuss the feasibility of renewing artificial lighting systems by referring an experiment performed in order to clarify how our perceived brightness

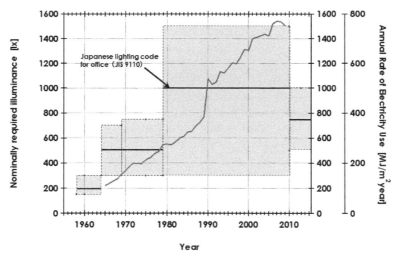

Fig. 5.17: Transition of nominally required illuminance and annual electricity use in office buildings in Japan.

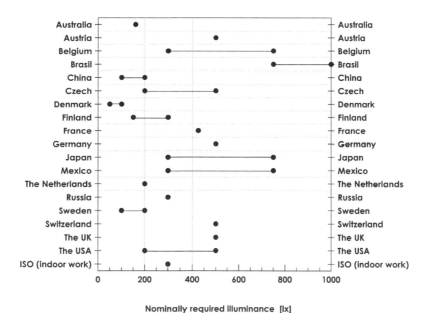

Nominally required illuminance [lx]

Fig. 5.18: Nominal values and ranges of required illuminance for office space in various countries (this is quoted from a graph originally made by Yamada 2011, based on the data compiled by Mills and Borg 1999).

is affected by the changes in illuminance in actual built environment (Kadokura and Shukuya 2013, Shukuya and Kadokura 2017).

This experiment was designed to invite 27 persons (fifteen men and twelve women, college students in early twenties) and ask them to visit 26 places and experience a variety of luminous environments during a one-hour period as summarized in Table 5.3. Among the 26 places, 19 are indoors and 7 are outdoors. Indoor luminous environment was provided either by daylighting or by electric lighting. In the course of visiting 26 places, the subjects were asked to vote their perceived brightness.

Each of the subjects was asked to wear photometers on the forehead during the period of tour as shown in Fig. 5.19a. Each forehead illuminance was measured at one-second intervals and recorded electronically to portable data loggers carried by each subject, while at the same time, the horizontal illuminance and the vertical illuminance of four directions were measured in the vicinity of the subjects at ten-second intervals with the sensors and recorders on a cart as shown in Fig. 5.19b.

Based on the respective perception of overall luminous environmental condition in each place, while they were seated and read newspaper or while they stopped for a moment in a corridor and so on as shown in Fig. 5.19c and d, the subjects made voting by choosing one of the five categories: "dark", "slightly dark", "bright enough", "slightly too bright" and "too bright" and checking one of the corresponding boxes (Maki and Shukuya 2009, 2012). The subjects voted 42 times as indicated in

Table 5.3: The route for experiencing a variety of luminous environments.

Places		Daylighting	Electric lighting	Posture	Vote Number
1 Office room	6F	Yes	No	seated	1
2 Corridor	6F	Yes	No	standing	2
3 Corridor	3F	Yes	Yes	standing	3
4 Class room	3F	Yes	No	seated	4, 5, 6, 7
5 Corridor	3F	No	Yes	standing	8
6 Class room	3F	No	Yes	seated	9, 10, 11, 12
7 Corridor	3F	No	Yes	standing	13
8 Class room	3F	No	Yes	seated	14, 15, 16
9 Corridor	3F	Yes	Yes	standing	17
10 Class room	3F	No	Yes	seated	18, 19, 20, 21, 22, 23, 24, 25, 26
11 Entrance hall	1F	Yes	No	standing	27
12 Courtyard	outdoors	-	-	standing	28
13 Courtyard	outdoors	-	-	standing	29
14 Courtyard	outdoors	-	-	standing	30
15 Entrance	1F	Yes	No	standing	31
16 Entrance	1F	Yes	No	seated	32
17 Corridor	1F	Yes	No	standing	33
18 Entrance	1F	Yes	No	standing	34
19 Courtyard	outdoors	-	-	standing	35
20 Courtyard	outdoors	-	-	standing	36
21 Courtyard	outdoors	-	-	standing	37
22 Courtyard	outdoors	-	-	standing	38
23 Corridor	1F	Yes	Yes	standing	39
24 Corridor	1F	Yes	No	standing	40
25 Corridor	6F	Yes	No	standing	41
26 Office room	6F	Yes	No	seated	42

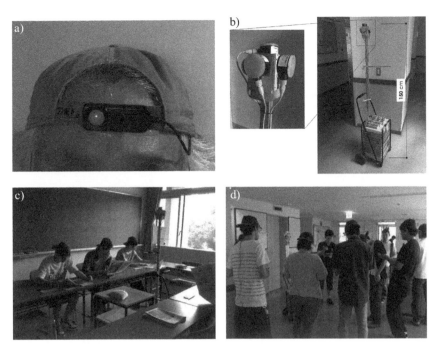

Fig. 5.19: Measurement of forehead, horizontal and vertical illuminances. (a) a sensor attached to a cap, (b) photometers for measuring horizontal and vertical illuminances on a cart, (c) a scene in a room with daylighting, (d) a scene in a corridor space.

Table 5.3. It took about twenty seconds to answer all of the questions at each of 42 times, which comprise not only the perceived brightness but also thermal perception and others.

The experiment was performed twice a day, one session from 10:30 to 11:30 and the other session from 12:30 to 13:30, for three days from 7th to 9th August, 2012. In each session, four or five subjects participated in and altogether 1134 votes from 27 persons with 42 times each were gathered.

Figure 5.20 shows two examples of measured forehead illuminance and the voting results of two subjects, who participated in the same session. Their forehead illuminance varies from 10 lx to 10000 lx for a one-hour period. Five categorical votes are indicated by closed circles. Although the variations of forehead illuminance look more or less the same for the two subjects, there are some differences in their respective votes.

The upper graph in Fig. 5.21 summarizes the whole profile of forehead illuminance measured at the times of voting. The total number of votes and their associated forehead illuminance is 1134 (= 27 subjects × 42 times/subject). The illuminance lower than 100 lx was 35% of the whole of measured illuminance values, that from 100 to 1000 lx was 45%, and that over 1000 lx was 20%. The lower graph in Fig. 5.21 shows the relationship between the forehead illuminance measured when the subjects were voting and the resulting votes.

As a whole, there is a tendency that the perceived brightness moves from the category of "too dark" to that of "too bright" as the forehead illuminance is

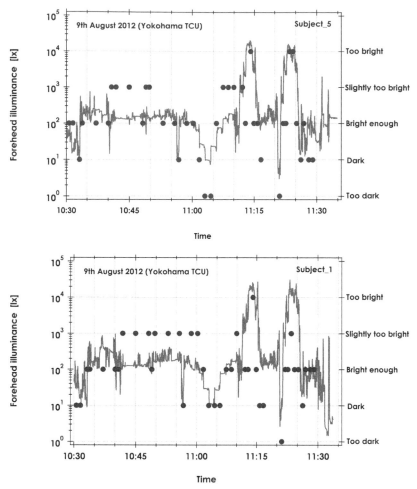

Fig. 5.20: Variations of forehead illuminance and perceived brightness of two subjects. The lines represent forehead illuminance and closed circular plots denotes the perceived brightness.

higher. But, there is a characteristic such that a certain value of illuminance does not necessarily correspond to a single category of perceived brightness but to a couple of categories. In other words, for one category of perceived brightness, its associated forehead illuminance can be either high or low. The vote of "dark" emerged as the forehead illuminance was higher than 2×10^2 lx, while on the other hand, the vote of "slightly too bright" emerged as the forehead illuminance was lower than 2×10^2 lx. The votes of "bright enough" are obtained from 10 lx to 2×10^4 lx. The ranges of forehead illuminance corresponding to "dark", "bright enough" or "slightly too bright" are quite wide and overlap each other. For example, the forehead illuminance of 100 lx could result in "dark", "bright enough" or "slightly too bright".

These results suggest that it should be possible to design indoor luminous environment in which sufficient brightness is with the requirement of small exergy input for lighting.

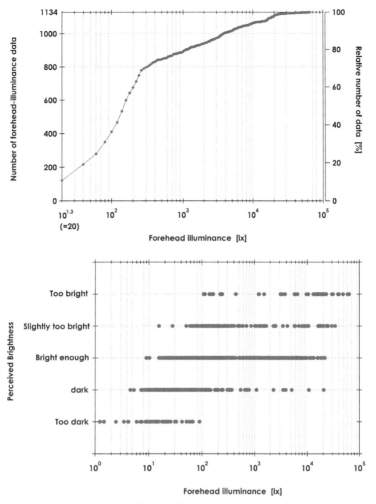

Fig. 5.21: The cumulative occurrence of forehead illuminance during the whole period of experiment and the relationship between forehead illuminance and perceived brightness.

5.4.2 Visible light exposed in the past affecting perceived brightness

The fact that a certain category of perceived brightness emerges depending on quite a wide range of forehead illuminance suggests that the perception of brightness is determined not only by the intensity of visible light coming into our eyes right at the moment when the perceived brightness is being asked, but also by the memory of luminous environment built up and etched in our brains as the relationship between the subjective brightness and the amount of visible light in the past.

For this reason, it is necessary to set up a reasonable variable that can reflect both effects of visible light at present and in the past. The variable should also reflect the general characteristic that an increase in the amount of visible light causes the perceived brightness to be brighter. Here we use the ratio of logarithm of illuminances at present and in the past (RIPP) defined as follows. Let us denote RIPP with symbol η.

$$\eta = \frac{\log_{10}\left(E_{\mathrm{fh},n}\right)}{\log_{10}\left(\overline{E_{\mathrm{fh},n}}\right)} \tag{5.3}$$

where $E_{\mathrm{fh},n}$ is forehead illuminance at present, n is integer representing the present time, and $\overline{E_{\mathrm{fh},n}}$ is the running mean of forehead illuminance, which is defined as follows.

$$\overline{E_{\mathrm{fh},n}} = \alpha E_{\mathrm{fh},n} + \left(1-\alpha\right)\overline{E_{\mathrm{fh},n-1}} \tag{5.4}$$

The running mean of forehead illuminance, $\overline{E_{\mathrm{fh},n}}$, is the weighted average of two illuminance values: one at present and the other the running mean of illuminance at the time, which is one interval of time before, that is, $(n-1)\Delta t$, with the weighting factor α. It may be regarded as the forehead illuminance in which a series of illuminances in the past is taken into account, expressed as follows.

$$\overline{E_{\mathrm{fh},n}} = \alpha \sum_{k=0}^{M}\left(1-\alpha\right)^{k} E_{\mathrm{fh},n-k} \tag{5.5}$$

The value of weighting factor, α, is somewhere between 0 and 1. The value of α being equal to unity implies that the present value of illuminance alone affects the perceived brightness, while on the other hand, α being smaller than unity implies that the illuminance in the past influences more or less the perception of brightness.

The relationship between RIPP as a stimulant and the relative number of subjects perceiving either "dark" or "too bright" as the response is, in general, to be expressed in the form of logistic function, which is known to be applicable to a variety of biological behaviours associated with the environmental stimulation. Their correlation was examined with respect to the logistic curves with a range of weighting factor, α. To do so, all votes are first sorted into the bins of RIPP, such as $0 \le \eta < 0.1$, $0.1 \le \eta < 0.2$ ..., with a certain assumed value of weighting factor, α, and then for each bin, the relative number of the subjects voting either "slightly dark" or "dark" is calculated. Next, the relative number of subjects voting either "slightly dark" or "dark" as the response was expressed as a logistic function of RIPP as the stimulant and the coefficient of determination (the square of correlation coefficient) were calculated for all logistic curves with the respective values of weighting factor, α, assumed. As the results of this series of analysis, it was found that the correlation was the highest with the largest coefficient of determination being 0.956 and its level of significance was less than 0.1% in the case of weighting factor, α, being equal to 0.02.

Figure 5.22 shows the relationship between the values of RIPP expressed by Eq. (5.3) and the relative number of the subjects. The closed circles are the relative number of those voting either "slightly dark" or "dark" corresponding to the respective bins of RIPP on the abscissa. The open circles represent the relative number of those voting either "slightly too bright" or "too bright" in the same manner as in the case of "slightly dark" or "dark"; the coefficient of determination was also the highest at 0.842 with the weighting factor, α, being equal to 0.02. Again, the level of significance was less than 0.1%.

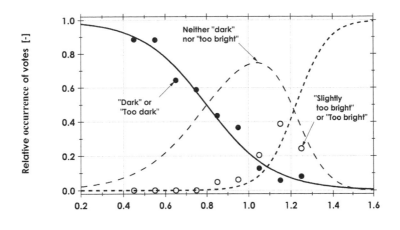

**Ratio of the logarithm of forehead illuminance at present
to that in the past (RIPP)**

Fig. 5.22: Relationship between RIPP and relative occurrence of votes.

The logistic curve with solid line is the regression of the case of either "slightly dark" or "dark" and that with dotted line the case of either "slightly too bright" or "too bright". Logistic curves are expressed in the following equations, respectively.

$$\ln\left(\frac{1-p_D}{p_D}\right) = 6.365\eta - 5.084 \tag{5.6}$$

$$\ln\left(\frac{1-p_{TB}}{p_{TB}}\right) = -13.461\eta + 16.495 \tag{5.7}$$

where p_D is the percentage of people who would vote either "slightly dark" or "dark", and p_{TB} is the percentage of people who would vote either "slightly too bright" or "too bright".

Figure 5.23 shows the relationship between the values of $\alpha(1-\alpha)^k$ appeared in Eq. (5.5), where $k = 0,1,2,3\cdots,M$, and the time towards the past, $k\Delta t$, where Δt is one second for the aforementioned experiment. It can be seen that the illuminance exposed in the past affects the numerator of RIPP for 240 seconds, that is, $M \leq 240$, but not more.

The dashed line in Fig. 5.22 indicates the percentage of people who vote neither for "slightly dark" and "dark" nor for "slightly too bright" and "too bright", that is, $1 - p_D - p_{TB}$ as a function of RIPP. Its peak is about 0.75 for RIPP = 1.04 on the abscissa. This suggests that, for example, if the whole lighting system of a room space together with its corridor space is designed so that the value of RIPP is about 1.04, then it would be likely that 75% of the occupants hardly perceive darkness in the room space. The value of RIPP should not go beyond about 1.1, since the number of people who would perceive neither "dark" nor "too bright" decreases.

Among all votes of "bright enough" shown in Fig. 5.21, those obtained under the condition of $1.0 \leq \eta < 1.1$ were found to be with the measured horizontal illuminance

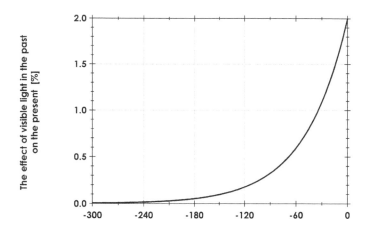

Fig. 5.23: Relationship between $\alpha(1-\alpha)^k$ and $k\Delta t$, where $k = 0,1,2,3\cdots,M$.

ranging from 100 lx to 1000 lx and the most occurred values within this set of horizontal illuminance were between 100 and 250 lx, while the second most were between 600 and 800 lx.

The horizontal illuminance from 100 to 250 lx corresponds to the lowest values of nominal required illuminance shown in Fig. 5.18 and also to the nominal required illuminance in late 1950s in Japan as can be seen in Fig. 5.17. The lighting designers, engineers, architects and others, who are used to high values of required illuminance such as 750 lx or 1000 lx, may regard that the horizontal illuminance between 100 and 250 lx may be too low, but with the experimental results discussed here in mind, it could be due to the fact that they themselves have been exposed to the luminous environment with very high illuminance.

Taking what has been described here in this section so far into consideration, let us imagine that the lighting system of an office space having the floor area of 50 m² including the lighting system of corridor space surrounding the office space is designed with RIPP being equal to 1.1. If the horizontal illuminance in the office space is assumed to be 300 lx, then the horizontal illuminance in the surrounding space turns out to be 180 lx. If it is 200 lx, then the horizontal illuminance in the surrounding space to be 125 lx.

Figure 5.24 shows a comparison of the whole exergy consumption patterns from the liquefied-natural-gas fired power plant to the wall and window of the assumed office space. As can be seen, a significant reduction in the exergy input should be possible: from 87 W/m² in the case of conventional required illuminance of 750 lx to 35 W/m² for 300 lx, or to 23 W/m² for 200 lx, with the assumption of a new lighting design taking the human adaptive nature into account.

If the half of all office buildings existing in Japan, the floor area of which is about 110 × 10⁶ m² (JREI 2015), adopt such a lighting system discussed here, then the total exergy input could reach the order of 3.5 GW, which is equivalent to three or four gigantic power plants, such as nuclear plants in Fukushima or Kashiwazaki-

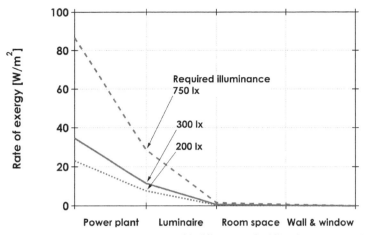

Fig. 5.24: Exergy consumption patterns with different required illuminance assumed.

Kariwa, to be replaced by more flexible and benign power supply systems in the future to come.

References

Architectural Institute of Japan (AIJ). 2005. Expanded AMeDAS Weather Data 1981–2000.

Asada H. and Shukuya M. 1995. An Exergy-entropy process of electric lighting using incandescent lamps. Proceedings of Annual Meeting of Architectural Institute of Japan, pp. 531–532 (in Japanese).

Asada H. and Shukuya M. 1996. Exergy-entropy process of electric lighting systems using fluorescent lamps. Journal of Architectural Planning and Environmental Engineering, Architectural Institute of Japan 483: 91–100.

Eloholma M. and Holonen L. 2008. Mesopic Vision and Photometry. Balkan Light 2008. Abstract paper (http://www.balkanlight.eu/abstract_pdf/ill.pdf).

Ikeda M. 1988. What are our eyes looking at?—Data processing of visual systems. Heibonsha. Tokyo Japan, pp. 125–126 (in Japanese).

Japan Real Estate Institute (JREI). 2015. http://www.reinet.or.jp/.

Kadokura S. and Shukuya M. 2013. Investigation on the luminous environment taking adaptive visual comfort into consideration. Proceedings of Annual Meeting of Architectural Institute of Japan, pp. 503–504 (in Japanese).

Maki Y. and Shukuya M. 2009. A field study on the visual and thermal comfort in a daylit room. Proceedings of Annual Meeting of Architectural Institute of Japan, pp. 421–422 (in Japanese).

Maki Y. and Shukuya M. 2012. Visual and thermal comfort and its relations to exergy consumption in a classroom with daylighting. International Journal of Exergy 11(4): 481–492.

Matthews G. G. 2001. Neurobiology—Molecules, Cells, and Systems—2nd ed. Blackwell Science Inc. Massachusetts USA.

Mills E. and Borg N. 1999. Trends in recommended illuminance levels: an international comparison. Journal of the Illuminating Engineering Society of North America, Winter Issue, pp. 155–163.

Naoi T., Wakatsuki T., Takeuchi A. and Shukuya M. 2003. An experimental study on brightness sensation acquired through experiences. Journal of Environmental Engineering, Architectural Institute of Japan 569: 55–60.

National Renewable Energy Laboratory (NREL). 2012. http://rrede.nrel.gov/solar/spectra/am 1.5/.

Parker A. 2003. In The Blink of an Eye—How Vision Sparked the Big Bang of Evolution. Basic Books. NY.

Shukuya M. 1993. Light and heat in the built environment—An approach by numerical calculation. Maruzen Publishers Ltd. Tokyo (in Japanese).

Shukuya M. 2013. Exergy—theory and applications in the built environment. Springer-Verlag London.

Shukuya M. and Kadokura S. 2017. An experimental investigation on the adaptive luminous comfort in the built environment. Proceedings of PLEA 2017 Edinburgh. Vol. II. pp. 3388–3395.

Yamada H. and Shukuya M. 2010. Entropy and exergy of light emitted by a fluorescent lamp and a LED lamp. Proceedings of Annual Meetings of Architectural Institute of Japan, pp. 145–146 (in Japanese).

Yamada H. 2011. Development of an Index for Luminous Comfort and Exergy Evaluation for Electric Lighting in Living Rooms. Ph.D. dissertation. Tokyo City University.

Appendix

5.A Estimation of outdoor and indoor daylight illuminances

5.A.1 Luminous efficacy of beam sunlight and diffuse sky light

Direct sunlight and diffuse sky light illuminances are usually not given as parts of the local weather data, but the rates of energy carried by direct and diffuse solar irradiances are usually available. Therefore, luminous efficacy of direct sunlight and diffuse sky light can be used for the estimation of direct sunlight and diffuse sky light illuminances. The luminous efficacy of direct sunlight and diffuse sky light given as respective VBA function codes shown in Tables 5.A.1 and 5.A.2 were used to produce the outdoor daylight illuminance (shown in Fig. 5.13) from the solar irradiance data in Yokohama.

5.A.2 Calculation of indoor daylight illuminance

The indoor daylight illuminance at a point on horizontal surface was calculated with the assumption that a window consisting of a transparent glass pane with an internal shading device and with an overhang and side fins is a uniform diffuse light source. The transmittance of transparent glass panes was given by a VBA code shown in Table 5.A.3. The shading effects of the overhang and the side fins were calculated

Table 5.A.1: VBA code for the estimation of luminous efficacy of direct sunlight.

```
Function Illum_DS(SH, IDN)
'
' This is a function program to convert the rate of energy delivered by direct
' solar radiation into the corresponding rate of sunlight [Lm/m^2].
' The estimation is based on an empirical formula made by Shukuya (1980).
'
' Input: SH -> sine value of solar altitude
'        IDN -> the rate of energy delivered by direct solar radiation [W/m^2]
'
E0 = 93.9: I0 = 1370#
eeta = E0 * ((6.25 * SH ^ 3 - 10 * SH ^ 2 + 3.94 * SH) * IDN / I0 + 0.983 * SH + 0.451)
Illum_DS = eeta * IDN
End Function
```

Table 5.A.2: VBA code for the estimation of luminous efficacy of diffuse sky light.

```
Function Illum_difS(SH, ISH)
'
' This is a function program to convert the rate of energy delivered by indirect
' solar radiation into the corresponding rate of diffuse sky light [Lm/m^2].
' The estimation is based on an empirical formula made by Shukuya (1980).
'
' Input: SH -> sine value of solar altitude
'       ISH -> the rate of energy delivered by indirect solar radiation [W/m^2]
'
E0 = 93.9
eeta = E0 * (3.375 * SH ^ 3 - 6.175 * SH ^ 2 + 3.4713 * SH + 0.7623)
Illum_difS = eeta * ISH
End Function
```

Table 5.A.3: VBA code for the calculation of transmittance of a transparent glass sheet.

```
Function staclr_T(ITT, CT)
'    This is a program to calculate the solar
'    transmittance of a clear glass pane of the thickness of 3, 6 or 12 mm
'Input: ITT -> thickness is given by the absolute value by 3, 6 or 12 mm;
'         if ITT>0 then the transmittance is for direct sunlight;
'         and if ITT<0 then for diffuse sky light.
'         CT-> cosine value of incident angle.
'
'         Data quoted from "Light and Heat in the Built Environment (Shukuya  1993)"
'                          25Feb2013 MS
    Dim CNT(4, 3), TF(3)
    CNT(1, 1) = 2.666504: CNT(2, 1) = -2.698242: CNT(3, 1) = 0.578125: CNT(4, 1) = 0.2900391
    CNT(1, 2) = 2.363159: CNT(2, 2) = -2.238281: CNT(3, 2) = 0.3154297: CNT(4, 2) =
0.3369141
    CNT(1, 3) = 1.845459: CNT(2, 3) = -1.557129: CNT(3, 3) = 0#: CNT(4, 3) = 0.3657227
    TF(1) = 0.762: TF(2) = 0.699: TF(3) = 0.577
    IT = Abs(ITT)
    If IT = 3 Then
          i = 1
    ElseIf IT = 6 Then
          i = 2
    ElseIf IT = 12 Then
          i = 3
    End If
    If ITT < 0 Then
      staclr_T = TF(i)
    Else
       T = 0
       For j = 1 To 4
       T = T + CNT(j, i) * CT ^ j
       Next j
       staclr_T = T
    End If
End Function
```

with the VBA code given in the appendices of Chapter 4. The indoor illuminance is the sum of two components: one is direct and the other indirect component. The direct component is calculated by multiplying the rate of daylight transmitted through the window and the configuration factor subtended by the window surface at a reference point indoors with the VBA code given in Table 5.A.4. The indirect component was calculated as the average illuminance of the whole interior surface of the room space as the result of inter-reflection.

Table 5.A.4: VBA code for the calculation of configuration factor subtended by a rectangular window.

```
Function UHV(a, b, c, IC)
    ' Configuration factor subtended by a rectangular surface at an infinitesimally small area,
    ' which is parallel or perpendicular to the surface.
    'Input:  a -> width
    '        b -> height
    '        c -> distance between the surface and the infinitesimally small surface
    '        IC ->   IC<=0 for "perpendicular"   and   0<IC for "parallel".
    '                                                   1992 and 2012  MS
    PAI = 3.14159265
    x = a / c
    y = b / c
    ry = Sqr(1 + y ^ 2)
    If IC <= 0 Then
        UHV = (Atn(x) - 1 / ry * Atn(x / ry)) / (2 * PAI)
    Else
      rx = Sqr(1 + x ^ 2)
      UHV = (x / rx * Atn(y / rx) + y / ry * Atn(x / ry)) / (2 * PAI)
    End If
    End Function
```

Chapter 6

Heat and Thermal Environment

6.1 Four paths of heat transfer—what happens in a candle?

Imagine that you light a candle and take a careful look at what goes on in the whole of the candle. It is quite interesting to observe as was wonderfully demonstrated in a series of lectures given by Faraday (1861, reprint in 2011). Even today when the modern theory of physics and chemistry has advanced so much, its concrete basis is nothing other than such phenomena in the flame of a candle light, which can be easily observed by anyone from ancient times to the present (Atkins 1991).

First, we find the shape of the whole flame as shown in Fig. 6.1: the middle and lower parts of the flame look thicker than the upper part and the bottom of the flame is in a little bluish colour and less bright, while on the other hand, the upper and middle parts are in yellowish colour and much brighter. This brightness is what you expect from the candle for lighting the surrounding space.

You also find that a small pool of liquefied wax has emerged on the top of the candle stick, the edge of which forms the shape of circle unless some wind blows away a portion of the liquefied wax. The position of the edge between liquid wax and solid wax is determined by how much of radiation is emitted downward by the flame and how much volume of surrounding air is induced by the hot flame, which is the ascending flow of the mixture of air and vaporized wax. The balance of the two phenomena, one called "radiation", as was already discussed in part especially on visible light in the previous chapter, and the other called "convection", which is the collective ascending flow of air mixed with the vaporized wax, affects the form, the size and the position of circular liquid-wax pool. Radiation and convection are two of the four paths of heat transfer.

In the very beginning before the flame starts having its stable form, there was no liquid-wax pool at all, but now there is. When you rub a match on a rough surface and thereby ignite the tiny amount of combustible matter on the top of match, you have a small flame. Then you put it on the top of wick not yet on fire. The flame grows at the top of the wick and starts transferring thermal radiation together with visible light and thereby the heat as the result of absorbing the portion of radiation conducts from the surface of solid wax into its inside because the temperature of the upper surface of solid wax increases as the radiation is absorbed and becomes higher than the temperature inside the solid wax. Due to their temperature difference, the

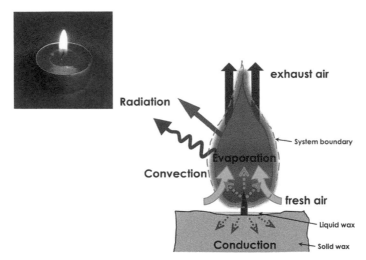

Fig. 6.1: A candle transferring heat into its environment by radiation, convection, evaporation and conduction.

heat conducts inside the solid wax. This is the third of the four paths of heat transfer, "conduction", which results in melting the solid wax into the liquid wax.

Without wick, a candle is not so easy to start burning. This suggests that the wick plays a crucial role in making the flame as a light source. Thermal radiation including visible light emitted by the flame is absorbed by the solid wax near the top surface of the candle, the portion of which is liquefied and thereby a tiny pool of liquid wax emerges as shown in Fig. 6.1. The wick soaks up the liquefied wax by so-called capillary effect. The liquefied wax, which ascends through the porous medium of wick due to the capillary effect, eventually evaporates at the top of the wick because of the high temperature. The vaporized wax is mixed together with an amount of air ascending along the wick, which is induced from the surrounding space of the candle stick, and thereby the hydrocarbon molecules and the oxygen molecules in the induced air meet and react vigorously with each other. This results in the formation of flame and the sustained emission of thermal radiation including visible light. The "evaporation" is considered to be the fourth path of heat transfer, since the evaporation occurs due to the effect of heating, which in turn results in keeping the temperature of matter remain stable where the evaporation takes place.

The major by-products of the whole process, which are carbon dioxide and water molecules, come out from the top of the flame as exhaust gas. The production of carbon dioxide can be confirmed by examining whether lime water turns milky as the exhaust gas goes through it, while on the other hand, the production of water can be confirmed by examining whether the condensation of water takes place at the bottom surface of a cold metal or glass sheet hung above the flame.

After a while when you have kept watching the candle, you will find that the length of the candle becomes shorter than before. The flame continues to emit light until all of the candle is gone unless some strong wind blows away the flame.

Conventionally, the set of three paths, radiation, convection and conduction, is called heat transfer and evaporation is called mass transfer. Nevertheless, the evaporation, or the condensation, which is the opposite phenomenon of evaporation, is involved, more or less, together with three paths of heat transfer; this is particularly so with respect to human body within the built environment. Therefore, heat transfer is re-defined here to consist of four paths: radiation, convection, conduction and evaporation.

The common feature of these four paths is that "dispersion" takes place due to the temperature or vapour pressure difference. As will be discussed in detail and clarified in Chapter 7, the dispersion proceeds so that the total of "energy" and "mass" involved is conserved while in due course the total of "entropy", which exactly quantifies dispersion itself, is increased.

6.2 What is heat? And what is temperature?

We are all very familiar with both "heat" and "temperature" in everyday life, from cooking to bathing, to wearing more clothes for warmth or less for coolness, to space heating and cooling. Probably because of such familiarity, we are not so conscious of their definition and when we are asked such questions formally as what the heat is or what the temperature is or even what the difference between heat and temperature is, we come to recognize that they are not so simple and not necessarily self-evident.

At what age do we usually come to know the two words: heat and temperature? It must be hard for most of you reading this treatise to recall exactly when these two words came into the part of your vocabulary that you use for everyday communications.

Suppose that there is an infant girl, who is hungry and wants to drink milk, and her mother just lets her drink milk from a nursing bottle. The mother carefully prepared the milk, but it is a little bit too hot. As soon as the infant girl starts drinking the hot milk, she starts crying because of sensing and feeling its hotness at her lips and tongue. Of course, her mother hushes her and speaks to her baby saying something, for example, "Oh, my poor little girl. I'm sorry. It's too hot, isn't it?"

The infant girl has been and will be exposed to a word "hot" that the mother just spoke to her and also some other associated words such as "cold", "warm" or "cool" again and again and in due course she acquires these words so that she can perfectly use them in the future. Repetition of similar experience now and then in various situations including such an experience mentioned above bring her to make a connection between her sensation and feeling of "hotness", "coldness" or others and their corresponding words. In such manners, all of us gradually build up the vocabulary. "Heat" and "temperature" are no exception.

The issue that we need to pay attention to is that the two words, "heat" and "temperature", are not necessarily self-evident and distinct. In our everyday conversations, for example, "heat" may imply the state of a body perceived as having a relatively high degree of warmth, the condition of being hot as described above, a perceived temperature higher than normal and so on. "Heat" may sound meaning "temperature" or vice versa.

To proceed scientific discussion with respect to heat transfer, it is necessary for us to distinguish the two words: "heat" and "temperature" clearly and make their definition more specific. To do so, let us think about a case of measuring the temperature of "hot" water contained in a vessel with a glass-tubed thermometer filled with ethanol (or kerosene), which is so-called alcohol thermometer, as shown in Fig. 6.2. When measuring the temperature of water or air with an alcohol thermometer, you need to bathe the thermometer in what is to be measured long enough before you read the value along the scale on the glass tube surface, to which the upper surface of alcohol sealed inside the glass tube matches.

Generally speaking, any matter, solid, liquid or gas, expands or shrinks depending on its state of being hot or cold. The alcohol inside the glass tube and the glass itself are no exception of course, but there is a significant difference between alcohol and glass regarding how easy the volumetric expansion is; alcohol is about forty times more capable to expand or shrink than glass is. Moreover, the alcohol is filled in a vertically very long and tiny space so that the relative expandability of alcohol to glass is considered to be much larger. Therefore, we can use such long glass tube filled with ethanol, kerosene, or mercury as thermometer.

The reason why we need to keep immersing the bottom of glass-tube thermometer for a while, where a relatively large amount of alcohol is stored, is that the temperature of thermometer itself, both of glass tube and alcohol in the bottom, is lower than the hot water whose temperature is to be measured. As shown in the middle case in Fig. 6.2, right after putting the thermometer in the hot water, the relationship between three temperatures that is, water temperature, t_{water}, glass-tube temperature, t_{glass}, and alcohol temperature, $t_{alcohol}$, is as follows: $t_{water} > t_{glass} > t_{alcohol}$. The level of alcohol inside the glass tube gradually increases and after a while it becomes unchanged. This is when we read the scale and equivalent to assuming that the water temperature, glass-tube temperature, and alcohol temperature are all equal to each other, that is, $t_{water} = t_{glass} = t_{alcohol}$. This condition of three temperatures being equal to each other is called "thermal equilibrium", which may be declared to be the Zeroth law of thermodynamics.

From the moment that the thermometer was immersed until the three temperatures become the same, what happens due to the temperature difference between high temperature and low temperature is called "heat" and we say that heat flows from

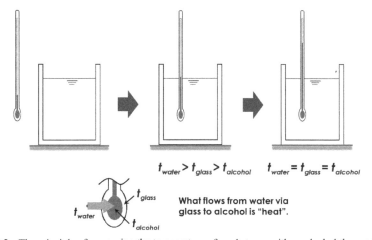

$t_{water} > t_{glass} > t_{alcohol}$ $t_{water} = t_{glass} = t_{alcohol}$

t_{glass}

t_{water}

$t_{alcohol}$

What flows from water via glass to alcohol is "heat".

Fig. 6.2: The principle of measuring the temperature of a substance with an alcohol thermometer.

where the temperature is high to where the temperature is low. Under the condition of thermal equilibrium, there is no heat flow because there is no temperature difference between the two surfaces or two points referred.

The discussion so far, I hope, has made the difference between "heat" and "temperature" clearer than in the very beginning, but you may still find it rather hazy. This is probably because what really flows due to the temperature difference between two points is not yet clarified but given the name "heat" alone.

In order to eliminate such haziness, let us next consider the relationship between molecular motion and the definition of temperature. Under contemporary educational systems, we usually learn in some classes of·science at junior or senior high school that any substance, not only surrounding us but also our living bodies themselves, consists of atoms and molecules. Most of us learn this principle as the basis of science and it is very rare that we suspect whether it is really true or not in the classes that elementary physics or chemistry is taught. If we simply believe the existence of molecules without any justification, our understanding would remain rather shallow and we have to admit that it is not scientific. Therefore, let us do confirm that the existence of molecules is true by observing what happens in a series of experiments as shown in Fig. 6.3.

We prepare two bottles: one containing an amount of water and the other an amount of fine sand grains, two ping-pong balls, and two copper balls. First, we drop a ping-pong ball and a copper ball in the bottle filled with water: the former floats because of its very low density and the latter sinks because of its much higher density than that of water.

We do the same for the bottle containing the fine sand grains. First, we drop a ping-pong ball and we find that it stays on the surface of sand grains. But, if we push down the ping-pong ball towards the bottom of the bottle while shuffling, then we can place the ping-pong ball near the bottom as can be seen in the top drawing in Fig. 6.3. Then we drop a copper ball. It sinks a little because of its heavy weight but stays almost as if it looks floating near the top surface of the whole of sand grains as also shown in the top drawing in Fig. 6.3. We see that the respective positions of a ping-pong ball and a copper ball in the water are opposite to those of the other ping-pong and copper balls in the sand grains. Suppose that two bottles are placed on respective flat plates as shown in Fig. 6.3, and the one on which the bottle filled with fine sand grains is placed is not mere a plate but a vibrator. Once we turn on this vibrator, then we can see that there emerges a seemingly random but still collective motion of the sand grains and soon some ascending movement of the sand grains emerges. In due course, as shown in the middle drawing of the bottle filled with sand grains, the ping-pong ball rises while on the other hand, the copper ball sinks and eventually the ping-pong ball reaches the surface of the whole of sand grains and the copper ball reaches the bottom. With the sand grains bouncing around, both the ping-pong ball and the copper ball eventually come to the same respective positions as those of the ping-pong ball and copper ball in the water.

This implies that liquid water molecules are all bouncing around each other, though we cannot see such behaviour of those water molecules directly with our own naked eyes, but with our mind eyes. We can now recognize that the concept of "temperature" is an indicator associated with how strong or weak the collective

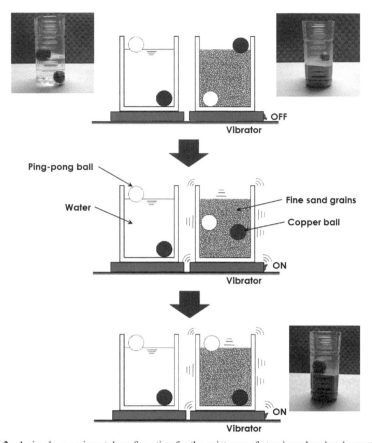

Fig. 6.3: A simple experimental confirmation for the existence of atomic and molecular particles.

but random movement of atoms and molecules is: this indicator is, as we all know through our everyday experience, being used thoroughly for all matters at the state of either solid, liquid or vapour.

Before the vibrator was turned on, none of the sand grains was moving at all; this corresponds to the state of ice because we can make a lump of ice with a ping-pong ball staying at the bottom and a copper ball on the upper surface of the real ice. In other words, each of the real water molecules as the state of ice stay within each of their tiny confined space, though each of them can vibrate within the respective tiny space. Once the ice melts and turns into liquid water, the water molecules go beyond their respective space to which they were confined to in the state of ice, and become freer to move, bounce around each other and thereby let the ping-pong ball float and the copper ball sink as we have discussed above.

The molecules of a body at high temperature are bouncing around more vigorously than those of the other body at low temperature. If these two bodies contact with each other, then the vibration of the molecules is transferred from the former at high temperature to the latter at low temperature. This is exactly what the heat transfer is; the "heat" is the transfer of randomly vibrating movement in the manner of dispersion, which is therefore irreversible. Such characteristic is quantified by the concept of entropy to be discussed in Chapter 7.

6.3 What happens as the whole of heat transfer

As mentioned earlier in 6.1, four paths of heat transfer takes place more or less simultaneously within the built environmental space and also within the occupants' body. This is considered to be very unique in thermal science associated with bio-climatology. In order to develop a holistic view of thermal features of the built environment both in winter and in summer, let us here discuss some typical results of a series of model-house experiments before learning about the various types of heat transfer: radiation, convection, conduction in the present chapter and evaporation in 8.5, Chapter 8.

6.3.1 Winter case studies

Figure 6.4 shows three scaled-down simple model houses, each of which is cubic-shaped with the size of 160 mm in external height, width and depth. Walls, roof and floor are made of hollow cardboard frame. The size of the window opening is 120 mm at the height and 120 mm at the width.

 Three houses are different with respect to thermal characteristics: "A" is the poorest among the three in thermal insulation and also with the least heat storing capacity; "B" has two pieces of bricks indoors for enhancing the heat storage capacity; "C" is identical to house "B" with respect to having two pieces of bricks for enhancing the heat storage capacity, but the frames of walls, roof and floor are filled with foam polystyrene. The frames of houses "A" and "B" are hollow inside.

 There is one transparent-plastic paned window in each house; houses "A" and "B" are single-paned and "C" is double-paned. There is no shading device over the windows in either of three houses. All windows are kept closed, since the experiment to be performed with these three houses is for winter.

 The experiment is performed in the following manner: first, an incandescent lamp as shown in the top of Fig. 6.4 is kept turned on for a period of fifteen minutes and then kept off for the next fifteen-minute period. Altogether, one thirty-minute period is taken to be a one-day cycle and then the experiment is repeated for three times, that is, to simulate a three-day period. In this series of experiment, the indoor space temperature is measured by alcohol thermometers inserted to the respective room space through the holes of the roofs as shown in the picture of experimental scene in Fig. 6.4. The bottom of the glass tube of each thermometer is placed around the centre of room space and exposed to the radiation coming from the interior surfaces of the room space and also to the room air. Therefore, the temperature value indicated by the thermometers is neither the pure average surface temperature nor the pure air temperature, but their combined average temperature. We call it here the indoor space temperature.

 Figure 6.5 demonstrates one example of measured indoor space temperature in the three model houses. As would be easily expected, while the lamp is on, that is, during daytime, the indoor space temperature gradually increases, and then while the lamp is off, that is, during night time, the indoor space temperature gradually

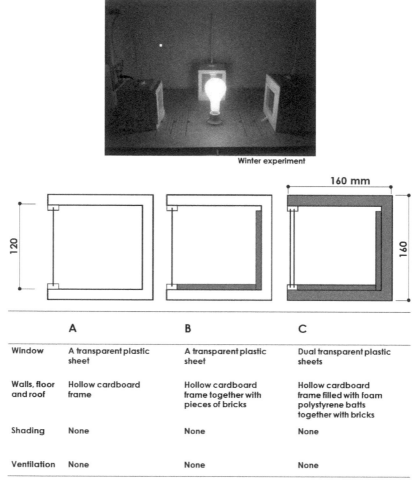

Fig. 6.4: An experiment for the comparison of indoor space temperature variations in three different model houses assuming winter settings.

decreases. The patterns of temperature variation in three houses are not different during the first daytime, but their difference gradually grows larger as the cycle of day and night proceeds.

During the whole period of experiment, the outdoor temperature for the three houses is almost constant at 22°C. The indoor space temperature reached at the end of third cycle of experiment in house "C" is about 5°C higher than that in house "A" or "B". This is because of the enhancement of thermal insulation in house "C" compared with houses "A" and "B". The difference in the highest and the lowest indoor space temperature in house "B" is slightly smaller than that in house "A". This is due to the effect of placing the bricks in house "B". It implies that massive substances, which can be indicated by the density, the mass per a unit volume, can contribute to attenuating the fluctuation of indoor space temperature, which should lead to creating a better indoor environment.

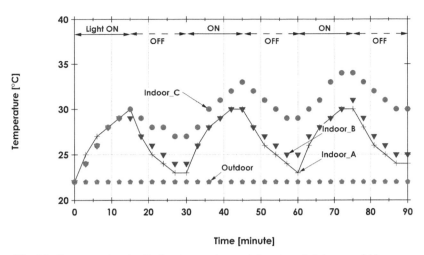

Fig. 6.5: Two example sets of indoor temperature variations in scaled-down model houses.

6.3.2 Summer case studies

Figure 6.6 shows three detached one-room houses prepared for summer-case experiment. All frames of three houses are filled with foam polystyrene boards. Houses "D" and "E" have no bricks but house "F" has. Three houses have different window settings from each other: in house "D", there is internal shading, which is fabric-made screen, while on the other hand in house "E", there is external shading called "sudare" screen, which is made of a bunch of dried reeds. The windows in house "D" and "E" are single paned. In house "F", the window is double paned and there is the same "sudare" screen over the external side of window as in house "E". The window with external shading in house "F" is kept open together with an opening in the upper part of rear wall as shown in Fig. 6.6. The experiment with these three houses was performed by following the same procedure as the winter case studies and the measurement of indoor space temperature was made with alcohol thermometers.

Figure 6.7 shows an example of the summer case experiment. The outdoor temperature during the whole experiment is between 24 and 25°C. There is, of course, a series of fluctuation in the indoor space temperature due to the turning on and off of a lamp as the simulated sun at every fifteen-minute interval as was observed in Fig. 6.5. There is a large difference in the values of temperature between houses "D" and "E". This is caused mainly by the respective positions of shading device hung along the window panes. The installation of external shading device has a significant effect on mitigating the indoor space temperature.

The difference between house "E" and house "F" is mainly due to the ventilation effect. It looks that ventilative cooling becomes very effective provided that the external shading is installed. At the end of experiment, there is nearly 4°C difference in indoor space temperature between house "E" and house "F".

After a whole of two sets of experimental investigation, one for winter and the other for summer, we confirm that solar radiation has heating effect, as we all know

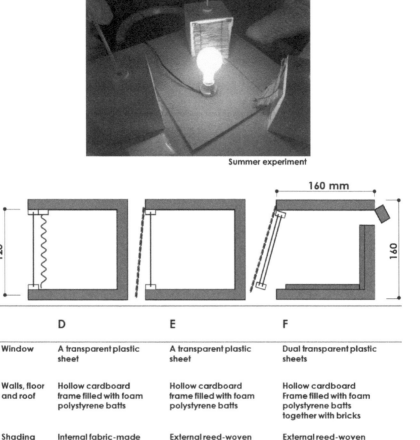

Summer experiment

Fig. **6.6:** An experiment for the comparison of indoor temperature variations in three different model houses assuming summer settings.

	D	E	F
Window	A transparent plastic sheet	A transparent plastic sheet	Dual transparent plastic sheets
Walls, floor and roof	Hollow cardboard frame filled with foam polystyrene batts	Hollow cardboard frame filled with foam polystyrene batts	Hollow cardboard Frame filled with foam polystyrene batts together with bricks
Shading	Internal fabric-made screen	External reed-woven screen	External reed-woven screen
Ventilation	None	None	Two openings

by everyday experience, which is preferable in winter, while on the other hand in summer, it is detrimental, thus to be avoided for creating thermally comfortable indoor environment.

Thermal insulation of the building envelopes, that is, to reduce the heat transfer by conduction, is essential to raise the indoor space temperature in winter. In order to increase solar control effect during summer time, the position of shading device relative to glass windows is of primal importance. The appropriate use of massive substances for indoor building envelopes should be effective in making the indoor space temperature fluctuation favourable for building occupants both in winter and in summer.

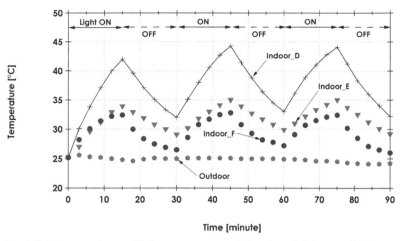

Fig. 6.7: Two example sets of indoor temperature variations in scaled-down model houses.

6.4 Basic characteristics of building materials

Here in this section, after knowing the conspicuous features of room space temperature in model houses in the previous section, let us learn the fundamental and quantitative characteristics in the three of four paths of heat transfer: radiation, convection and conduction; the fourth path, evaporation, will be discussed in Chapter 8.

6.4.1 Radiation

As shown in the upper drawing of Fig. 6.8, let us prepare a plastic bottle filled with hot water. One half of the round-shaped side is covered by a matte black thin film and the other half by an aluminium thin film. The plastic material used in this bottle together with these two types of cover is thin enough so that the surface temperature of the bottle can be assumed to be very close to the temperature of hot water.

Then, we measure the surface temperature with a so-called infrared thermographic camera, which allows us to measure the temperature of surfaces without touching them by deducing the values of temperature corresponding to the rate of black-body radiation that equals the rate of radiation received by the sensor of the infrared thermographic camera.

As can be seen in the pictures of Fig. 6.8, the surface, which looks black to our eyes, has the surface temperature much higher than the other surface, which looks shiny to our eyes; the former is 48°C and the latter 33.2°C. This is because that the matte black surface emits much more radiation than the shiny aluminium surface does.

The rate of energy emitted by black-body radiation is known, thanks to Stefan (1835–1893) and Boltzmann (1844–1906), to be proportional to the fourth power of absolute temperature of the black-body surface; this is expressed as

$$R_B = \sigma T_B^{\,4},\tag{6.1}$$

where R_B is the rate of energy emitted by black-body radiation [W/m²], σ is the proportional constant (= 5.67 W/(m²K⁴)), and T_B is the absolute temperature of the black-body surface [K], which is equal to the Celsius temperature plus 273.15.

The value of temperature indicated by the infrared thermometer is the temperature of an imaginary "black-body" surface emitting the same rate of long-wavelength radiation, which is the sum of both emitted and reflected radiation. In the case of matte black surface, which can be regarded to be very close to the ideal black-body surface, the value obtained by the infrared thermometer is quite close to the actual surface temperature. This implies that in the example shown in Fig. 6.8, the temperature of hot water is close to 48 or 49°C.

On the other hand, the surface covered by the aluminium film is significantly reflective so that the measured temperature is not the target surface temperature itself, but some temperature value including the effect of reflection from the surrounding space. In the case shown in Fig. 6.8, the measured temperature is 33.2°C, clearly lower than the actual surface temperature. The measured value with respect to reflective surface is considered to indicate a kind of weighted average of the target

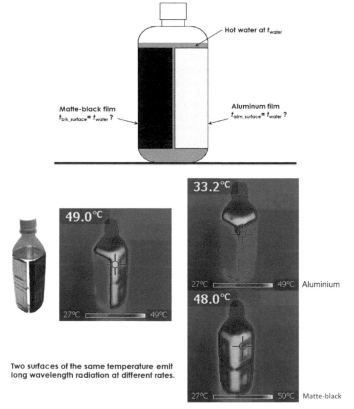

Fig. 6.8: Measurement of the surface temperature of a plastic bottle filled with hot water by a thermographic camera, which detects long-wavelength radiation emitted and reflected by the target surfaces. Measured temperatures of matte-black and aluminium surfaces become different from each other due to the difference in respective optical properties.

surface temperature and the surrounding average surface temperature. The rate of radiant energy detected by the sensor of the infrared thermographic camera matches the sum of emitted and reflected rates of radiant energy by the target surface. This relationship can be expressed as follows.

$$R_B = \sigma T_B{}^4 = \varepsilon \cdot \sigma T_{target}{}^4 + (1-\varepsilon)\sigma T_{surround}{}^4 \qquad (6.2)$$

where ε is the emittance of the target surface, which is at some value between 0 and 1.0, in which the unity is the emittance of the ideal black-body surface. Note that Eq. (6.2), with ε being 1.0, becomes exactly as Eq. (6.1). The emittance is the ratio of the rate of emitted radiation from a surface in question to the maximum rate of radiation to be emitted with the same surface temperature, that is, black-body radiation. T_{target} is the absolute temperature of the target surface [K] and $T_{surround}$ is the average absolute temperature of the whole surface surrounding the target surface [K].

(a) Long-wavelength emittance and solar absorptance

Most building materials have high values of their emittance, well above 0.9, as shown in Fig. 6.9. The exception is metal surface such as aluminium film; very newly polished one in particular has the emittance value of around 0.1 and the used one around 0.2. Since the materials shown in Fig. 6.9 are thick, they do not transmit long wavelength radiation; this means that their transmittance is null. Therefore, the value obtained by subtracting the emittance value from the unity represents the ability of reflection, which is called reflectance; this is the factor expressed within the bracket in the second term of the right-hand side of Eq. (6.2).

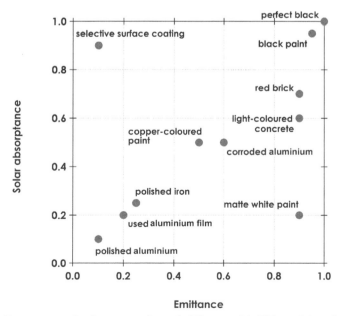

Fig. 6.9: Emittance versus solar absorptance of some building materials. This graph is produced using the set of data given by Watanabe (1965) and AIJ (1978).

Shown together with the emittance in Fig. 6.9 are the values of solar absorptance. If both emitted and absorbed radiations are in the same range of wavelengths, the emittance is exactly the same as the absorptance; this is called Kirchhoff's law, which states that the ratio of the emitted radiation to black-body radiation is identical to the ratio of the absorbed radiation to black-body radiation, provided that the wavelength of absorbed radiation is the same as that of emitted radiation. This law is required by the second law of thermodynamics, which will be discussed in Chapter 7. Kirchhoff (1824–1887) was one of the scientists in the field of thermodynamics and coined the name black-body radiation that will be described later in this section. Therefore, Fig. 6.9 is the comparison of two kinds of absorptance: one for long wavelength radiation and the other for solar radiation, which is short wavelength radiation.

Ideal perfect black surface locates the edge of the upper and right corner, both the emittance and solar absorptance being unity. Any other actual surfaces have lower emittance and lower solar absorptance. Very close to perfect black is matte black paint. Polished aluminium is the most opposite to the perfect black or matte black paint. The characteristics indicated either in upper-left corner or in lower-right corner may be called selective, since they represent either lower emittance with higher solar absorptance or vice versa.

The surfaces with so-called selective surface or selective absorber is appropriate for the purpose of solar thermal energy collection because of having high solar absorption while at the same time having low emittance for minimizing the long wavelength radiation dissipating into the surrounding space. Matte white paint is also selective in the opposite implication, that is, very effective in reducing the absorption of solar radiation, while on the other hand, very effective in dissipating the long wavelength radiation into the cold medium such as the clear sky vault, the temperature of which is much lower than the ground surface temperature. This is probably one of the reasons why vernacular buildings in Mediterranean islands are painted matte white as was shown in Fig. 2.4 in Chapter 2.

We may use Eq. (6.2) for making a rough estimation of the average surface temperature surrounding the bottle shown in Fig. 6.8 by substituting the value indicated by the infrared thermographic camera at the aluminium film surface, 306.35 K (= 33.2°C), to T_B and assuming that T_{target} can be given approximately by the value indicated by the infrared thermographic camera viewing the matte black surface, 321.15 K (= 48°C), and also assuming that ε is equal to 0.2, referring to Fig. 6.9. By solving Eq. (6.2) for the single unknown variable, $T_{surround}$, we get the value of $T_{surround}$ being equal to 302.25 K (= 29.1°C).

(b) Spectral distribution of radiation

Although the rate of radiant energy can be given by Eq. (6.1), which is solely the function of surface temperature, the reason why there exists such differences in the characteristics of material surface as shown in Fig. 6.9 is that the emissive or reflective characteristic of the material surfaces depends highly on the wavelength and on the spectral intensity of radiation, which shifts from one range to another if the difference in source temperature is very large. Such characteristics can be theoretically determined by the perfect-black-body radiation formula, thanks to

Planck (1858–1947), who was one of the students of Kirchhoff mentioned above. The perfect black-body radiation formula coded by VBA is given in 6.A.

Figure 6.10 demonstrates five cases of the black-body surface temperature from –20 to 60°C. The wavelength ranges from 3 to 60 µm and this range of wavelength is called "long". The peak of the spectral radiant energy emission is positioned around 10 µm, to which the sensors used in the infrared thermometers and thermographic cameras are usually sensitive, and amounts to 12 to 55 W/(m²µm),

For comparison with the spectral distribution of "long" wavelength radiation shown in Fig. 6.10, the spectral distribution of solar radiation at extra-terrestrial surface normal to the Sun and also that incident on the ground surface at typical atmospheric condition of air mass 1.5, which corresponds to solar altitude of 41.8°, is presented in Fig. 6.11 (NREL 2012). Shown together are the two cases of black-body radiant distribution to be received at the extra-terrestrial surface from the black-body sphere at 5500 and 5700°C, which are assumed to be at the same position as the actual Sun.

The overall spectral distribution of extra-terrestrial solar radiation is very close to those of the black-body radiation at 5500 and 5700°C. The difference between the actual extra-terrestrial solar radiation and the black-body radiation must be caused by the dynamics of solar wind, Earth's magnetosphere and others. The range of wavelength in solar radiation, which is from 0.3 to 3 µm, is much shorter, one tenth to one twentieth, than long wavelength radiation demonstrated in Fig. 6.10. Therefore, solar radiation is often called "short" wavelength radiation in particular for its clearly distinctive range of wavelength compared to so-called "long" wavelength radiation. Note also that the maximum rate of solar radiation is 40 to 100 times larger than those of "long" wavelength radiation.

It is known that the Universal space surrounding the Earth has the temperature at almost 3 K, which was first detected by Penzias (1933 ~) and Wilson (1936 ~), in fact accidentally, in 1964 while they were performing radio astronomical measurement

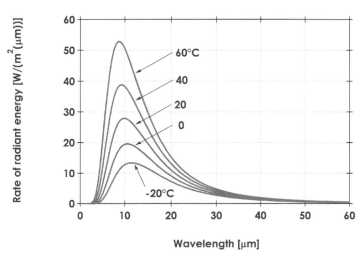

Fig. 6.10: Spectral distribution of radiant energy emitted by black-body surface at the temperature from –20 to 60°C.

with huge speaker-like shaped antenna. Since then, there have been a variety of measurements in terms of the Universal background radiation and at present the temperature is known to be exactly at 2.725 K due to the fact that the measured spectral radiant energy, which was detected by extra-terrestrial satellites such as COBE (NASA 2016), fits very well to the black-body radiation at 2.725 K.

Figure 6.12 shows the background radiation in the Universe. We can say that the Universal space is filled with the cosmic background radiation at 2.725 K. This background radiation is considered to have originated from the emergence of Universe some 13.7 billion years ago according to the so-called Big-Bang theory (Silk 1994).

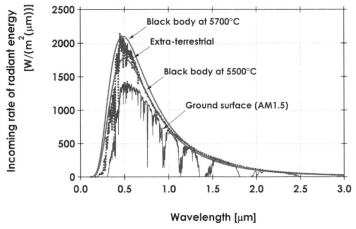

Fig. 6.11: Comparison of the spectral distribution of the incoming solar radiation at an extra-terrestrial surface and the ground surface together with the black-body radiation at 5500 and 5700°C to be incident on the extra-terrestrial surface.

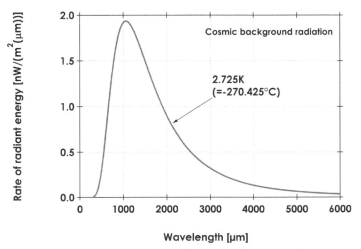

Fig. 6.12: The spectral distribution of black-body radiation at 2.725 K, which is known to fit very well to the measured rate of energy detected by COBE that explored the radiant characteristics in the Universe (NASA 2016).

The range of wavelength of the background radiation, which is from 300 to 6000 μm, is in the order of 100 times longer than the "long" wavelength radiation shown in Fig. 6.10. We can say that the "long" wavelength radiation, to which we are always being exposed indoors and outdoors, has its range of wavelength in the middle of "super long" and "short". Note that, not only the range of wavelength, but also the maximum rate of radiant energy is so different among "short", "long" and "super long" wavelength radiations: from 2×10^3 W/(m²μm), via 50×10^0, to 2×10^{-9} W/(m²μm). The maximum rate of energy in the background radiation is only 10^{-12} times of solar radiation: this implies that radiant environmental space in the Universe is super cold. Note that the Universal space is filled not only with such super-cold radiation, but also with a lot of ionizing radiation that is intrinsically harmful to biological systems. More on ionizing radiation will be discussed in Chapter 12.

(c) Spectral transmission

Using an infrared thermometer or thermographic camera, we can confirm that a sheet of glass used for windows has the characteristic of selective transmission. Suppose that you look at a glass window, through which you can see some trees, the objects in the garden, the adjacent buildings and others. As we can see all those things with our naked eyes, can the sensor of an infrared thermographic camera detect them through the window glass? The answer may seem yes, if we consider that what our eyes detect and what an infrared thermographic camera detects are anyway both radiations emerged based on the electro-magnetic phenomena. On the other hand, it may also seem no, if we imagine that there may be some similar characteristics in terms of transmission referring to the aforementioned selective characteristics on the emittance of long wavelength radiation and the solar absorptance shown in Fig. 6.9.

Figure 6.13 demonstrates an example of the right answer by comparing two pictures of a window in a residential building: one taken by an ordinary camera, and the other by an infrared thermographic camera on a sunny summer day in Yokohama, Japan. The left-hand side picture is the ordinary picture of a window glass that matches how our eyes can see and the right-hand side is the picture taken by an infrared thermographic camera, which allows us to know the distribution of surface temperature of the targeted window.

One half of the window, the left-hand side, is open for natural ventilation except the insect-preventing screen and the other half, the right-hand side, with double panes of glass is kept closed. In the outdoor side of this window, there is an external shading device hung from the edge of the roof. As can be seen in the thermographic picture, the left-hand side of the window, where there is no glass pane but insect-preventing screen, is at high temperature well over 40°C, while on the other hand, the right-hand side, where the glass pane exists, is at lower temperature well below 30°C. This implies that the insect-preventing screen transmits both solar radiation transmitted through the external shading and also the long wavelength radiation mainly emitted by the external shading, the temperature of which is very high because of the absorption of solar radiation, while on the other hand, the glass pane does not transmit long wavelength radiation at all, but visible light.

Figure 6.14 shows the spectral characteristics of a transparent glass pane, the thickness of which is 3 mm, over the range of wavelength from 0.3 to 100 μm. The

Fig. 6.13: Confirmation of whether or not the long wavelength radiation transmits a window paned by sheets of glass.

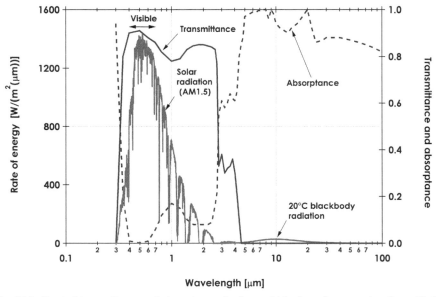

Fig. 6.14: Spectral transmittance and absorptance of a 3 mm thick clear glass pane together with the spectral distribution of solar radiation at AM1.5 and the black-body radiation at 20°C.

spectral transmittance is very high, well over 0.8 in the range of wavelength from 0.3 to 3 μm, and it is the highest especially in the visible range. On the other hand, the glass pane is totally opaque in the range of wavelength from 5 to 100 μm and the spectral absorptance, which is exactly spectral emittance, is higher than 0.8 and its average is close to the emittance of black paint shown in Fig. 6.9. Because of the characteristic shown in Fig. 6.14, we can say that a glass sheet is, in general, selectively transparent. This characteristic is optimal both for daylighting discussed in Chapter 5, and for the purpose of passive solar heating. It is also optimal for cooling, provided that external shading devices are utilized together as will be demonstrated later in this chapter.

(d) Coldness to be given by radiation

As everyday experiences, people are very familiar with the warmth or hotness given by solar radiation or by a frying pan or an oven. Therefore, the word "radiation" is very likely to call the image of "warmth" or "hotness" in mind while it is hard to call an image of "coldness" or "coolness" to be provided by long wavelength radiation. But long wavelength radiation emitted by the surface having lower temperature than our body can surely cause "coldness" or "coolness". Long wavelength radiation causing the perception either of warmth or coolness is more or less in the same range of spectral radiant distribution as was demonstrated in Fig. 6.10.

Figure 6.15 demonstrates the existence of long wavelength radiation causing "coldness" or "coolness". As can be seen in the left-hand side ordinary picture, a glass filled with a couple of ice cubes is hung above an aluminium bowl. Although the average temperature of the bowl as a whole is 28 to 29°C, the surface temperature detected by the infrared thermographic camera is 9.7°C. This thermographic picture assures that the long wavelength radiation emitted by the bottom of the glass is very well reflected by the internal surface of the bowl, which has low emittance as an aluminium film. The internal surface of the bowl having low emittance, that is, having high reflectance, reflects the long wavelength radiation coming from the bottom surface of the glass filled with ice and from other surrounding surfaces. Therefore, the reflected radiation originating from the glass filled with ice has an effect of lowering the surface temperature of the matter exposed to this radiation. If the matter is our skin or clothes, then this radiation causes the associated thermal sensation and then the resulting perception of "coldness" or "coolness".

Once we come to be conscious to such effect with respect to radiation described above, how we see our nature with our mind eyes become richer, I think. The sky is cold, though it is not so noticeable during daytime because of the dominating warming effect of solar radiation, but noticeable during night time because of the cooling effect of the sky radiation originating from the upper atmosphere, the temperature of which is kept always very low for the primary cooling due to the Universe being filled with the cosmic background radiation demonstrated in Fig. 6.12.

Fig. 6.15: Long wavelength radiation emitted by a cold surface, being reflected by a lustrous surface at a much higher temperature.

6.4.2 Convection

Convection takes place almost everywhere, along our body surface, wall, floor, window and ceiling surfaces. The cause of convection, as we have come to recognize through a discussion on the flame of a candle being formed by the collective movement of an amount of air, is a difference in temperature between the solid or liquid surface and the amount of air existing in its vicinity.

As we discussed what heat is in 6.2, the molecules of air are always bouncing around each other at a very high speed and changing their directions randomly within extremely short distances. Higher speed corresponds to higher temperature and lower to lower. With such nature of air in mind, if there is a temperature difference between one part of air which is very close to a solid surface, to the other part of air, which is relatively remote from the solid surface, and the air molecules involved can move collectively, then the convection as a unique heat transfer path emerges; this is the heat transfer due to the collective movement of lumps of air near the surfaces, whose temperature is different from that of air temperature.

Since it is a certain movement of mass of air from one position to another position, the convection can be regarded as a kind of "work" generated by a temperature difference between the air and the surface nearby. That is, it is a heat engine producing convective air movement as "work" in the course of "heat" transfer; such relationship between "work" and "heat" will be discussed more in detail in Chapter 7.

The process of convection and its associated heat transfer in general may be schematically drawn as shown in Fig. 6.16. Let us focus on an imaginary enclosed but open space denoted by the closed dashed line, which is assumed to be a system that transfers heat by convection due to the collective movement of lumps of air, along a vertical wall surface. Suppose that the wall surface temperature, T_s, is lower than the surrounding air temperature, T_a. Because of the temperature difference, an amount of air existing very close to the wall surface is cooled, its density becomes larger and thereby it sinks. Although the actual movements of air is complex, we assume that an amount of air flows into the imaginary system crossing its boundary surface denoted by the dashed line, while at the same time the same amount of air flows out from the system crossing its boundary.

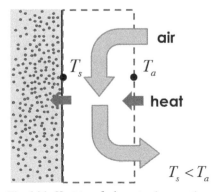

Fig. 6.16: Heat transfer by natural convection.

In general, the larger the temperature difference between the wall surface and the air in its vicinity, the greater the rate of heat transfer; this relationship may be expressed as the rate of heat transfer being proportional to the temperature difference as follows.

$$q_{cv} = h_c\,(T_a - T_s), \tag{6.3}$$

where q_{cv} is the rate of heat transfer in the unit of W/m², and h_c is convective heat transfer coefficient in the unit of W/(m²K) or W/(m² °C). Such an expression as Eq. (6.3) for convection was first given by Fourier (1768–1830) (Bejan 2004). The values of h_c varies according to a given set of temperature, T_s, and T_a. A variety of empirical formulae and also nominal values are given in various textbooks such as Holman (1981), Bejan (2004), Turns (2006), Shukuya (1993) and others. With respect to the units used here, "W" and "K", some basic explanation will be given in the sections and chapters that follow.

Figure 6.17 shows the values of convective heat transfer coefficient which may appear indoors; they were obtained from empirical formulae for a vertical or horizontal flat surface as a function of temperature difference between the surface and the air in the vicinity, the VBA codes for which are given in 6.A.

As already mentioned above, the values of convective heat transfer coefficient becomes larger as the temperature difference becomes greater; in ordinary built-environmental space, the temperature difference is usually smaller than 4°C so that the convective heat transfer coefficient values are considered to range from 0.5 to 2.5 W/(m²K). In the convection along vertical surface, the values of heat transfer coefficient are determined simply by the temperature difference; the lines for the case of surface temperature being lower than air temperature look symmetrical to the lines for the case of surface temperature being higher than air temperature.

In the case of floor surface, whether the surface temperature is higher or lower than air temperature brings about a different value of heat transfer coefficient. For example, in the case of floor heating system, with which the floor surface temperature is higher than air temperature, the convective heat transfer coefficient becomes 2 to 3 W/(m²K), while on the hand, in the case of floor cooling, with which the floor surface temperature is lower than air temperature, it becomes 0.5 to 1 W/(m²K).

Such a difference similar to floor surface can also be seen in the case of ceiling surface. Low-temperature ceiling surface tends to bring about higher convective heat transfer coefficient than high-temperature ceiling surface does.

Convective heat transfer coefficient may also be given as a function of air velocity as shown in Fig. 6.18 instead of temperature difference. The solid line is the convective heat transfer coefficient along a flat vertical or horizontal surface as a function of air velocity (Kreith and Bohn 1986), which was obtained from another VBA code given in 6.A; it ranges from 0.5 to 2.5 W/(m²K).

Shown together with a dashed line is the average convective heat transfer coefficient around the human-body surface; this was also obtained from an empirical formula developed by Gagge (1986), the VBA code for which is given in 6.A. The air velocity near the body surface is usually less than 0.2 m/s within indoor spaces having still air for the windows and doors being closed. On the other hand, the air velocity fluctuates very much, most likely between 0.2 and 1.2 m/s, within indoor

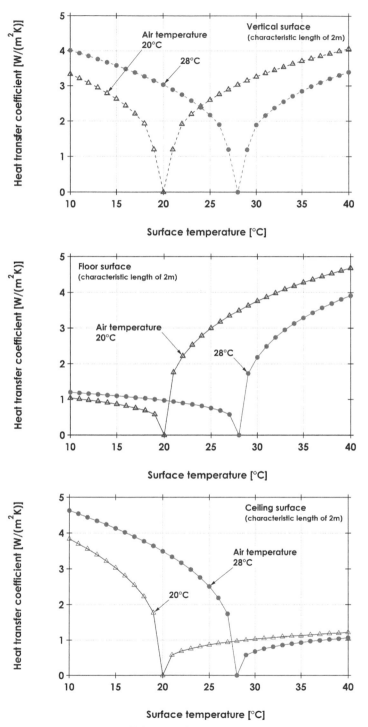

Fig. 6.17: Convective heat transfer coefficient as a function of temperature difference between vertical wall, floor and ceiling surfaces and air nearby.

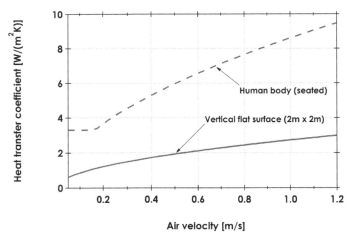

Fig. 6.18: Convective heat transfer coefficient along vertical surface as a function of air velocity nearby together with the average values of convective heat transfer coefficient around the human body.

spaces, where cross natural ventilation is being made by keeping the windows and doors open; more on this will be discussed in Chapters 10 and 11. In the latter cases in particular, we can see that the heat transfer coefficient increases significantly as the air velocity becomes higher. Therefore, we perceive the breeze provided by natural ventilation for the enhancement of convection, if solar radiation and long wavelength radiation incident on the human body surface is properly controlled (Shukuya 2015).

The air velocity is one physical quantity that relates to convection as demonstrated in Fig. 6.18, but the origin of air movement that is characterized by air velocity is necessarily the temperature difference as we discussed earlier in this section. In the case of human body, even if there seems no air movement, that is, air velocity is less than 0.1 m/s or so, the convective heat transfer coefficient becomes about 3 W/(m²K). This corresponds to the convective heat transfer coefficient given in Fig. 6.17 in the case of temperature difference being 4 to 6°C. The core temperature of human body is, as described in Chapter 3, at about 37°C and our surrounding air temperature is usually some values between 15 and 30°C. Therefore, even if the indoor air is kept as still as possible, there exists spontaneous and thus unavoidable current of air caused by the temperature difference.

The wind in the outdoor environment to be used for natural ventilation for indoor air conditioning is also brought by temperature difference, which emerges necessarily between the bottom of atmospheric air, which is exposed to the surfaces of the ground and the sea, and the upper atmospheric air, which is always cooled by the Universe filled with the cosmic background radiation. This temperature difference provides us, who live near the ground surface, with high and low atmospheric pressure patterns all over the Earth. The wind, which is the bulk movement of lower atmospheric air from a high pressure zone to a low pressure zone, is provided by a combined effect of the "hotness" given by solar radiation and the "coldness" given by the Universe.

Selective utilization of convection provided by both large scaled or small scaled heat engines within the built environmental space indoors and outdoors is thus to be recognized as important for built-environmental conditioning.

6.4.3 Conduction

Imagine how you change your clothes depending on the seasons throughout one year; in summer we wear less and in winter we wear more. This is because there is always some temperature difference between the human body and the surrounding space and it is necessary for us to control the rate of heat transfer from our body, through the clothes, to the environmental space depending on how high or low the environmental temperature is. As we all know through our daily experiences, the thicker the clothes are, the less is the rate of heat transfer. A cloth thicker than another cloth is thermally more resistant than the latter. Such a relationship applies also to the heat transfer through solid matters, that is, conduction.

Thermal resistance and thickness of a material is considered to be in a proportional relationship. Figure 6.19 demonstrates quantitatively the relationships between the thermal resistance of some typical building materials and their respective thickness. As expected, the thicker a material is, the more is its resistance. Thermal resistance shown in the vertical axis is the quantity that indicates how much the product of the surface area of a material in question and the temperature difference across the material is required for one unit of thermal conduction to take place within one unit length of time. The amount of heat is measured in "Joule" and the time in "second" here. The quantification of heat was originally made with the unit of "calorie", which is defined to be the amount of heat that raises by one degree Celsius ($1°C$) in one gram (1 g) of water. One calorie of heat is exactly equal to 4.186 J as will be discussed later in Chapter 7 and in contemporary science, the unit of "Joule", abbreviated as "J", is used thoroughly. In Fig. 6.19, the surface area is assumed to be 1 m² as shown in the drawing at the top. Therefore, the number in the vertical axis indicates how much of temperature difference is required for the rate of heat flowing at 1 W (= J/s).

As can be seen in Fig. 6.19, the slopes of the lines representing respective materials are different. From a gentle to steep slope, it is in the order of concrete, glass, plywood, cedar, cellulose fibre, and finally vacuum insulated panel (VIP). Thermal resistance of plywood (hard wood) at the thickness of 100 mm is about 0.55 m²K/W. With the same thickness, cellulose fibre board requires 2.7 m²K/W. This implies that cellulose fibre board is nearly five times more thermally resistant.

Glass and concrete, both of which have been used much in contemporary buildings, are thermally not resistant at all. The reason why such materials as glass and concrete are thermally least resistant is that they are very massive and there is little air space within them. In other words, light materials having more or less tiny air spaces within them are thermally more resistant than such materials as glass and concrete. Cellulose fibre materials contain a lot of tiny air space in between the slender filaments so that the corresponding slope is very steep.

The VIP is a highly sealed panel in which air is kept scant. Because of less air, the convection taking place within the space becomes less and most of the heat transfer inside the panel takes place by long wavelength radiation in the hollow spaces and by conduction through filaments; the radiation can be minimized with the use of reflective films as discussed earlier in this section. This is why the line of VIP is much steeper than cellulose fibre. In reality though, note that the size of VIP is limited because it has to be pre-manufactured in factories so that one may not be able to choose its size freely.

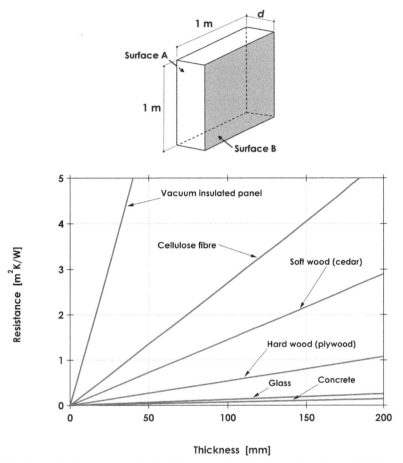

Fig. 6.19: Thermal resistance of typical building materials being proportional to their thickness; the conduction of heat assumed from surface A to surface B.

Let us express the relationship between the thickness and the thermal resistance as follows:

$$R = rd, \tag{6.4}$$

where R is thermal resistance [m²K/W], d is thickness [m], and r is the proportional constant, which characterizes each of the solid materials, usually called "thermal resistivity", whose unit is mK/W (= (m²K/W)/m).

Inversely proportional to thermal resistance, R, is thermal conductance, C, whose unit is W/(m²K), is expressed as follows and its relation to thickness can be demonstrated as shown in Fig. 6.20.

$$C = \frac{1}{R} = \frac{1}{rd} = \frac{\lambda}{d}, \tag{6.5}$$

where C is thermal conductance [W/(m²K)], λ is the inversely proportional constant and is called "thermal conductivity", whose unit is W/(mK).

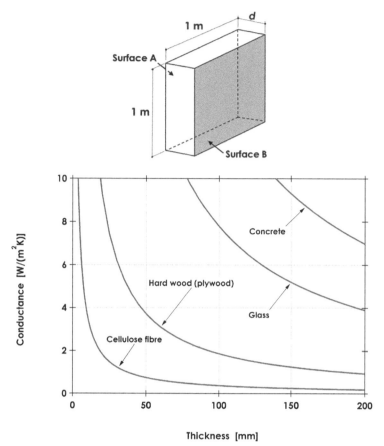

Fig. 6.20: Thermal conductance of typical building materials being inversely proportional to their thickness; the conduction of heat assumed from surface A to surface B.

The rate of heat flow for a unit surface area, q [W/m²], is expressed as

$$q = C (T_A - T_B),$$

(6.6)

where T_A and T_B are the temperatures of two boundary surfaces of a solid material shown in Fig. 6.20.

Table 6.1 lists thermal resistivity and conductivity of building materials demonstrated in Figs. 6.19 and 6.20 respectively and also those of other materials as well for comparison. Three other columns show specific heat capacity, volumetric heat capacity and density, respectively, to be discussed later in this sub-section. The materials are tabulated in the order of the thermal resistivity ascending from the top towards the bottom except the two materials in the bottom, water and air. As we have already discussed, the slope of a line in Fig. 6.19 corresponds to thermal resistivity; for example, thermal resistance of soft wood (cedar) is 1.45 m²K/W at the thickness of 100 mm and we can calculate the slope, which is thermal resistivity, as 14.5 mK/W by dividing 1.45 m²K/W by 0.1 m (= 100 mm). You can also find the corresponding value (14.49 mK/W) in Table 6.1.

Table 6.1: Thermal characteristics of building materials.

Material	Thermal resistivity [mK/W]	Thermal conductivity [W/(mK)]	Specific heat capacity [J/(kgK)]	Volumetric heat capacity [kJ/(m³K)]	Density [kg/m³]	Note
aluminium	0.005	210.0	879.1	2373.5	2700	
mortar	0.661	1.512	795.3	1590.7	2000	
concrete	0.717	1.396	879.1	1933.9	2200	
plaster	1.264	0.791	837.2	1632.5	1950	
glass	1.283	0.779	753.5	1913.8	2540	
adobe	1.457	0.686	879.1	1125.2	1280	
fibre reinforced plastic	3.839	0.261	1172.1	1875.3	1600	
linoleum	5.374	0.186	1172.1	1465.1	1250	
plywood	5.374	0.186	1297.7	713.7	550	hard wood
gypsum board	5.732	0.174	1130.2	1028.5	910	
tatami mat	6.614	0.151	1255.8	288.8	230	
cypress	7.692	0.130	800.0	519.2	649	
carpets	12.28	0.081	795.3	318.1	400	
cedar	14.49	0.069	1300.0	390.0	300	completely dry, soft
cellulose fibre insulating material	25.00	0.040	1260.0	31.5	25	
wood fibre insulating batt	25.64	0.039	2100.0	105.0	50	
expanded polystyrene	26.87	0.037	1255.8	35.2	28	
vacuum insulated panel (VIP)	125.0	0.008	1000.0	20.0	20	availability in size limited, the heat capacity and density assumed
water	1.678	0.596	4186.0	4186.0	1000	the values for liquid phase at 25°C
air	38.46	0.026	1007	1.18	1.184	the values at 25°C

Looking at Table 6.1, we find that the thermal resistivity of aluminium is remarkably small, in other words, remarkably conductive. Such a material is usually highly conductive for electric current, while on the other hand, highly reflective to electro-magnetic radiation and tends to be thermally highly conductive, that is, thermally less resistant. In Fig. 6.19, if the line for aluminium is drawn, it is almost on the horizontal axis. On the other hand, remarkably high resistivity is of vacuum insulated panel as already explained, since the convection and radiation are suppressed inside the panel and the heat has to be transferred only through fine structural members of the panel.

The materials having thermal resistivity higher than 25 mK/W or thermal conductivity lower than 0.04 W/(mK) are usually called thermally insulating materials. Table 6.1 shows that the resistivity of air is 38.5 mK/W, higher than 25 mK/W. This value is for the case where there is no convection at all and the heat is transferred only by conduction. As mentioned above, the reason why so-called insulating materials such as cellulose fibre, wood fibre, or expanded foam polystyrene are highly resistant is that they are porous and composed of tiny air spaces so that the convection hardly occurs. Figure 6.21 shows thermal conductance of air space, whose height is much longer than the depth, filled with air together with the hypothetical thermal conductance assuming no convection to occur. In this example, the temperature difference between two vertical surfaces is assumed to be 10°C. Up until the depth of 16 mm or so, the thermal conductance of free air

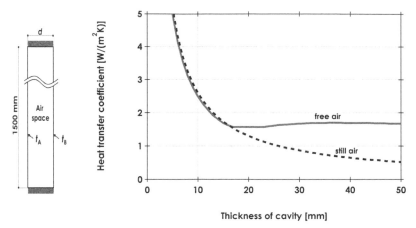

Fig. 6.21: Thermal conductance of air space due to the combination of conduction and convection (the temperature difference between two vertical surfaces assumed to be 10°C).

is almost the same as that of still air. But, with the depth larger than 16 mm, the convection emerges and thereby the values of heat transfer coefficient increases. At the depth of 50 mm, that is the depth is one third of the height in this example, the heat transfer by conduction diminishes to 30%, but there emerges the rest of heat transfer, 70%, owing to convection. Air can be a good insulating material provided that it is confined to a small space.

In the heat transfer by conduction, one other important characteristic in addition to thermal resistivity or conductivity is that how fast or slow the temperature of a material increases or decreases. As we observed in a series of model experiments discussed in 6.3, indoor space temperature increases and decreases gradually. How fast or slow the increase of indoor space temperature was dependent on the kinds of materials used. This is characterized by the concept of heat capacity, which is the amount of heat to be stored for a lump of matter to raise its temperature by 1°C (= 1 K).

Figure 6.22 shows the relationship between the heat capacity and the thickness of four typical building materials, assuming the surface area to be 1 m². The heat capacity increases proportionally depending on the thickness. This proportional relationship can be expressed as

$$Q = (c\,\rho)d, \tag{6.7}$$

where Q is the heat capacity of a material having the surface area of 1 m² and the thickness of d. The unit of Q is J/(m²K) and that of d is metre. The product of c and ρ, $c\rho$, is the proportional constant for the respective lines in Fig. 6.22 and called volumetric heat capacity. Denoted by c is specific heat capacity, which is the heat capacity per one unit of mass, 1 kg, that is J/(kgK), and ρ is the density of material in the unit of kg/m³. The respective values of $c\rho$, c and ρ are shown from the third to the fifth column in Table 6.1.

Shown together in Fig. 6.22 is the heat capacity of water. The slope of water is very steep; this means that the specific heat capacity of water is very large. This is

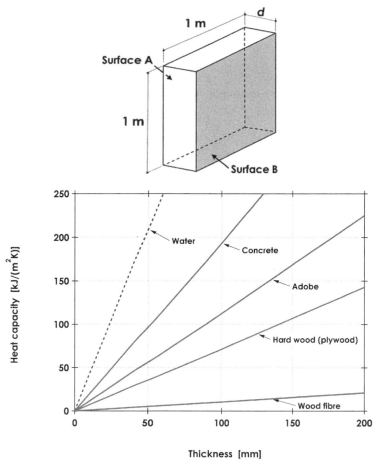

Fig. 6.22: Heat capacity for one-squared-metre surface area of typical building materials being proportional to their thickness; the conduction of heat assumed from surface A to surface B.

due to the fact that water consists of huge number of polar molecules, hydrogen side being positively charged while oxygen side being negatively charged; this implies that those water molecules are not so easy to move around if compared with neutrally charged molecules and because of this a lot of heat can be transferred by evaporation. Water cannot be used directly as a building material, since it is at liquid phase in the range of temperature within ordinary environmental space but because of this characteristic, it can be a good carrier of heat provided that it is filled, sealed inside pipe space and thereby circulated between two remote places.

6.5 Overall heat transfer characteristics of building envelopes

Applying all together the fundamental characteristics of heat transfer hitherto described, we can set up a physical and mathematical model of a wall, window, roof and others as a system and thereby calculate the space-wise distribution and the time-wise variation of temperature within the system. Making use of the values of temperature obtained, we can also perform further analyses with respect to exergetic feature of the system. Here in the last section of the present chapter, some of the examples in these regards are discussed and thereby extract the important features for building envelope systems to be equipped in order to create an environmental space for the occupants' well-being.

How to set up the mathematical models focusing on a variety of components and their aggregate is described in Chapter 9 and also in 7.6, Chapter 7.

6.5.1 The effect of thermal insulation

Figure 6.23 shows a comparison of the distribution of temperature across a 120 mm thick concrete wall without any insulating material attached and that across the same concrete wall with a 100 m thick foam polystyrene board. The assumption for the calculation are as follows: it is night time with outdoor sky radiant temperature at –15°C, ground surface temperature at 5°C, and outdoor air temperature at 5°C. Average indoor radiant temperature is 15°C for the not-well insulated wall and 19°C for the insulated wall, while the indoor air temperature is 20°C for both cases.

Provided that the energy balance equation set up for three systems: (1) exterior surface system; (2) internal boundary system where two materials meet each other; and (3) interior surface system, all of which are connected in series without any heat storage in between are solved with the assumed outdoor and indoor sets of temperature as the boundary condition, then the values of temperature shown in Fig. 6.23 are obtained. With the pieces of information assumed beforehand and also calculated thereafter, we can further determine the flows of energy, entropy and exergy through both walls and also the generation of entropy and the consumption of exergy. Figure 6.23 presents exergy inflow, outflow and consumption rate at the system of interior surface of the walls (Shukuya 2013, 2015). Two wave-like and two bold straight arrows represent "warm" radiant exergy and "warm" exergy transfer by convection and conduction, respectively. The numbers shown in rectangles are exergy consumption rates.

Comparing the two walls, one not-well insulated and the other well insulated, with each other, the most significant difference that we can see is the interior surface temperature. With the thickness of 100 mm of foam polystyrene board, the interior surface temperature can be raised more than 8°C. This is due to the decrease of thermal conduction; the exergy flow into the wall by conduction is decreased by 72% from 990 to 275 mW/m². This results in increasing "warm" radiant exergy emission by seven times more from 234 to 1699 mW/m². As a whole, with the enhancement of thermal insulation, the exergy consumption rate occurring at the interior surface decreases remarkably from 225 to 3 mW/m², more than 99% reduction.

The difference in temperature between interior surface and air is decreased from 10°C to 1.4°C by thermal insulation. This results in the diminishing convective heat transfer and thereby "warm" exergy transferred into the interior surface in the case of well-insulated wall becomes 36% of the case of not-well insulated wall, that is, 190 versus 531 mW/m². This is consistent with avoiding draft, unfavourable air movement in winter, which should be minimized as much as possible. Imagine wearing a muffler while walking outdoors; this is to avoid air movement around our

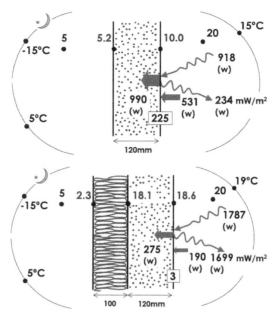

Fig. 6.23: Comparison of not-well and well-insulated walls in a wintry night time with respect to the temperature distribution and thermal exergy balance at the interior surface.

necks. Appropriate thermal insulation is for minimizing convection to cause draft while at the same time maximizing radiant exergy emission.

Figure 6.24 shows the results of a similar calculation made for daytime in summer. The assumption for the boundary conditions is as follows. Outdoor radiant sky and ground surface temperatures are 20°C and 36°C, respectively. There is solar radiant energy at the rate of 300 W/m² incident on the exterior surface having the solar absorptance of 0.5. Average indoor radiant temperature is 31°C in the case of not-well insulated wall while on the other hand, 28°C in the case of well insulated wall. Indoor air temperature is assumed to be 28°C in both cases.

The interior surface temperature is lowered by 5.9°C by insulation. Without insulation, "warm" exergy is transferred by conduction towards the interior surface through the inside of the concrete so that the interior surface temperature cannot be kept lower than outdoor air temperature and thereby "warm" radiant exergy is emitted at 50 mW/m². But with insulation, "warm" exergy is completely consumed inside the foam polystyrene board and thereby "cool" exergy flows by conduction from the interior surface into the concrete wall inside, towards exterior surface, while at the same time "cool" radiant exergy emission emerges from the interior surface of the wall. Exergy consumption rate at the interior surface is significantly reduced from 102 to 2 mW/m² by 98%. Exergy transfer by convection without insulation is 144 mW/m² for "warm" exergy towards indoor space, while on the other hand, it is 16 mW/m² for "cool" exergy with insulation. The absolute value of exergy which turns out to be only about 11% of the former is of course important, but switching from "warm" exergy to "cool" exergy is very much worth noting.

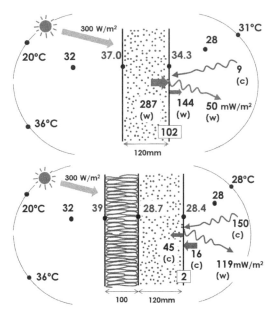

Fig. 6.24: Comparison of not-well and well-insulated walls in a summer daytime with respect to the temperature distribution and thermal exergy balance at the interior surface.

6.5.2 The effect of heat capacity

Figure 6.25 shows the variation of interior surface temperature (upper plate) and "warm" radiant exergy emission rate (lower plate) of the same two walls shown in Fig. 6.23 for a wintry four-day period in Yokohama. This temperature variation was obtained from the calculation of thermal energy balance equations set up for the whole of the walls taking the effect of heat capacity into consideration. As we observed in the series of model experiment discussed in 6.3, the heat capacity of the walls acts as a kind of damper to soften the fluctuation of temperature. As can be seen in the upper plate of Fig. 6.25, although the outdoor radiant temperature and air temperature fluctuate quite sharply, the interior surface temperature remains stable, that is, the fluctuation is much less. This is the effect of heat capacity.

In addition to such effect of heat capacity, the level of interior surface temperature is raised almost by 10°C with the effect of thermal insulation. This brings about a significant increase in "warm" radiant exergy emission from the interior surface as shown in the lower plate of Fig. 6.25. Filling the indoor environmental space with such sufficient rates of "warm" radiant exergy emission should be the key for the occupants to perceive warmth indoors during winter period.

Such enrichment realized by the appropriate combination of thermal insulation and heat capacity in fact dramatically decreases the exergy consumption rate within the wall. Figure 6.26 shows this effect, as an example, in the node having the heat capacity closest to the interior wall surface (about 15 mm inside). The exergy consumption rate ranges from 200 to 400 mW/m² in the case of poor insulation, but it becomes smaller than 10 mW/m² or almost nothing in the case of insulation

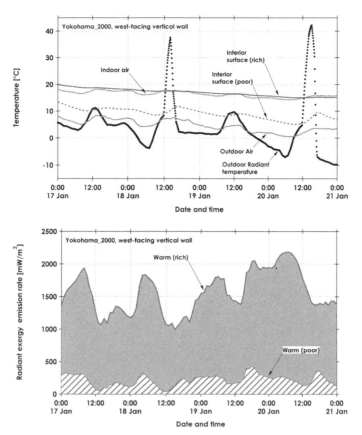

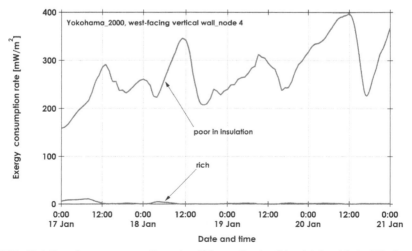

Fig. 6.25: Heat capacity together with appropriate thermal insulation providing with an increase of warm radiant exergy emission in winter.

Fig. 6.26: Variation of exergy consumption rate within a wall not well insulated and that within the other wall with better insulation in winter.

enriched. This implies that the avoidance of unnecessary exergy consumption for poor insulation results in a significant increase in "warm" radiant exergy emission as demonstrated in Fig. 6.25.

Figure 6.27 demonstrates the effect of heat capacity for summer in the case of flat roof, again in Yokohama. As can be seen in the top plate, the interior surface temperature fluctuates rather sharply, if there is no adequate insulation at all, but it remains rather unchanged provided that the effects of insulation and heat capacity are combined.

The lower two plates show the time series of "warm" and "cool" radiant exergy emission in the cases of not well insulated (left-hand side) and well insulated (right-hand side), respectively. We can see that without insulation, "warm" radiant exergy emission tends to be extremely large and "cool" radiant exergy emission to be small. But with insulation, "warm" radiant exergy emission can be significantly reduced and instead, "cool" radiant exergy emission is increased.

This increase of "cool" radiant exergy has become possible, since the exergy consumption rate within the roof material was reduced dramatically by appropriate thermal insulation as shown in Fig. 6.28. In the case of poor insulation, large exergy consumption tends to take place right before "cool" radiant exergy shown in Fig. 6.27 starts to emerge. On the other hand, in the case of insulation enriched, the exergy consumption rate tends to decrease dramatically as a whole and also the

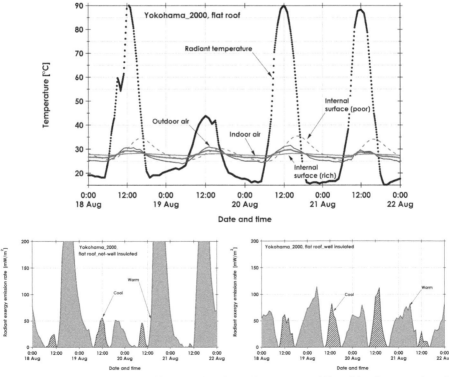

Fig. 6.27: Heat capacity together with appropriate thermal insulation providing with an increase of "cool" and decrease of "warm" radiant exergy emissions in summer.

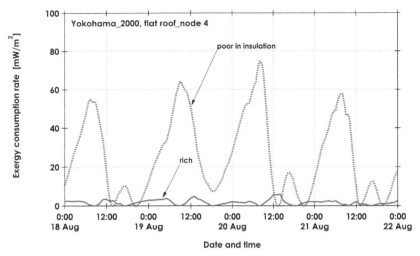

Fig. 6.28: Variation of exergy consumption rate within a wall not well insulated and that within the other wall with better insulation in summer.

swinging nature of exergy consumption rate looks almost the opposite to that in the case of poor insulation. At the time when the exergy consumption rate becomes the largest in the case of poor insulation, that in the case of enriched insulation tends to become the smallest. Such differences are considered to result in the increase of "cool" radiant exergy emission in the case of enriched insulation.

The examples shown in Fig. 6.27 suggest that appropriate thermal insulation together with the use of heat capacity is the key to produce the "cool" radiant exergy emission, which plays crucial role in providing the occupants with sufficient thermal comfort during summer seasons with rational consumption of both non-renewable and renewable exergy sources.

6.5.3 The effects of insulation and solar control on the window performance

Figure 6.29 shows a comparison of two double-glazed windows: one without shading and the other with internal shading. Since the glass panes and shading material are partly transparent, reflective and absorptive against solar radiation, the effect of absorption has to be taken into account when calculating each of the window components: (1) external glass pane; (2) internal glass pane; and (3) internal shading. Their solar optical properties: transmittance, reflectance, and absorptance, are assumed to be 0.8, 0.1, and 0.1 for glass pane and 0.5, 0.2, and 0.3 for internal shading material, respectively. The glass panes and shading material are so thin that their heat capacity is negligible. Therefore, the calculation of temperature for windows is basically the same as for that of walls demonstrated in Figs. 6.23 and 6.24, except the consideration of solar absorption by each of transparent window components.

As shown in both plates of Fig. 6.29, solar exergy of 275 W/m² is incident on the window surface. Other outdoor and indoor conditions assumed for the calculation

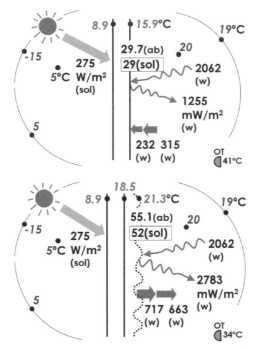

Fig. 6.29: Comparison of two double glazed windows: one without and the other with internal shading in terms of temperature distribution across the window components and exergy balance at the inner most surface in a wintry daytime.

of window systems are the same for those of the walls discussed with the examples shown in Fig. 6.23.

The value of temperature shown at a small hemisphere with a symbol OT (abbreviation of operative temperature) at the bottom right of each plate is the temperature of a hypothetical black hemisphere without heat capacity which was determined from transmitted solar radiation together with the internal surface temperature of window and room air temperature assumed. This is a kind of indicator of how much of transmitted solar radiation could influence the internal environment. It is 41°C without shading and 34°C with internal shading. This implies that in the case of no shading, much of solar radiation is transmitted. The solar radiant exergy absorbed at the internal glass pane is 29.7 W/m² and its 98%, 29 W/m², is consumed, while on the other hand the solar exergy absorbed at the internal shading is 55.1 W/m² and its 94%, 52 W/m², is consumed.

The difference between solar exergy absorbed and the associated consumption is "warm" exergy to be transferred into the surrounding space outwards and inwards by long wavelength radiation and convection. The temperature of internal shading at 21.3°C is 5.4°C higher than the internal glass surface temperature of the window without shading at 15.9°C because of the internal shading device absorbing much of solar radiation and thereby resulting in the lower OT than the case without shading.

"Warm" radiant exergy being emitted by internal shading at 2783 mW/m² is more than twice larger than that without shading at 1255 mW/m². This "warm" radiant exergy emission from the window with internal shading is also much larger than that from the interior surface of the well-insulated wall demonstrated in Fig. 6.25. This increase of "warm" radiant exergy emission is owing to the absorption of solar radiation by internal shading.

By convection, "warm" exergy of 232 mW/m² is transferred into the interior surface of glass pane without shading, but from the internal shading, 663 mW/m² is transferred into the room air. The directions of "warm" exergy transfer by convection in the two cases are opposite to each other. For this reason, the internal shading device may be regarded as acting both as a radiant heating panel and convective air heater during daytime while keeping the privacy of occupants indoors.

Figure 6.30 is a comparison of two double-glazed windows: during a summer daytime, one window has internal shading, which is the same as the one shown in the lower plate of Fig. 6.29, and the other has external shading, the optical property of which is exactly the same as the internal shading. The solar radiant exergy at the rate of 456 mW/m² is incident on the window. Other outdoor conditions are the same as those for the walls demonstrated in Fig. 6.24. The indoor conditions are 30°C for average radiant temperature and 28°C for air temperature.

The solar radiant exergy absorbed at the inner most window component is 127 W/m² in the case of internal shading and its 98%, 124 W/m², is consumed in the course of absorption, while on the other hand in the case of external shading, it is 6 W/m² and again 98%, 5.9 W/m², is consumed. There is no difference in the ratio of exergy to solar radiant exergy absorption between the internal shading and

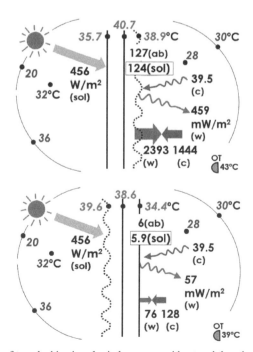

Fig. 6.30: Comparison of two double glazed windows: one without and the other with internal shading in terms of temperature distribution across the window components and exergy balance at the inner most surface in a summer daytime.

the inner most glass pane, but the rate of exergy consumption in the latter is much smaller than the former, 5.9 W/m^2 versus 124 W/m^2. This large difference brings about a large difference in surface temperature: the temperature of internal shading at 38.9°C is 4.5°C higher than the temperature of interior-side surface of internal glass pane at 34.4°C. Since the outdoor and indoor conditions and also the materials used are the same, this difference is brought by the location of shading device relative to glass panes, which have selective transmitting characteristic with respect to the wavelength of radiation as described in 6.4. Note that there is a large difference in the emission of warm radiant exergy towards indoor space; internal shading device emits "warm" radiant exergy at 459 mW/m^2, which is eight-times larger than 57 mW/m^2 emitted by the internal glass pane with external shading device.

Furthermore, along the internal shading device, there is "warm" exergy flowing inward at the rate of 2393 mW/m^2 by convection, which meets with "cool" exergy being transferred by convection from indoor space at the rate of 1444 mW/m^2 towards the internal shading. Their sum, 3837 (= 2393 + 1444) mW/m^2, is totally consumed and turns into nothing. The existence of such a large rate of exergy consumption implies that a large rate of "cool" exergy is required to be supplied into the indoor space.

The phenomenon, in which "warm" exergy and "cool" exergy transferred by convection meet with each other and thereby both are consumed, happens also along the surface of the interior glass pane of the window having external shading, but the rates of "warm" and "cool" exergy transferred by convection are much smaller as shown in the lower plate of Fig. 6.30. Their sum, 204 (= 76 + 128) mW/m^2, is totally consumed, is only 5% in the case of internal shading. Installation of external shading device is thus confirmed to be crucially important in order to provide the occupants with a comfortable summer indoor environment.

References

Architectural Institute of Japan (AIJ). 1978. Collection of the Information for Architectural Design—I. Environment. Maruzen Publisher (in Japanese).

Atkins P. W. 1991. Atoms, Electrons, and Change. Scientific American Library.

Bejan A. 2004. Convective Heat Transfer. Third Ed. John Wiley & Sons, Inc.

Faraday M. 1861. The Chemical History of a Candle—A reprint version edited and introduced by James F.A.J.L (2011) Oxford University Press.

Gagge A. P., Foblelets A. P. and Berglund L. G. 1986. A standard predictive index of human response to the thermal environment. ASHRAE Transactions 92(3B): 709–731.

Holman J. P. 1981. Heat Transfer. 5th ed. McGraw-Hill.

Kreith F. and Bohn M.S. 1986. Principles of Heat Transfer. 4th ed. Harper & Row Publishers.

National Aeronautics and Space Administration (NASA). 2016. Cosmic Background Explorer (COBE). http://science.nasa.gov/.

National Renewable Energy Laboratory (NREL). 2012. http://rrede.nrel.gov/solar/spectra/am1.5/.

Shukuya M. 1993. Light and heat in the built environment—An approach by numerical calculation. Maruzen Publishers Ltd. Tokyo (in Japanese).

Shukuya M. 2013. Exergy—Theory and applications in the built environment. Springer-Verlag London.

Shukuya M. 2015. Indoor-environmental requirement for the optimization of human-body exergy balance under hot/humid summer climate. PLEA2015 conference, Bologna, 9–11th September.

Silk J. 1994. A short history of the universe. Scientific American Library.

Turns S. R. 2006. Thermal-fluid Science—An Integrated Approach. Cambridge University Press.

Watanabe K. 1965. Principles of architectural planning. Maruzen Publisher (in Japanese).

Appendix

6.A Characteristics of black-body radiation and convective heat transfer

6.A.1 Calculation of the spectral emission rate of black-body radiation

The spectral distribution curves demonstrated in Figs. 6.10 and 6.12 were obtained from a VBA function code shown in Table 6.A.1. The data used for Fig. 6.11 were also based on the results of calculation obtained from the VBA code in Table 6.A.1, but they are the ones converted so that they represent the radiation received at the extra-terrestrial surface by considering the relationship between the diameter of the Sun, $d_{sun} = 1.39 \times 10^9$ m, and the distance between the Sun and the Earth, $D_{SE} = 1.5 \times 10^{11}$ m. The factor for conversion is $d_{sun}^2 \big/ \left(4D_{SE}^2\right)$.

Table 6.A.1: VBA code for the calculation of spectral black-body irradiance.

```
Function planck_energy(lamda, TK)
'
'This is a function program to calculate the spectral emission rate of
'Planck's black-body radiation.
'input: lamda -> wavelength [nm]
'       TK -> black-body temperature in Kelvin
'Output: planck_energy -> the rate of energy is in the unit of W/(m^2)/micro-meter
'
'                09September2016 MS
'
h_plk = 6.626 * 10 ^ (-34): c_light = 2.998 * 10 ^ 8: k_boltzmann = 1.381 * 10 ^ (-23)
pai = 3.1415926
LM = lamda * 10 ^ (-9)
EE = h_plk * c_light / (LM * k_boltzmann * TK)
CH = 2 * pai * h_plk * c_light ^ 2 / (LM ^ 5)
planck_energy = (CH * (1 / (Exp(EE) - 1))) / 10 ^ 6
End Function
```

6.A.2 Empirical formulae to estimate convective heat transfer coefficient

Convective heat transfer coefficient is characterized as the relationship between Nusselt number, which is the ratio of heat transfer coefficient to the corresponding thermal conductance of the layer where convection takes place, and a couple of other dimensionless numbers such as Grashof number, Prandtl number, and Reynolds number.

For natural convection, Rayleigh number, the product of Grashof number and Prandtl number, is correlated to Nusselt number. Grashof number is a dimensionless number determined by gravitational acceleration, the volumetric expansion coefficient, the temperature difference, the characteristic length, and the kinematic viscosity. Prandtl number is a dimensionless number determined by kinematic viscosity and thermal diffusivity. Tables 6.A.2 and 6.A.3 show the VBA function codes for the calculation of convective heat transfer coefficient along a vertical wall and a horizontal (floor or ceiling) surface under the condition of natural convection, respectively. The values demonstrated in Fig. 6.17 were obtained from these VBA function codes.

For forced convection, Reynolds number is correlated mainly to Nusselt number and slightly to Prandtl number. Reynolds number, which is determined by the velocity, the characteristic length and the kinematic viscosity, plays a key role in the case

Table 6.A.2: VBA code for the estimation of heat transfer coefficient due to natural convection along a vertical wall surface.

```
Function h_ncv_v(DH, TS, TA)
'
'    This is a function sub-program to calculate the convective
'    heat transfer coefficient under the condition of free(natural)
'    convection for the heat flow across vertical surface
'
'                    16Mar1989 MS,   05Sep2016 MS
'
'Input:  DH -> the characteristic length [m]
'    TS -> surface temperature [degree C]
'    TA -> air temperature [degree C]
'
    G = 9.8: GPCHK = 1 * 10 ^ 9
    '
    CKS = 0.56: CKL = 0.13
    '
    TM = (TS + TA) * 0.5
    XNY = (0.138 + TM * (0.000891667 + TM * (-5.09317E-11 + _
        TM * (0.0000000208324 + TM * 3.55271E-15)))) * 0.0001
    GR = G / (TM + 273.16) * Abs(TS - TA) * DH * DH * DH / XNY / XNY
    ATM = (0.0689 + TM * (0.000487083 + TM * (0.000000927068 + _
        TM * (-0.0000000177091 + TM * 0.000000000182306)))) / 3600#
    Pr = XNY / ATM
    GP = GR * Pr
    If GP < GPCHK Then
        XNU = CKS * GP ^ 0.25
    Else
        XNU = CKL * GP ^ 0.333
    End If
    XLM = (0.0207 + TM * (0.0000716666 + TM * (-0.0000000208381 + _
        TM * (-0.00000000416685 + TM * 5.20841E-11)))) * 1.163
    h_ncv_v = XNU * XLM / DH
End Function
```

Table 6.A.3: VBA code for the estimation of heat transfer coefficient due to natural convection along a horizontal surface such as floor or ceiling.

```
Function h_ncv_h(DH, UD, TS, TA)
'
'    This is a function sub-program to calculate the convective
'    heat transfer coefficient under the condition of free(natural)
'    convection for the heat flow across horizontal surface
'                                11December2016 MS
'Input: DH -> the characteristic length [m]
'      UD -> UD>=0 for upward surface and UD<0 for downward surface
'      TS -> surface temperature [degree C]
'      TA -> air temperature [degree C]
'

    G = 9.8: GPCHK = 1 * 10 ^ 7
    CKS = 0.54: CKL = 0.15: CKSL = 0.27
    '
    TM = (TS + TA) * 0.5
    XNY = (0.138 + TM * (0.000891667 + TM * (-5.09317E-11 + _
        TM * (0.0000000208324 + TM * 3.55271E-15)))) * 0.0001
    GR = G / (TM + 273.16) * Abs(TS - TA) * DH * DH * DH / XNY / XNY
    ATM = (0.0689 + TM * (0.000487083 + TM * (0.000000927068 + _
        TM * (-0.0000000177091 + TM * 0.000000000182306)))) / 3600#
    Pr = XNY / ATM
    GP = GR * Pr
    If UD >= 0 Then            'upward surface
      If TS > TA Then          'upward heat flow
        If GP < GPCHK Then
          XNU = CKS * GP ^ 0.25
        Else
          XNU = CKL * GP ^ 0.333
        End If
      Else                     'downward heat flow
          XNU = CKSL * GP ^ 0.25
      End If
    '

    Else                       'downward surface
      If TS < TA Then          'upward heat flow
        If GP < GPCHK Then
          XNU = CKS * GP ^ 0.25
        Else
          XNU = CKL * GP ^ 0.333
        End If
      Else                     'downward heat flow
          XNU = CKSL * GP ^ 0.25
      End If
    End If
    XLM = (0.0207 + TM * (0.0000716666 + TM * (-0.0000000208381 + _
        TM * (-0.00000000416685 + TM * 5.20841E-11)))) * 1.163
    h_ncv_h = XNU * XLM / DH
End Function
```

of forced convection instead of Grashof number in the case of natural convection. Table 6.A.4 is a VBA function code for the calculation of convective heat transfer coefficient under the condition of forced convection. Table 6.A.5 is a VBA code for the calculation of average heat transfer coefficient along human-body surface as a function of air velocity in the vicinity of human body. The values demonstrated in Fig. 6.18 were obtained from the VBA function codes in Tables 6.A.4 and 6.A.5.

Table 6.A.4: VBA code for the estimation of heat transfer coefficient due to forced convection along a flat surface.

```
Function h_fcv(DH, TS, TA, VA)
'
'    This is a function sub-program to calculate the convective
'    heat transfer coefficient under the condition of forced
'    convection for the heat flow across vertical surface
'    or for heat flow across horizontal surface.
'
'                      16Mar1989 MS,   05Sep2016 MS
'
'Input: DH -> the characteristic length [m]
'    TS -> surface temperature [degree C]
'    TA -> air temperature [degree C]
'    VA -> air velocity [m/s]
'
    G = 9.8: RECHK = 5 * 10 ^ 5
    '
    CKS = 0.664: CKL = 0.036
    '
    TM = (TS + TA) * 0.5
    XNY = (0.138 + TM * (0.000891667 + TM * (-5.09317E-11 + _
         TM * (0.0000000208324 + TM * 3.55271E-15)))) * 0.0001
    GR = G / (TM + 273.16) * Abs(TS - TA) * DH * DH * DH / XNY / XNY
    ATM = (0.0689 + TM * (0.000487083 + TM * (0.000000927068 + _
         TM * (-0.0000000177091 + TM * 0.000000000182306)))) / 3600#
    Pr = XNY / ATM
    RE = VA * DH / XNY
    If GP < RECHK Then
         XNU = CKS * (RE ^ 0.5) * (Pr ^ 0.333)
    Else
         XNU = CKL * (RE ^ 0.5) * (Pr ^ 0.333)
    End If
    XLM = (0.0207 + TM * (0.0000716666 + TM * (-0.0000000208381 + _
         TM * (-0.00000000416685 + TM * 5.20841E-11)))) * 1.163
    h_fcv = XNU * XLM / DH
End Function
```

Table 6.A.5: VBA code for the calculation of heat transfer coefficient along human-body surface.

```
Function hcv_g(v, Met)
'
'Calculation of convective heat transfer coefficient
'Equations used by Gagge et al. (1986)
'It is also used in human-body exergy balance calculation.
'
  qmet = Met * 58.2
  QQ = qmet / 58.2 - 0.85
  If QQ < 0 Then
    hc1 = 0
  Else
    hc1 = 5.66 * QQ ^ 0.39
  End If
  hc2 = 8.6 * v ^ 0.53
  hcv_g = hc1
  If hc1 < hc2 Then hcv_g = hc2
End Function
```

Chapter 7

Thermodynamics

7.1 Work, heat and energy

A variety of artificial systems driving our societies at large including contemporary buildings and cities are realised by the development of physical science and its engineering applications. The physical science taught in modern academia has been usually considered to consist of the following three branches: mechanics, electromagnetics and thermodynamics; we can say so looking at the curricula arranged for science and engineering education in universities.

Thermodynamics stands unique compared to mechanics and electromagnetics since it focuses directly on "dispersion" that necessarily takes place in actual macroscopic phenomena. In other words, as far as the mechanics and electromagnetics confine their focus on the behaviour of celestial bodies or charged particles, they may look having nothing to do with thermodynamics.

Nonetheless, all of the mechanical and electromagnetic phenomena that we experience in everyday life can never escape from thermodynamic phenomena, that is, the dispersion of heat and matter involved. The branches of mechanical and electrodynamic sciences, having been taught as major courses, tend to ignore the consequence of dispersion and thereby seem to result in letting those students ignore the thermodynamic consequence of actual macroscopic phenomena; therefore, they should be further re-developed taking thermodynamics into consideration so that one can develop his or her own holistic and connective ways of thinking in learning mechanics and electromagnetics. This is, I think, important in particular for the contemporary technology to advance further towards more environmentally-benign and better solutions.

In Chapter 2, subsection 2.5.2, we already discussed how a heat engine can produce "work" by a series of thought experiment. We also already made clear what the "heat" and "temperature" are in Chapter 6, section 6.2. Here in this section, let us first make the definition of work clearer and then discuss the relationship between "work", "heat" and "temperature".

"Work" is defined quantitatively as the product of "force", which is the input rate of momentum, as will be further described in Chapter 9, necessary to move a lump of matter and its resulting "distance" that it moves. For example, suppose that a matter having its mass of 100 kg is moved from the ground surface to a floor surface at the height of 10 m. The force exerted on the matter by the Earth is 100 kg multiplied by the gravitational acceleration rate of 9.8 m/s², that is 980 N (= 100 kg × 9.8 m/s²). Therefore, the minimum work required to lift the matter for 10 m becomes 9800 J (= N·m) by multiplying 980 N and 10 m. As mentioned in Chapter 6, the unit of "Joule", the product of force and distance, is used not only for "work" itself but also for "heat" in contemporary scientific and engineering discussion.

Note that the amount of work required for a mass of 1000 kg moved for 1 m is exactly the same as a mass of 100 kg for 10 m or 10 kg of matter for 100 m. Simply complying with the definition of work, we get the same figures, but as a common naïve sense, if we imagine how we perceive them when carrying respective matters for respective distances, it is hard for us, I think, to convince ourselves that they are really equivalent to each other.

The reason why we use "Joule" both for work and heat is according to the fact that we can have hot water not only by making use of "heat", for example, as putting a kettle containing water above the fire on an oven as shown in Fig. 7.1, but also by making use of "work", for example, as stirring water vigorously with the blades as shown in Fig. 7.2. Namely, there are two ways to raise water temperature: one is "heat" and the other is "work".

It may sound self-evident, but it was not. Scientists in early to mid-nineteenth century struggled with whether and how the "work" and "heat" are related to each other until J. P. Joule (1811 ~ 1889) found that there is exact relationship between "work" and "heat" leading to the conception of "energy", for which three scientists, not only Joule but also J. R. von Mayer (1814 ~ 1878) and H. Helmholtz (1821 ~ 1894) are considered as the foremost contributors.

What Joule did was a series of experiments on the quantitative relationship between work and heat using several devices, one of which is exactly as the one schematically shown in Fig. 7.2. The weight has a potential to do work before falling down, but it does not after. On the other hand, the water in the pot has no influence of heat before the weight has fallen down, but it has a certain influence as the rise of water temperature after the weight has fallen down. Suppose that first of all the water temperature was at t_0[°C] in the beginning, and it eventually reached at t_1 [°C]. The friction between the water and the blades results in an increase of water temperature from t_0 to t_1. Here we assume that the friction occurs only between the water and the blades, but not the pulleys and the rope so that all of heat caused by the work exerted emerges inside the water tank. Note that the water temperature never decreases spontaneously by such an apparatus as the one shown in Fig. 7.2. Again it may sound too obvious, but is worth keeping in mind, in association with what follows in this chapter.

Figure 7.3 exemplifies what Joule did, though he and his contemporaries did not show experimental results as shown in this figure (Harman 1982). Each plot represents one case of experimental study with an amount of work by the fall of a weight and its resulting amount of heat. In short, what he found is that the amount of work done is always proportional to the amount of heat available and the ratio of work to heat is always constant. In other words, with a certain amount of work given, there is no way to generate more heat or less heat. The slope is known to be 0.2389 cal/J or its reciprocal 4.186 J/cal today. One calorie (cal) is the amount of heat

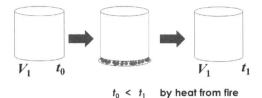

$t_0 < t_1$ **by heat from fire**

Fig. 7.1: Raising the water temperature by heating.

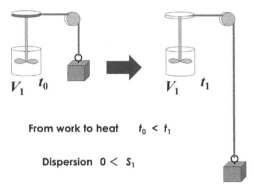

From work to heat $t_0 < t_1$

Dispersion 0 < S_1

Fig. 7.2: The friction between the blades and water, which is brought by "work" performed by a weight falling down, emerges "heat".

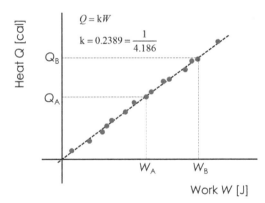

Fig. 7.3: Relationship between work and heat established in the mid-19th century by Joule and others.

to raise the temperature of 1 gram of water for 1°C. The relationship so far described can be written as

$$Q = kW \tag{7.1}$$

where Q is heat in the unit of calorie, k is the proportional constant connecting work with heat (= 0.2389 cal/J) and W is work in the unit of Joule.

The concept, which can connect two concepts so far used, "heat" and "work", is "energy", whose essential characteristic is to be conserved, since it is impossible to generate either more heat or less heat with a certain amount of work exerted according to the experimental evidence exemplified in Fig. 7.3. This is called the law of "energy conservation" or the first law of thermodynamics. Because of heat being exactly proportional to work as described above, the units of heat, work, and energy can be unified into one single universal unit, that is, "J", which comes from the name of Joule as already mentioned before.

With the concept of energy, let us explain again what was described above. We first consider an amount of "energy" being latently held by the weight; it is called the potential "energy" held by weight. It is converted first in the course of the weight falling down into "work" as one of the two ways of "energy" transfer, next into "heat" as the other of "energy" transfer, which emerges due to the friction between the blades and water, and finally into thermal "energy" held by the water in the tank. Either the potential energy held by the weight or thermal energy held by water should be recognized as the quantity of "state", which is in contrast to the quantity of "transfer" or "flow" such as work and heat, as their implication will be discussed later in this chapter and also in Chapter 9.

7.2 Dispersion, entropy and absolute temperature

As we discussed in 6.2, we know that things are made of molecules. All things around you and even your body yourself are made of molecules, which are composed of atoms. With this evidence in mind, the difference between potential energy and thermal energy can be explained as follows. The former is in an ordered form, while on the other hand, the latter is in a dispersed form. The characteristic of "work" is that all molecules making up the shape of weight move altogether maintaining their relative positions (Shukuya 2013). On the other hand, what is realized as the water temperature rises is that the water molecules in the pot turn into their collective state as more vigorous and random motions than before having been stirred by the blades.

The concept of temperature is the primary index to signify the vigour with randomness of the molecular motion within a certain body, whether it is either in solid, liquid or gas state. As discussed in Chapter 6, heat flows from a body whose temperature is higher to the other body whose temperature is lower. This implies in fact that the whole of molecular vibrating motions with a certain vigour and randomness existing in a body at higher temperature transfers into the other body at lower temperature, whose molecular vibrating motions as a whole are with less vigour but more randomness. What happens in the course of energy transfer from work to heat and also from heat at higher temperature to heat at lower temperature is "dispersion". All natural phenomena involving a huge number of molecules cannot escape from such course of dispersion.

In addition to energy transfer, it is also true that there is dispersion in the course of "mass" transfer as was introduced in the beginning of Chapter 6 and is also to be discussed later in Chapter 8. Therefore, we need to quantify how much or little of dispersion emerges in a variety of natural phenomena, with which all of energy and matter involved are necessarily conserved.

The quantity specific to dispersion is called "entropy", which was originally conceived by R. Clausius (1822 ~ 1888), a German scientist, one of the two foremost contributors of the 2nd law of thermodynamics, while the other is W. Thomson, a

Scottish scientist (1824 ~ 1907), whose lord name, "Kelvin", is used as the unit of thermodynamic temperature to be introduced later in this chapter. In any of natural processes, an amount of entropy is necessarily and inevitably generated. Therefore, this is also called the law of "entropy generation" in contrast to the law of "energy conservation". In other words, the concept of entropy is to quantify the irreversibility within the frame of energy and mass conservation.

Let us quantify the entropy as we did for heat, work, and energy. To do so, we first need to confirm that the concept of "entropy" is one of the extensive quantities such as length, area, volume, mass, momentum and energy. The most fundamental characteristic of these extensive quantities is their additivity; for example, the distance between two places A and B can be added to that between B and C and thereby the total distance between A and C is obtained. The same is true in such quantities as area, volume, mass, momentum and energy. In order to confirm that such a characteristic also inheres in the concept of entropy, let us make a thought experiment on how the dispersion takes place in a series of natural phenomena starting from what was already shown in Fig. 7.2, together with what happens in the course of phenomena shown in Figs. 7.4 and 7.5.

In Fig. 7.2, as we already described, an amount of work brings about an increase of water temperature from t_0 to t_1 in a tank having the volume of V_1 [m³]. We do the same with a larger tank, whose volume is V_2 [m³], as shown in Fig. 7.4. If the amount of work in Fig. 7.4 is the same as that in Fig. 7.2, then the final water temperature, t_2, is lower than t_1. Of course, either in the case of Fig. 7.2 or that of Fig. 7.4, dispersion takes place. Denoting the amount of dispersion, that is entropy, to occur in Fig. 7.2 to be S_1 and that in Fig. 7.4 to be S_2, let us ask the following question: which case generates more dispersion, or more precisely, whether it is $S_1 < S_2$, $S_1 = S_2$ or $S_1 > S_2$. The answer is $S_1 < S_2$. This is due to the fact that t_2 is lower than t_1. This can be confirmed, as shown in Fig. 7.5, by thinking about mixing the volume of water, V_1, at the water temperature of t_1, with the volume of water, $V_2 - V_1$, at t_0 and thereby generating the volume of water, V_2, at t_2. The relationship between the volumes and temperature can be expressed as follows, according to the law of energy conservation, applying the concept of heat capacity described in Chapter 6.

$$W = Q = c\rho V_1(t_1 - t_0) = c\rho V_2(t_2 - t_0), \tag{7.2}$$

where W is work exerted [J], Q is heat generated [J], c is specific heat capacity of water (= 4186) [J/(kg·K)], and ρ is the density of water (= 1000) [kg/m³]. Equation (7.2) can be rewritten as

$$t_2 = \frac{V_1}{V_2}t_1 + \left(1 - \frac{V_1}{V_2}\right)t_0. \tag{7.3}$$

Equation (7.3) affirms that the water temperature reached after mixing is the weighted average of respective water temperature of two volumes, V_1 and V_2, before mixing. Since $V_1 < V_2$, $t_0 < t_2 < t_1$. Note that the temperature is one of the typical intensive quantities, whose characteristic is neither additive nor subtractive but related to multiplication and division, such as density, pressure, concentration and others. The right hand side of Eq. (7.3) exactly shows that such nature exists in the concept of temperature.

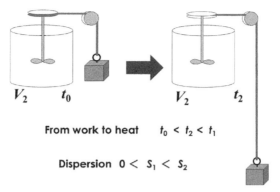

From work to heat $t_0 < t_2 < t_1$

Dispersion $0 < S_1 < S_2$

Fig. 7.4: Increase of water temperature by friction due to the work transferred by the fall of weight in the case of tank volume larger than that in Fig. 7.1.

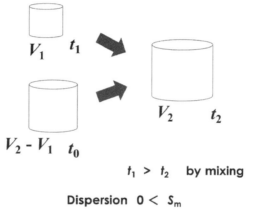

$t_1 > t_2$ **by mixing**

Dispersion $0 < S_m$

Fig. 7.5: Mixing of two volumes of water whose temperatures are different from each other is also a typical phenomenon of dispersion.

Mixing of two volumes of water at different temperature values is one of the typical thermal phenomena in which dispersion takes place, so that let us denote the corresponding entropy by S_m. Since the final result of mixing illustrated in Fig. 7.5 is exactly the same as what is illustrated in Fig. 7.4, with the entropy of S_2, the following equation should stand valid. That is,

$$S_1 + S_m = S_2. \tag{7.4}$$

Equation (7.4) clarifies that the concept of entropy has the characteristic of additivity. This is why "entropy" is one of the extensive quantities.

The next to do is to derive the formula of entropy that allows us to perform numerical calculation of entropy together with that of energy. To do so, let us imagine that there is a huge tank as shown in Fig. 7.6 containing a large amount of water, whose temperature hardly changes by the exertion of work because of heat capacity being so large. Suppose, for example, that the tank is the size of a swimming pool, 15 m wide, 50 m long, and 2.5 m deep. The heat capacity of water contained in the

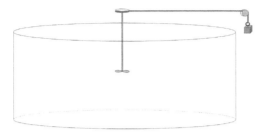

Fig. 7.6: Water temperature in a huge tank hardly increases by the input of a finite amount of work.

tank is 4186 J/(kg·K) × 1000 kg/m³ × (15 m × 50 m × 2.5 m) = 7.8 × 10⁹ J/K. If you let 100 kg of weight fall down for the height of 10 m, the work to be provided is 100 kg × 9.8 m²/s × 10 m = 9800 J. This implies that the water temperature increases only by 1.26 × 10⁻⁶ K (= 9800 J/(7.8 × 10⁹ J/K)). That is, we may assume that the water temperature can be held constant if the volume of a tank is sufficiently huge.

Suppose that we use weight A for an amount of heat, Q_A, to emerge and weight B for Q_B to emerge. Again, since the fall of weights A and B also makes the dispersion emerge, we denote the corresponding amounts of entropy as S_A and S_B, respectively. Their relationship can be expressed schematically as shown in Fig. 7.7a and also written mathematically in the following equation, since we now know that entropy is one of the extensive quantities.

$$\frac{Q_B}{Q_A} = \frac{kW_B}{kW_A} = \frac{S_B}{S_A} \tag{7.5}$$

The slope of the line, ς, shown in Fig. 7.7a, can be expressed as

$$\varsigma = \frac{S_A}{Q_A} = \frac{S_B}{Q_B} \tag{7.6}$$

The slope ς has an important implication as will be discussed in what follows, so let us give it the name "dispersality".

Next, take a look at Fig. 7.7b. There are two dashed lines in addition to the line originally shown in Fig. 7.7a. One dashed line has a steeper slope and the other a gentler slope. In the case of steeper sloped line, the amount of heat Q results in a larger entropy S^* than S in the original case, while on the other hand, in the case of gentler sloped line, the amount of heat Q results in a smaller entropy S^{**} than S. That is,

$$S^{**} < S < S^*. \tag{7.7}$$

Whether the entropy becomes larger or smaller as the same amount of heat flows into the water is associated with the temperature level of water. If the water temperature in the tank shown in Fig. 7.6 is lower, the entropy to emerge shall be larger and if the water temperature is higher, the entropy to emerge shall be smaller. Therefore, the slope becoming steeper, as shown in Fig. 7.7b, corresponds to the water temperature being lower and the slope being gentler corresponds to the water temperature being higher. At their extremes, one is the steepest slope at 90° against horizontal axis and the other is the gentlest slope at 0°. The slope at 90° implies

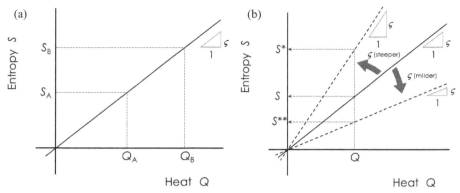

Fig. 7.7: Relationship between heat and entropy: (a) entropy is proportional to heat; (b) entropy becomes larger if the slope is steeper and becomes smaller if it is milder.

that infinitely large dispersion emerges with a finite amount of heat Q and that of 0° implies that no dispersion emerges.

There is, of course, no more beyond infinitely large dispersion so that the slope at 90° corresponds to the lowest possible temperature, while on the other hand, no dispersion at all corresponds to the highest possible temperature. Therefore, let us conceive a temperature scale, T, to be determined absolutely by complying with the nature of the slope, ς. The relationship between ς and T may be expressed schematically as shown in Fig. 7.8 and is written mathematically as follows.

$$\varsigma = \frac{1}{T} \tag{7.8}$$

The temperature, T, in inversely proportional relation to dispersality, ς, is called absolute temperature. This temperature scale defined by Eq. (7.8) is exactly the same as thermodynamic temperature that Thomson, the other of the two founders of second law of thermodynamics as mentioned earlier, found through his in-depth investigation based on the ideal heat engine conceived by S. Carnot (1796 ~ 1832), a French scientist, whose research work is regarded to be the foundation of thermodynamics (Carnot 1824). Commemorating Thomson's discovery of thermodynamic temperature, which is absolutely independent of any thermo-physical properties of matter and henceforth universal, the unit of absolute temperature is Kelvin, which originated from his name as Lord.

The derivation of absolute temperature so far reached in this section is rather similar to what Lewis and Randall explained in their treatise of chemical thermodynamics (Lewis and Randall 1961), though their explanation is so brief that it would be hard for the beginners to grasp the essence of absolute temperature and entropy.

Combining the relationship expressed by Eq. (7.8) with that by Eq. (7.6), we find the following relationship.

$$S = \left(\frac{1}{T}\right)Q. \tag{7.9}$$

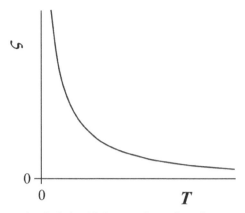

Fig. 7.8: Inversely proportional relationship between thermodynamic temperature and dispersality.

Equation (7.9) shows that entropy is determined as the product of "dispersality" and "heat". Note that dispersality, which is the amount of entropy per one unit of heat, has the nature of intensive quantity, since it is purely equal to the inverse of absolute temperature. Also note that dispersality is defined as the ratio of two extensive quantities, entropy, S, and heat, Q; this can be confirmed by dividing both sides of Eq. (7.9) by Q and having the relationship, $S/Q = 1/T$.

7.3 Heat engine, exergy and environmental temperature

As pointed out in the previous section, a certain amount of dispersion inevitably emerges in the course of energy transfer from work to heat. In other words, it is not possible for heat to turn itself wholly into work. The same is true with respect to heat itself. Heat can only flow from high temperature to low temperature, but never from lower temperature to higher temperature. The opposite of dispersion never happens spontaneously.

Nonetheless, it does not imply that the production of work from heat or the delivery of heat from where the temperature is low to where the temperature is high is impossible. The driving agents of contemporary societies such as automobiles, air planes, motor boats, electric power plants, refrigerators and heat pumps for space heating and cooling, with which we are familiar, function by making use of work produced from heat, of course without violating the laws of entropy generation and also energy conservation.

Here we confirm how they function by taking a careful look at how an imaginary heat engine works and its utmost efficient one known as Carnot engine produces the maximal limit of available work. This is to deepen our understanding of what we discussed in 2.5 and also to introduce the concept of "exergy". Exergy is exactly the concept that allows us to discuss so-called energy and environmental issues with the least vagueness and thereby leads us to conceiving rational systems for lighting, heating, cooling, and ventilating systems for the built environment (Shukuya 2013). Once we come to understand the cyclic operation of heat engine that can keep producing work from the flow of heat, we come to recognize that both "exergy consumption" and "entropy disposal" are inevitable and rather necessary for a system to keep functioning (Tsuchida 1992, Shukuya 2013).

Let us start the discussion with the following two assumptions originally given by Carnot. First is that some amount of work can be extracted from an amount of heat flowing from a source with high temperature at T_H [K] to a sink with lower temperature T_L [K]. This implies that it is essential to place a heat engine between heat source and heat sink to produce work. Second is that the extraction of work is made by the volumetric change of a working fluid under the isothermal condition. This implies that the heat engine is operated in a way, on the one hand, to maximize the amount of work, while on the other hand, to minimize the heat flow due to space-wise temperature difference within the engine system. This leads to maximizing the extraction of work from the heat flowing into and flowing out the engine system. Note that we assume here that the size of both source and sink are huge enough that their temperatures, T_H and T_L, are held constant even if heat keeps flowing out or in.

Based on the two assumptions mentioned above and the heat source and sink at respective constant temperature values making the heat flow emerges only with an infinitesimally small temperature difference that is provided by the infinitesimally slow expansion of a working fluid allows a heat engine to produce the maximal limit of work. Such an imaginary ideal heat engine is exactly what Carnot thought as of primal importance and is therefore called "Carnot engine". The cyclic process within Carnot engine is called "Carnot cycle".

In order to understand what the Carnot engine is, let us first take a look at Fig. 7.9. Suppose that there is a cylindrical vessel with a lid, which can move upwards or downwards freely without friction against the wall surface of the vessel, while at the same time the lid is perfectly tight so that there is neither infiltration nor exfiltration.

The volume of space, which is initially V_0 [m^3], is surrounded by the lid and the vessel wall is filled with a gaseous matter as the working fluid for the engine. There is a weight on the lid, which initially balances the internal pressure of the fluid, P_0 [Pa]. The initial temperature of this fluid is assumed to be T_H [K] by placing the bottom of this vessel right above a huge heat source at constant temperature, T_H. The lid is moved upwards extremely slowly so that little temperature difference emerges, then the weight is raised by the expansion of fluid inside the vessel. Such ideal expansion with perfectly no temperature change is called isothermal expansion. If the fluid temperature is held constant, the amount of work exerted to move the weight

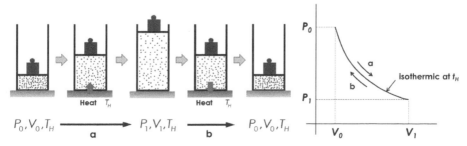

Fig. 7.9: Heat generating work to raise a weight and work by the weight squeezing heat.

upwards becomes exactly the same as that of heat flowing into the fluid at constant temperature, T_H, since there is no thermal energy left inside the fluid contained by the vessel because of no change in fluid temperature. After this isothermal expansion, the working fluid turns into the condition of the pressure and volume at P_1 [Pa] and V_1 [m³], respectively, with the temperature at T_H, and thereby the position of weight changes so that it holds an amount of potential energy.

The potential energy held by the weight can be used to compress the fluid isothermally, that is, without any change of temperature of the fluid, by discarding the corresponding amount of heat into the sink whose temperature is T_H, as shown from the state of P_1 and V_1 to that of P_0 and V_0. Thus one cycle completes because the state of the fluid returns to its initial state. Since all of the potential energy held by the weight is used up to lower the weight and let the working fluid be again in the same state as the initial state, there is no energy left to be extracted from the cycle in Fig. 7.9. This proves that the cycle shown in Fig. 7.9 is not useful at all and thereby suggests that a certain temperature difference is required to extract an amount of work from the engine. As can be seen in the pressure-volume relationship illustrated as a diagram attached in Fig. 7.9, the process from "a" to "b" and that from "b" to "a" are exactly on the same path.

Let us now take a look at Fig. 7.10 for a further discussion on what would happen if the temperature of the sink is lowered compared to that of the source. We assume that the heat sink is also as huge as the heat source that the temperature remains constant. This time, we need to make the expansion and compression consist of a series of isothermal and adiabatic processes, respectively, since there is a temperature difference between the heat source and the heat sink. In isothermal process, as used in the cycle shown in Fig. 7.9, there is no temperature change in the

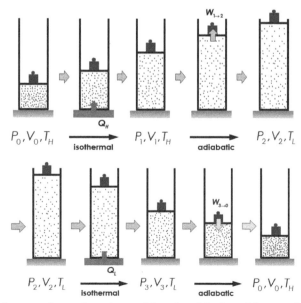

Fig. 7.10: Heat, Q_H, generating work to raise a weight and work by the weight squeezing heat, Q_L, which is smaller than Q_H.

working fluid, while on the other hand, in adiabatic process, in which the working fluid is expanded or compressed, no heat is transferred either into or out from the working fluid contained by the vessel. The cycle completes taking four steps: the first two steps are isothermal and adiabatic expansion processes and the following two are isothermal and adiabatic compression.

With the series of expansion, the heat amounting to Q_H flows into the working fluid, while at the same time the work exactly corresponding to the amount of $(-Q_H)$ for raising the weight and then the work amounting to $W_{1\rightarrow2}$ flows out so that the weight holds its potential energy being equal to $-Q_H + W_{1\rightarrow2}$.

Then the two types of compression, isothermal and then adiabatic, follow. In the isothermal compression, the heat amounting to Q_L flows out from the working fluid as the result of the work amounting to $(-Q_L)$ exerted into the working fluid and then the work amounting to $W_{3\rightarrow0}$ is adiabatically utilized in order to return the working fluid to the initial state, where the fluid temperature is T_H. For these series of compression, the change of potential energy held by the weight is $-Q_L + W_{3\rightarrow0}$.

Since the whole amount of energy involved has to be conserved as the first law of thermodynamics requires, and also, after completing the four steps of expansion and compression, all of the fluid temperature, pressure and volume return to their initial values, the sum of energy input to the working fluid, $Q_H + W_{3\rightarrow0}$, and the sum of energy output from the working fluid, $Q_L + W_{1\rightarrow2}$, has to be equal to each other.

$$Q_H + W_{3\rightarrow0} = Q_L + W_{1\rightarrow2}. \tag{7.10}$$

Equation (7.10) can be rewritten as follows.

$$Q_H = W + Q_L, \tag{7.11}$$

where

$$W = W_{1\rightarrow2} - W_{3\rightarrow0}. \tag{7.12}$$

Equation (7.11) is energy balance equation for the heat engine that we are focussing on; it implies that a portion of thermal energy supplied to the engine, Q_H, from the heat source at the absolute temperature of T_H is converted into the amount of work, W, while at the same time discarding the rest of heat, Q_L, into the heat sink at the absolute temperature of T_L. The abstract of this relationship is schematically presented in Fig. 7.11a; this is what most of thermodynamic textbooks present in the very beginning and tends to let the readers get into a dense fog, which usually intimidates them. Figure 7.11b shows a case that there is no heat engine between the heat source and the heat sink so that heat flows spontaneously from high to low temperature. Referring to Eq. (7.11) and Fig. 7.11a, we may say that energy with the amount of Q_H is transferred into the heat engine, while at the same time energy with the amount of W and Q_L is transferred out from the heat engine.

While the two diagrams shown in Fig. 7.11 may look too abstract, but having an image of an actual heat engine as shown in Fig. 7.12a together in parallel must help confirm the usefulness of these diagrams. This is a model of an engine which makes use of air as the working fluid, originally invented by Stirling in 1820s (Cardwell 1971). This tiny model engine functions in the flow of heat from the heated floor surface at about 27°C with a floor heating system to the room space at about 23°C as

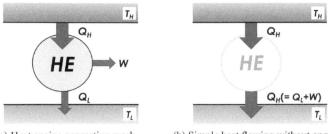

(a) Heat engine generating work   (b) Simple heat flowing without engine

Fig. 7.11: Heat flow and extraction of work from a heat engine and simple heat flow without producing work.

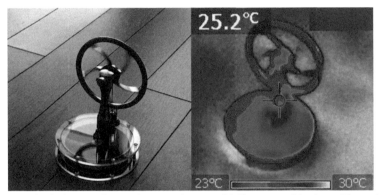

(a) The wheel of 4 cm radius turning around   (b) Thermo-graphic view of the engine

Fig. 7.12: A small model Stirling engine functioning on a heated floor as its heat source and room space as its heat sink.

shown in Fig. 7.12b. The wheel, the radius of which is 4 cm, keeps turning around as long as the temperature difference is kept larger than 5°C; in the case of Fig. 7.12, the temperature difference is about 7°C. A volume of air as the working fluid filled inside a flat cylinder below the wheel repeats expansion and contraction and in due course, work is extracted at a certain rate so that the wheel can keep turning around. This tiny Stirling engine corresponds to the heat-engine diagram shown in Fig. 7.11a and the heat flow alone from the floor surface to the room space corresponds to the diagram without heat engine shown in Fig. 7.11b.

As we discussed in the previous section, "work" is not dispersed at all, but "heat" is dispersed more or less depending on its associated absolute temperature so that we can say that a heat engine is a device extracting non-dispersed energy, that is "work", from the dispersive course of energy transfer, that is, "heat" flowing from high temperature to low temperature.

Following what we have discussed and reached so far in the previous section, the inflow of entropy can be expressed as the product of heat flowing in, Q_H, and its corresponding dispersality, $\frac{1}{T_H}$, and also the outflow of entropy as the product of heat flowing out, Q_L, and its corresponding dispersality, $\frac{1}{T_L}$. The work extracted, W, is not dispersed at all so that, as we discussed with Figs. 7.7 and 7.8, its corresponding dispersality is zero, that is $\varsigma = \dfrac{1}{\infty} = 0$. This implies that work is equivalent to heat at absolute temperature of infinity (Yamamoto 2008).

Since there may be some dispersion that emerges within the whole of cyclic operation shown in Fig. 7.10, let us assume that it amounts to S_g. Based on what has been so far defined in terms of entropy, we can set up the following equation.

$$\left(\frac{1}{T_H}\right)Q_H + S_g = \left(\frac{1}{\infty}\right)W + \left(\frac{1}{T_L}\right)Q_L. \tag{7.13}$$

Equation (7.13) is entropy balance equation to be considered in parallel to energy balance equation expressed as Eq. (7.11). In a similar manner to what we read in the implication of energy balance equation, let us read the implication of entropy balance equation as follows: the entropy with an amount of $\left(\dfrac{1}{T_H}\right)Q_H$ is transferred into the heat engine from the heat source and, in due course, the entropy with an amount of S_g is generated within the heat engine and then the entropy with an amount of $\left(\dfrac{1}{T_L}\right)Q_L$ is transferred out from the heat engine.

Rewriting the relationship expressed by Eq. (7.11) as $Q_L = Q_H - W$ and then substituting it into Eq. (7.13) yields the following equation.

$$\left(1 - \frac{T_L}{T_H}\right)Q_H - S_g T_L = W, \tag{7.14}$$

Equation (7.14) is called "exergy" balance equation. We can explain what the heat engine does with this exergy balance equation as follows: the heat engine feeds on an amount of exergy expressed as $\left(1 - \dfrac{T_L}{T_H}\right)Q_H$ from the heat source and consumes its portion amounting to $S_g T_L$ within the whole process of heat engine cycle, and eventually produces exergy output amounting to W. Exergy consumption, $S_g T_L$, is proportional to entropy generation S_g with the proportional constant of T_L.

It is important for us to recognize that there is the term that exactly expresses "consumption" as can be seen in Eq. (7.14), while on the other hand there is no such term in energy balance equation expressed as Eq. (7.11). Such an expression as "energy consumption" has been widely used not only in everyday conversation but also in scientific and engineering discussions, but it is, strictly speaking, not correct, since the concept of energy has the characteristic of conservation, not of consumption. Therefore, "exergy" is the legitimate concept, which can articulate precisely and accurately what the consumption is (Shukuya 2013). More on the implication of energy, entropy and exergy balance equations will be given in Chapter 9.

It is also important to recognize that the exergy to be supplied into the engine is determined not only by the temperature of heat source, T_H, but also by the temperature of heat sink, T_L. The heat sink may be regarded as the environment for the heat engine. This implies that how much of exergy we can have as resource is dependent not only on the temperature of heat source but also on the environmental temperature both for the engine as a system and the resource. A schematic drawing shown in Fig. 1.3 in Chapter 1 was in fact the reflection of such exergetic view with respect to resource, system, and environment. Therefore, let us summarize what has been described as "exergy consumption theorem" as shown in Fig. 7.13.

If the lid of the vessel shown in Fig. 7.10 moved so slowly that the uniformity of temperature and pressure with the working fluid is never broken, the dispersion would be perfectly avoided. The imaginary heat engine functioning in such a limited condition is Carnot engine, in which the entropy generation, S_g, turns into null and thereby the available work is maximized to be W_{max}. For Carnot engine, Eq. (7.14) is reduced to the following equation.

$$\left(1-\frac{T_L}{T_H}\right)Q_H = W_{max}.$$ (7.15)

The maximum available work, which is exactly exergy itself, can never be fully available in reality, since any actual heat engines have to function faster than infinite slowness and thus its portion has to be consumed more or less.

The entropy balance equation to be consistent with Eq. (7.15) is written as follows.

$$\left(\frac{1}{T_H}\right)Q_H = \left(\frac{1}{T_L}\right)Q_L.$$ (7.16)

Equation (7.16) indicates that the entropy flowing into the heat engine, $\left(\frac{1}{T_H}\right)Q_H$, outflows the heat engine as $\left(\frac{1}{T_L}\right)Q_L$; this is considered to be crucial, since the entropy amounting to the entropy input itself has to be discarded even if there is no entropy generation at all inside the Carnot engine. This confirms why "entropy disposal" is important. The model heat engines that we discussed in section 2.5, Chapter 2, could keep functioning provided that an amount of heat was necessarily discarded into the environment. That is in fact the entropy disposal, which is essential for a system to sustain its activity.

Exergy to be extracted from Carnot engine is the absolute maximum, which can never be exceeded. But, is it really so? Isn't it possible to conceive a heat engine

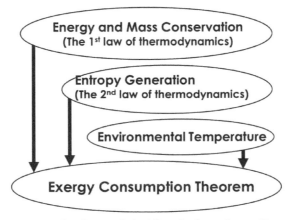

Fig. 7.13: The exergy consumption theorem derived from the first and second laws of thermodynamics and environmental temperature.

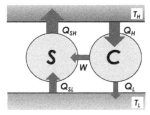 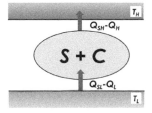

(a) A super engine working with Carnot engine   (b) Net energy transfer through the two engines

Fig. 7.14: Combination of a Carnot engine and a Super Carnot engine. This thought experiment clarifies that the amount of work to be generated by the Carnot engine is the maximum.

exceeding Carnot engine? Let us examine this by raising a question: if there were a super heat engine being more efficient in producing work than Carnot engine would, what could have happened?

To answer this question, let us think about a combination of a super engine and a Carnot engine as shown in Fig. 7.14a. The super engine is assumed to function as a heat pump that draws an amount of heat, Q_{SL}, from the heat sink at the absolute temperature T_L and deliver further the amount of heat, Q_{SH}, into the heat source at the absolute temperature T_H by the exertion of work, W, produced by the Carnot engine. Since the super engine is assumed to exceed the performance of Carnot engine, the amount of heat drawn from the heat sink, Q_{SL}, should be larger than the heat discarded by Carnot engine into the heat sink, Q_L, while on the other hand, the heat pumped up into the heat source by the amount of Q_{SH} would also be larger than the heat flowing into Carnot engine, Q_H. This implies that, as shown in Fig. 7.14b, the net flow of heat would have to be from low temperature to high temperature. This contradicts what the nature allows to happen so that we have to admit that the assumption of a super engine performing better than Carnot engine was false. Namely, Carnot engine provides the maximum available work.

7.4 Ideal-gas and absolute temperatures

For a heat engine to function properly, it is essential to use a matter whose range of expansion and compression is sufficiently wide. So-called ideal gas is a hypothetical matter of such kind to which the ambient air surrounding us is very close. The model Stirling engine shown in Fig. 7.12 uses air as working fluid and it really functions very well. The characteristics of ideal gas, which was very well studied and established by Towneley (1629–1707), Boyle (1627–1691), Hooke (1635–1703), Mariotte (1620–1684), Charles (1746–1823) and Gay-Lussac (1778–1850), can simply be expressed as

$$PV = nR\Theta, \tag{7.17}$$

where P is pressure [Pa], V is volume [m^3], and n is the number of molecules [mol]; one mol corresponds to 6.02×10^{23} molecules. R is gas constant (= 8.314) [J/(mol·K)], which is the proportional constant that relates the product of pressure and volume containing one mol of gaseous molecules to the ideal gas temperature, Θ. The ideal gas temperature, Θ, is known to be expressed as

$$\Theta = t + 273.15 \tag{7.18}$$

where t is temperature in Celsius [°C]. More on how the characteristics of ideal gas to be expressed by Eqs. (7.17) and (7.18) will be given in Chapter 8.

As described above, the Carnot engine would produce the maximum work. It is so with any type of working fluid. This means that it is also true in the case of Carnot engine, in which the ideal gas is used as the working fluid. With this in mind, we should be able to find the relationship between absolute temperature and Celsius temperature via ideal gas temperature.

Figure 7.15 shows the relationship between pressure and volume, often called P-V diagram, of air as an ideal gas used as the working fluid in a small Carnot engine, whose heat source is at 80°C and heat sink is 20°C as an example set of temperature. In the isothermal expansive process from state "0" to "1", the following relationship is valid since the temperature of the working fluid is held constant at Θ_H.

$$P_0 V_0 = P_1 V_1 = nR\Theta_H. \tag{7.19}$$

In adiabatic process that goes from state "1" to "2", the following Eq. (7.20) can be derived by combining the relationship expressed in Eq. (7.17) and the relationship, $-PdV = nC_V d\Theta$, which is based on the law of energy conservation under the constraint of adiabat,

$$\Theta_L = \Theta_H \left(\frac{V_1}{V_2} \right)^{R/C_V} \tag{7.20}$$

where C_V is molar heat capacity of air at constant volume (= 20.8) [J/(mol·K)]. Note that natural logarithm has to be applied to mathematical operation to reach Eq. (7.20). For those who need to know more about natural logarithm, refer to Chapter 9.

Taking the same manner from state "2" to "3" and from "3" to "1", like Eqs. (7.19) and (7.20), the following equations can be derived,

$$P_2 V_2 = P_3 V_3 = nR\Theta_L \tag{7.21}$$

$$\Theta_L = \Theta_H \left(\frac{V_0}{V_3} \right)^{R/C_V}. \tag{7.22}$$

Comparison of Eqs. (7.20) and (7.22) lets us know that $V_0/V_1 = V_3/V_2$. The amounts of heat, Q_H and Q_L, are quantified as follows, again by using the ideal-gas characteristics expressed in Eq. (7.17) and the law of energy conservation,

$$Q_L = nR\Theta_L \ln\frac{V_0}{V_1}, \quad Q_H = nR\Theta_H \ln\frac{V_3}{V_2}. \tag{7.23}$$

where symbol "ln" represents natural logarithm. Again, those who are not familiar with natural logarithm are invited to take a look at its explanation in Chapter 9. Combining two equations in Eq. (7.23) and taking the relationship with respect to volumes, $V_0/V_1 = V_3/V_2$, into consideration, we finally find the following relationship.

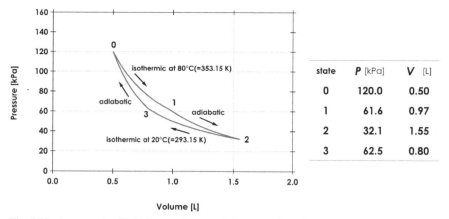

Fig. 7.15: An example of P-V diagram for a small Carnot engine using dry air as the working fluid.

$$\left(\frac{1}{\Theta_H}\right)Q_H = \left(\frac{1}{\Theta_L}\right)Q_L . \tag{7.24}$$

Equation (7.24) has the same form as that of Eq. (7.16). Therefore, we may take an arbitrary proportional constant, α, connecting absolute temperature and ideal-gas temperature as $T = \alpha\Theta$. If we choose α being equal to unity for making it the simplest, then absolute temperature becomes exactly the same as ideal-gas temperature. That is,

$$T = \Theta = t + 273.15. \tag{7.25}$$

Because of this relationship, the unit of ideal-gas temperature is the same as that of absolute temperature, Kelvin.

The sets of pressure and volume in cyclic operation taken from state "0" via "1" via "2" to "3" form a closed line, whose internal area corresponds to the amount of work extracted from the Carnot engine and in this particular case, it is 40 J; Carnot engine functions taking infinitely a long period of time for extracting the maximum work of 40 J but its rate of work is 0 ($= 40/\infty$) W. In reality, what we need is to have a certain rate of work. This implies that the exergy consumption expressed as $S_g T_L$ in Eq. (7.14) can never be zero. In other words, a certain rate of exergy consumption is necessary even if it should be minimized.

7.5 Warm and cool exergies contained by a stationary system

The discussion in 7.2 to derive entropy and absolute temperature was based on the case of a huge tank containing a large volume of water whose temperature is held constant. Here we extend the concept of entropy to be applicable to a variety of cases, in which the temperature does not necessarily remain constant.

First, instead of considering a huge tank, we consider a case of a finite volume of tank as shown in Fig. 7.16. A weight falls for an infinitesimally small distance, dh, with which the water temperature increases only scantly for an infinitesimally small value, dT. This is equivalent to no increase in water temperature in a huge

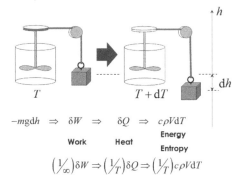

$$-mg\mathrm{d}h \Rightarrow \delta W \Rightarrow \delta Q \Rightarrow c\rho V\mathrm{d}T$$

Work   Heat   Energy
                            Entropy

$$\left(\tfrac{1}{\infty}\right)\delta W \Rightarrow \left(\tfrac{1}{T}\right)\delta Q \Rightarrow \left(\tfrac{1}{T}\right)c\rho V\mathrm{d}T$$

Fig. 7.16: An infinitesimally small work exerted to the water resulting in an infinitesimally small heat causing an infinitesimal increase of energy and entropy.

tank so that the relationship between heat and absolute temperature that we reached in 7.2 can be applied. The infinitesimally small amount of work, δW, is given as the result of the corresponding infinitesimally small change of potential energy held by a weight as $-mg\,\mathrm{d}h$, where m is mass of the weight [kg], g is gravitational acceleration rate [m/s^2], and the negative sign in front of the mass is to denote a decrease in height. This infinitesimal work, δW, results in an infinitesimally small energy transfer by heat, δQ, with its associated dispersality as $\tfrac{1}{T}$. That is, the entropy amounts to $\left(\tfrac{1}{T}\right)\delta Q$. Note that there are two types of symbol representing infinitesimal smallness, "d" and "δ", the former is to denote an infinitesimal quantity of "state" and the latter an infinitesimal "flow".

Focussing on the weight as a system and following the general expression of energy balance as [Energy input] = [Energy stored] + [Energy output], there is no input of energy, the energy stored due to an infinitesimally small increase in the height, $\mathrm{d}h$, is $mg\mathrm{d}h$ and the output as an infinitesimally small work, $\mathrm{d}W$, as the result of the weight falling the infinitesimally small distance, $\mathrm{d}h$, we may write as

$$0 = mg\,\mathrm{d}h + \delta W. \tag{7.26}$$

The infinitesimally small work, δW, turns out to be the infinitesimally small input of heat, δQ, flowing into the water contained by a tank, whose temperature is T [K], by the friction between the blades and water. Then, for the blades surrounded by the water as a system, energy balance equation as the work being equal to the heat and entropy balance equations as [Entropy input] + [Entropy generation] = [Entropy output] can be written as follows, respectively, based on what we discussed in 7.3.

$$\delta W = \delta Q. \tag{7.27}$$

$$\left(\frac{1}{\infty}\right)\delta W + \mathrm{d}'S_g = \left(\frac{1}{T}\right)\delta Q. \tag{7.28}$$

where $\mathrm{d}'S_g$ is an infinitesimally small generation of entropy; "d'" denotes infinitesimal smallness of generation as "d" and "δ" denote for the change of "state" and "flow". "d'" is also used to represent an infinitesimally small amount of exergy consumption.

In similar manner to Eqs. (7.27) and (7.28), energy and entropy balance equations for the water as another system in the tank can be written as

$$\delta Q = c\rho V \mathrm{d}T. \tag{7.29}$$

$$\left(\frac{1}{T}\right)\delta Q = \mathrm{d}S. \tag{7.30}$$

Note that how to set up such balance equations just described above is more thoroughly given later in Chapter 9.

Combining all equations from Eq. (7.26) to Eq. (7.30) and assuming the environmental temperature of the tank to be T_o [K], the following exergy balance equation can be derived.

$$-mg\,\mathrm{d}h - \mathrm{d}'S_g T_o = c\rho V \mathrm{d}T - T_o \frac{c\rho V \mathrm{d}T}{T}. \tag{7.31}$$

Equation (7.31) implies the following: an infinitesimally small exergy input provided by the weight falling down for an infinitesimally small distance is partly consumed by an infinitesimally small amount of exergy, $\mathrm{d}'S_g T_o$, and thereby infinitesimally small amount of thermal exergy, which is expressed in the right-hand side of the equation, is generated. If the weight falls from the height of h to that of h_0, then the entropy generation is summed up to S_g and thereby the water temperature increases from T_0 to T. The whole of this process is expressed in the integrated form as follows.

$$-mg\int_h^{h_0}\mathrm{d}h - T_o\int_0^{S_g}\mathrm{d}'S_g = c\rho V\int_{T_0}^T\mathrm{d}T - T_o c\rho V\int_{T_0}^T\frac{\mathrm{d}T}{T}. \tag{7.32}$$

Integral operation of Eq. (7.32) provides us with the following exergy balance equation.

$$mg\left(h-h_0\right) - S_g T_o = c\rho V\left\{\left(T-T_0\right) - T_o\ln\frac{T}{T_0}\right\}, \tag{7.33}$$

where symbol "ln" is natural logarithm (see Chapter 9). We read this equation as follows: the change in potential exergy held by the weight from h to h_0 lets the blades rotate, generate heat in part as exergy consumption as $S_g T_o$ and entropy generation as S_g, and thereby raise the water temperature from T_0 to T to let the water contain thermal exergy expressed by the right-hand side of Eq. (7.33).

Figure 7.17a shows numerical examples of exergy contained by 1 m^3 of water with its temperature from 0 to 40°C, assuming three cases of environmental temperature: 5, 20, 35°C. The heat capacity of water was assumed to be 4186 J/(kg·K) and the density to be 1000 kg/m^3. Thermal exergy never becomes negative. Therefore, we call thermal exergy in the case of $T > T_o$ as "warm" exergy and that in the other case of $T < T_o$ as "cool" exergy.

When the water temperature is higher than environmental temperature, the molecules of water are more vigorous than the molecules of air in the environmental space surrounding the system of water as a whole. This causes dispersion in the direction from the system to the environment. Under such a condition of the system relative to the environment, there exists "warm" exergy. On the other hand, when the

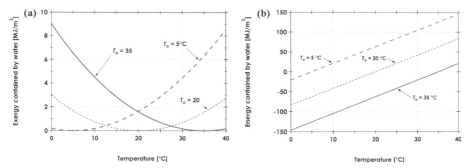

Fig. 7.17: (a) Thermal exergy contained by a unit volume of water. In the case of water temperature higher than environmental temperature, there is "warm" exergy, while on the other hand, in the case of lower than environmental temperature, there is "cool" exergy; (b) Thermal energy contained by the same volume of water. Negative values indicate a lack of energy relative to the environment.

water temperature is lower than environmental temperature, the molecules of water are less vigorous than the molecules of air in the environmental space surrounding the system of water as a whole. This causes dispersion in the direction from the environment to the system. Under such a condition of the system relative to the environment, "cool" exergy is considered to exist in the system.

"Warm" exergy may be regarded as the capability of energy in the system to disperse into the environment and "cool" exergy as a capability of the system, which allows the thermal energy in the surrounding space to disperse into the system due to its lack of energy. Figure 7.17b shows thermal energy contained by the same volume of water. There is a lack of energy in the cases of water temperature lower than environmental temperature. In summer case, cold water is useful because of the lack of thermal energy, for which the concept of exergy can present a positive value that indicates the true usefulness to be consumed for the purpose of cooling.

Looking at Fig. 7.17a, we can find that a volume of water at a certain temperature has different values of thermal exergy depending on the environmental temperature. For example, a volume of water at 35°C, which should be quite useful for the purpose of space heating, has 6.3 MJ/m³ (= kJ/L) of "warm" exergy under a winter condition at environmental temperature of 5°C, but it has none at all under a summer condition at the environmental temperature of 35°C. Under the same summer condition at 35°C, a volume of water at 20°C has 1.58 MJ/m³, and that of 10°C has 4.49 MJ/m³ of "cool" exergy. They are useful for the purpose of space cooling.

Such a characteristic that appears as "warm" or "cool" depending on the relationship between the system temperature and the environmental temperature is unique in the concept of exergy. Because of this characteristic, "warm" exergy and "cool" exergy should be applied to explain how the passive heating measures function with respect to the storage of "warmth" of wall and floor materials, and how the passive and active cooling measures function with respect to the storage of "coolness" under the ground and others (Li et al. 2014, Kazanci et al. 2016a, b, Menberg et al. 2017).

The characteristic of thermal exergy in the shape of quadratic curve as presented in Eq. (7.33) lets us notice one other important feature. If an amount of thermal energy alone is concerned, the amount of energy to be added to a certain medium for carrying an amount of thermal energy, for example, from 20 to 30°C is exactly the same as that from 30 to 40°C as can be expected from Fig. 7.17b, but from the viewpoint of exergy, they are different. In winter, for the environmental condition of 5°C, the difference in "warm" exergy contained by a unit volume of water between 40°C and 30°C is about 4.07 kJ/m³ and that between 30°C and 20°C is about 2.8 kJ/m³. The former is about 1.45 times larger than the latter for the quadratic nature exists in thermal exergy as the quantity of state.

This suggests that it is important to use such heat sources having as low temperature as possible for a heating system to require as less exergy consumption as possible. The same applies to "cool" exergy. The choice of higher-temperature levels is attractive in reducing the amount of "cool" exergy supply and its consumption, as much as possible. This is exactly why the development of low-exergy heating and cooling systems is important (Li et al. 2014, Kazanci et al. 2016a, b, Menberg et al. 2017).

7.6 Warm and cool exergy transfer

As we discussed in Chapter 6, within building envelopes, the heat transfer process by radiation, convection and conduction is always taking place, more or less. Here in this section, let us introduce how the thermal exergy transfer by thermal radiation, convection, and conduction is described (Shukuya 2013). Understanding the whole heat transfer from the exergetic viewpoint is, as demonstrated in the last section of Chapter 6, of vital importance in order to plan, design, and realize low-exergy systems for human thermal health and comfort to be sought in the built environment.

7.6.1 Radiation

Thanks to a series of intensive research done by Stephan (1835–1893), Boltzmann (1844–1906), Planck (1858–1947) and others during a couple of decades at the turn of century from 19th to 20th, we are able to calculate precisely the rates of energy and entropy of thermal radiation expressed by the following equations, respectively.

$$R_b = \int_0^\infty R_{b\lambda} \mathrm{d}\lambda = \sigma T^4, \tag{7.34}$$

$$S_b = \int_0^\infty S_{b\lambda} \mathrm{d}\lambda = \frac{4}{3}\sigma T^3. \tag{7.35}$$

where R_b is radiant energy emission rate [W/m²]; $R_{b\lambda}$ is spectral radiant energy emission rate per one meter of wavelength [(W/m²)/m]; λ is the wavelength [m]; σ is Stephan-Boltzmann constant (= 5.67×10^{-8} W/(m²K⁴)); T is absolute temperature of the black-body surface [K]; S_b is radiant entropy emission rate [W/(m²K)]; and $S_{b\lambda}$ is spectral radiant entropy emission rate per one meter of wavelength [(W/(m²K))/m]. The values of $R_{b\lambda}$ and $S_{b\lambda}$ are obtained from the formulae given by Planck (Shukuya 2013).

Figure 7.18 shows thermal energy and entropy emission rates as a function of black-body surface temperature calculated from Eqs. (7.34) and (7.35). Both rates increase as the surface temperature rises but their ascending natures are different due to their respective exponent values.

Based on these energetic and entropic characteristics of black-body radiation, let us derive thermal radiant exergy equation to be used for the evaluation of built environment. Suppose that there is an external wall in a room as shown in Fig. 7.19. The internal surface of this wall, whose temperature is T_1 [K], is surrounded by the ceiling, interior wall surfaces and floors, whose overall average surface temperature is T_2 [K]. Environmental temperature for this room is T_o [K].

Focusing on a very thin layer including the interior surface of this external wall, we first set up energy and entropy balance equations. With the assumption that the layer is so thin that there is neither energy nor entropy stored, we can make [Energy input] be equal to [Energy out], according to the law of energy conservation, and [Entropy input] + [Entropy generation] be equal to [Entropy output], according to the law of entropy generation.

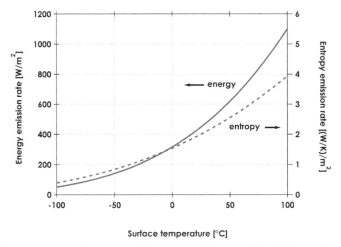

Fig. 7.18: Thermal energy and entropy emission rates as a function of black-body surface temperature.

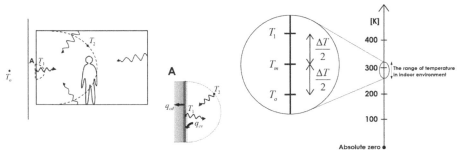

Fig. 7.19: A room with an external wall, whose interior surface temperature is T_1, assumed for theoretical consideration of radiant exergy calculation within indoor space. The whole of the room is assumed to be surrounded by the environmental space at the temperature of T_o.

[Energy input] consists of two kinds of heat transfer: one is the rate of energy delivered by the incoming long wavelength radiation from the opposite surfaces of the room; the other is the rate of energy transferred by convection between the room air and the internal surface that we are focussing on. Similarly, [Energy output] consists of two kinds of heat transfer: one is the rate of energy discharged by the surface towards the room space by long wavelength radiation and the other is the rate of energy flowing inside the external wall by conduction.

Similarly to energy input and output, [Entropy input] consists of radiation and convection and [Entropy output] consists of radiation and conduction. [Entropy generation] occurs due to the absorption of long wavelength radiation coming from the whole of opposite surfaces in the room.

Assuming that the absorptance for long wavelength radiation is unity, what has been described so far above can be written as follows

$$\varepsilon \sigma T_2^4 + q_{cv} = \varepsilon \sigma T_1^4 + q_{cd}, \tag{7.36}$$

$$\varepsilon \left(\frac{4}{3} \sigma T_2^3 \right) + \left(\frac{1}{T_1} \right) q_{cv} + s_{g\_rad} = \varepsilon \left(\frac{4}{3} \sigma T_1^3 \right) + \left(\frac{1}{T_1} \right) q_{cd}. \tag{7.37}$$

where ε is the emittance of wall surfaces, which is usually very close to unity, say 0.95 in the cases of ordinary building walls, as described in Chapter 6 (here, we assume the emittance of wall 1 and wall 2 to be equal to each other as ε); q_{cv} is the rate of energy delivered by convection [W/m²]; q_{cd} is the rate of energy delivered by conduction [W/m²]; and S_{g_rad} is the rate of entropy generation [(W/K)/m²]. In setting up Eqs. (7.36) and (7.37), the effect of mutual reflection between the wall surfaces is neglected since the emittance of ordinary building wall surfaces are close to unity as mentioned above.

The rates of radiant energy input and output which appeared in Eq. (7.36) and those of radiant entropy input and output which appeared in Eq. (7.37) are expressed as the whole rates summed up from the absolute zero to the respective absolute temperature; this means that they include the effect of long wavelength radiation at environmental temperature. Therefore, we rewrite Eqs. (7.36) and (7.37) by introducing the rates of radiant energy and entropy at T_o to both left-hand and right-hand sides of these equations so that the equality of both sides remains true. Namely,

$$\varepsilon \sigma \left(T_2^4 - T_o^4 \right) + q_{cv} = \varepsilon \sigma \left(T_1^4 - T_o^4 \right) + q_{cd}, \tag{7.38}$$

$$\varepsilon \left(\frac{4}{3} \sigma \right) \left(T_2^3 - T_o^3 \right) + \left(\frac{1}{T_1} \right) q_{cv} + s_{g\_rad} = \varepsilon \left(\frac{4}{3} \sigma \right) \left(T_1^3 - T_o^3 \right) + \left(\frac{1}{T_1} \right) q_{cd}. \tag{7.39}$$

We can drive the exergy balance equation as follows: first, multiply environmental temperature to both sides of Eq. (7.39), then subtract left-hand side and right-hand side of the resulting equation from the respective sides of Eq. (7.38). This procedure is to come up with the exergy balance equation on the basis of the principle shown in Fig. 7.13. The resulting exergy balance equation is

$$x_{rad\_2} + \left(1 - \frac{T_o}{T_1} \right) q_{cv} - x_{c\_rad} = x_{rad\_1} + \left(1 - \frac{T_o}{T_1} \right) q_{cd}. \tag{7.40}$$

where
$$x_{c\_rad} = s_{g\_rad} \cdot T_o, \tag{7.41}$$

$$x_{rad\_2} = \varepsilon\sigma\left\{\left(T_2^4 - T_o^4\right) - \frac{4}{3}\left(T_2^3 - T_o^3\right)\right\}, \tag{7.42}$$

$$x_{rad\_1} = \varepsilon\sigma\left\{\left(T_1^4 - T_o^4\right) - \frac{4}{3}\left(T_1^3 - T_o^3\right)\right\}. \tag{7.43}$$

We can read the implication of Eq. (7.40) as follows: radiant exergy input, x_{rad_2}, and convective exergy input, $\left(1 - \frac{T_o}{T_1}\right)q_{cv}$, are supplied, their portion, x_{c_rad}, is consumed, and thereby, two exergy outputs are produced: one by radiant exergy, x_{rad_1}, outgoing into the room space, and the other by conductive exergy transfer into the wall inside, $\left(1 - \frac{T_o}{T_1}\right)q_{cd}$. Exergy consumption emerges due to the absorption of radiant exergy. This is because, as explained above, we are focusing on a very thin layer including the interior surface as a thermodynamic system, in which there can be no exergy consumption due to either convection or conduction.

What follows in the rest of this sub-section explains the characteristics of radiant exergy and then the next two sub-sections 7.6.2 and 7.6.3, will explain the detail of exergy transfer by convection and conduction.

The temperature level that we encounter within the built environment ranges from −10 to 40°C and the difference in temperature between the environment and the wall surfaces is usually small, say 20°C at the largest, if compared with the values of absolute temperature. These two facts allow us to use simplified form of Eqs. (7.42) and (7.43) that indicates explicitly the characteristic of radiant exergy (Shukuya 2013). Such simplified form of x_{rad_2} is expressed by

$$x_{rad\_2} = \varepsilon h_{rb} \frac{\left(T_2 - T_o\right)^2}{T_2 + T_o} \tag{7.44}$$

where h_{rb} is radiative heat transfer coefficient of black-body surface, which can be approximated to be $4\sigma T_m^3$. T_m is the average of T_2 and T_o and can be taken as a constant assuming a nominal value; for example, $4\sigma T_m^3 \simeq 5 \sim 6\text{W}/(\text{m}^2\text{K})$. Replacing T_2 in Eq. (7.44) with T_1 gives us the simplified formula for x_{rad_1}. Note that the error induced by the use of Eq. (7.44) against the accurate calculation to be made by Eq. (7.42) is within 3% for the ordinary indoor and outdoor conditions.

Due to the fact that $0 < \varepsilon$, $0 < h_{rb}$, $0 < T_2 + T_o$, and $0 < (T_2 - T_o)^2$, radiant exergy never becomes negative. For the cases where $T_o < T_2$, there is "warm" radiant exergy and for the cases where $T_2 < T_o$, there is "cool" radiant exergy.

Figure 7.20 shows two numerical examples of warm and cool radiant exergies calculated from Eq. (7.44) for two outdoor conditions, winter and summer. In the former, the environmental temperature is assumed to be 278.15 K (= 5°C) and in the latter, it is 303.15 K (30°C). The values of radiative heat transfer coefficient, εh_{rb}, are assumed to be 5.3 W/m²K for winter and 6.3 W/m²K for summer, respectively.

In winter, the interior surface temperature of a wall with a scant thermal insulation board may become 10°C or even below, under the condition of outdoor

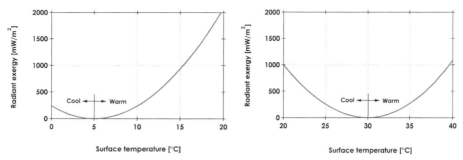

Fig. 7.20: Radiant exergy emission rate in winter (left) and in summer (right).

air temperature at 5°C, as described in 6.5. In such a case, "warm" radiant exergy available is about 250 mW/m². From such a poor insulation level on, the thickness of thermal insulation board is increased, the "warm" radiant exergy available increases in a quadratic manner and it can reach over 1600 mW/m² for the wall surface temperature at 18°C. This is more than six-times larger than the case of a wall with poor insulation. The improvement of thermal insulation level of building envelopes is to increase the availability of "warm" radiant exergy as was thoroughly demonstrated in 6.5. It becomes the prerequisite for a low-exergy heating system to be designed.

In summer, the rate of "warm" or "cool" radiant exergies is usually much smaller than "warm" radiant exergy available in winter. This is due to a much smaller temperature difference between indoors and outdoors in summer than in winter. Interior surface temperature of a window with an internal shading device tends to be very high because of the absorption of solar exergy incident on the shading device. It may reach 35°C or higher. This results in "warm" radiant exergy emission at the rate of about 250 mW/m². But, if an external shading device is installed, then the interior surface of glass windows may stay almost the same as or only slightly higher than outdoor air temperature and thereby "warm" radiant exergy becomes negligibly small or none. These features were also demonstrated in 6.5.

In addition to installing such external shading device, if the interior surface temperature of walls and floors stays a little lower than outdoor air temperature, then some rate of "cool" radiant exergy becomes available. For example, "cool" radiant exergy is emitted from the surface whose temperature is 27°C at the rate of about 100 mW/m². "Cool" radiant exergy emission rate even smaller than 100 mW/m² can help realize a low-exergy cooling system (Shukuya 2013). Thermal perception of human beings seems to be related very much to radiant exergy emission rate; this issue will be discussed more in Chapter 10.

7.6.2 Convection

Here we describe how to express the exergy transfer by convection more in detail. The exergy transfer by convection, which takes one of the exergy inputs as $\left(1 - \dfrac{T_o}{T_1}\right)q_{cv}$ in Eq. (7.40), was based on the assumption that the interior surface temperature of the

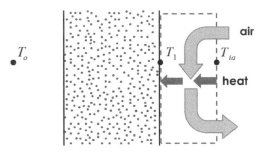

Fig. 7.21: Heat transfer by convection taking place within the boundary layer of room air near the interior side of an external wall.

external wall, T_1, is lower than room air temperature, T_{ia}, under a winter condition; that is, $T_o < T_1 < T_{ia}$.

Let us consider an air layer, where convection takes place along the wall surface as shown in Fig. 7.21. Inside this air layer as a system, thermal energy transferred from the room air, q_{cv_ia}, is balanced with that transferred into the interior wall surface, q_{cv}, assuming that no thermal energy is stored within this system of air layer. That is

$$q_{cv\_ia} = q_{cv}, \tag{7.45}$$

where
$$q_{cv\_ia} = h_{cv}(T_{ia} - T_1). \tag{7.46}$$

The proportional factor to be multiplied by the temperature difference which appeared in Eq. (7.46), h_{cv}, is convective heat transfer coefficient [W/(m²K)], which was explained in 6.4.2. Its values are considered to be in the range from 0.5 to 4 W/(m²K) along the interior surface of building envelopes as was demonstrated in Fig. 6.17.

The entropy balance equation to be expressed in parallel to Eq. (7.45) is

$$\left(\frac{1}{T_{ia}}\right)q_{cv\_ia} + S_{g\_cv} = \left(\frac{1}{T_1}\right)q_{cv}, \tag{7.47}$$

where S_{g_cv} is the rate of entropy generation due to convection taking place within the boundary layer [W/(m²K)].

Combining Eqs. (7.45) and (7.47) together with environmental temperature brings about the following exergy balance equation for the air layer as

$$\left(1 - \frac{T_o}{T_{ia}}\right)q_{cv\_ia} - x_{c\_cv} = \left(1 - \frac{T_o}{T_1}\right)q_{cv}, \tag{7.48}$$

where
$$x_{c\_cv} = S_{g\_cv} \cdot T_o. \tag{7.49}$$

We read Eq. (7.48) as follows: thermal exergy at the rate of $\left(1 - \frac{T_o}{T_{ia}}\right)q_{cv\_ia}$ is transferred into the boundary air layer from room air, its portion, x_{c_cv}, is consumed and thereby thermal exergy at the rate of $\left(1 - \frac{T_o}{T_1}\right)q_{cv}$ is produced and transferred into the interior surface of the external wall. Note that the exergy consumption by convection occurs within the boundary air layer shown in Fig. 7.21, while on the other hand as described in 7.6.1, the exergy consumption by long wavelength radiation occurs exactly at the surface by absorption. Thus, we come to recognize that there is an intrinsic difference between convection and long wavelength radiation in terms of their respective manners of dispersion.

Substituting the relationship given by Eqs. (7.45) and (7.46) into Eq. (7.47), and a little bit of algebraic operation, brings about

$$S_{g\_cv} = \frac{h_{cv}(T_{ia} - T_1)^2}{T_{ia} \cdot T_1} \geq 0. \tag{7.50}$$

Since $h_{cv} > 0$, $(T_{ia} - T_1)^2 > 0$, $T_{ia} > 0$ and $T_1 > 0$, Eq. (7.50) proves that entropy is necessarily generated at a certain rate as far as there is a temperature difference and the corresponding amount of exergy is consumed at the rate given in Eq. (7.49).

By denoting the rate of exergy transfer by convection into the boundary air layer as $x_{cv\_ia} = \left(1 - \dfrac{T_o}{T_{ia}}\right) q_{cv\_ia}$, and substituting q_{cv_ia} expressed as Eq. (7.46) into this equation, we find the following formula.

$$x_{cv\_ia} = h_{cv}(T_{ia} - T_1)\left(1 - \frac{T_o}{T_{ia}}\right). \tag{7.51}$$

Taking a look at Eq. (7.51), the rate of exergy transfer by convection can be either positive or negative depending on the relationship between three temperature values: T_o, T_1 and T_{ia}.

Because of three variables, there are six combinations, each of which is different in terms of whether it implies "warm" or "cool" and also whether it implies "incoming" or "outgoing" for the system shown in Fig. 7.21. Table 7.1 shows these six combinations for the exergy flow of x_{cv_ia} given by Eq. (7.51). In Table 7.1, the sign of Carnot factor denoted by b determines "warm" or "cool" exergy: positive is "warm" exergy and negative is "cool" exergy. The other factor denoted by a is the difference in temperature between two surfaces and the sign of the product, $a \cdot b$, determines "inflow" or "outflow": positive is inflow and negative outflow, with respect to the boundary layer as the system. The diagram attached to Table 7.1 shows respective six ranges determined by the combination of three temperatures, T_o, T_1 and T_{ia}.

The exergy output from the boundary air layer, which was exactly one of the exergy input to the wall surface by convection as expressed in Eq. (7.40), can be rewritten as follows by tracing the manner that we took for having Eq. (7.51).

$$x_{cv} = h_{cv}(T_{ia} - T_1)\left(1 - \frac{T_o}{T_1}\right). \tag{7.52}$$

Again because of three variables, there are six combinations, which are different in terms of whether it implies "warm" or "cool" and also whether it implies "outgoing" or "incoming". Table 7.2 shows these six combinations for the exergy flow of x_{cv} given by Eq. (7.52). How to read Table 7.2 is the same as Table 7.1. The

Table 7.1: Thermal exergy going into the boundary layer by convection, x_{cv_ia}.

	Temperature	$a = T_{ia} - T_1$	$b = 1 - \dfrac{T_o}{T_{ia}}$	$a \cdot b$	Warm/Cool	In/Out
I	$T_o \leq T_1 < T_{ia}$	+	+	+	Warm	In
II	$T_o < T_{ia} \leq T_1$	-	+	-	Warm	Out
III	$T_1 < T_o \leq T_{ia}$	+	+	+	Warm	In
IV	$T_{ia} \leq T_o \leq T_1$	-	-	+	Cool	In
V	$T_1 \leq T_{ia} < T_o$	+	-	-	Cool	Out
VI	$T_{ia} < T_1 < T_o$	-	-	+	Cool	In

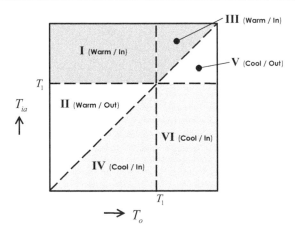

sign of Carnot factor c determines whether it is "warm" or "cool" exergy and the sign of the product, $a \cdot c$, determines whether it is "outflow" or "inflow".

Figure 7.22 demonstrates how the rates of exergy input, consumption and outputs as the surface temperature, T_1, varies. The upper graph is for a winter case, in which indoor air temperature, T_{ia}, and environmental temperature, T_o, are assumed to be 293.15 K (= 20°C) and 273.15 K (= 0°C), respectively; the lower graph is for a summer case, in which indoor air temperature, T_{ia}, and environmental temperature, T_o, are assumed to be 301.15 K (= 28°C) and 306.15 K (= 33°C), respectively.

In winter, since both the surface temperature, T_1, and indoor air temperature, T_{ia}, are assumed to be higher than environmental temperature, T_o, the input and output are both necessarily "warm" exergy. The exergy consumption rate increases quadratically as the surface temperature is either higher or lower than the indoor air temperature. Such characteristic is what we can expect from the form of Eq. (7.50).

Table 7.2: Thermal exergy coming out the boundary layer by convection, x_{cv}.

	Temperature	$a = T_{ia} - T_1$	$c = 1 - \dfrac{T_o}{T_1}$	$a \cdot c$	Warm/ Cool	In/ Out
I	$T_o \leq T_1 < T_{ia}$	+	+	+	Warm	Out
II	$T_o < T_{ia} \leq T_1$	−	+	−	Warm	In
III	$T_1 < T_{ia} \leq T_o$	+	−	−	Cool	In
IV	$T_{ia} \leq T_o < T_1$	−	+	−	Warm	In
V	$T_1 < T_o < T_{ia}$	+	−	−	Cool	In
VI	$T_{ia} \leq T_1 \leq T_o$	−	−	+	Cool	Out

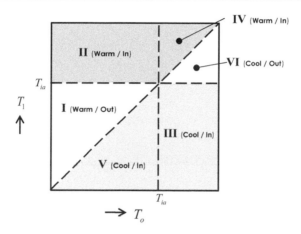

While the exergy inflow increases more and more as the surface temperature decreases, the exergy outflow looks reaching a plateau as the surface temperature decreases. This is due to quadratic increase of the exergy consumption rate. In the case of the surface temperature being higher than room air temperature, both exergy inflow and outflow become negative. This means that "warm" exergy comes out from the interior surface of the wall and goes into the air layer, while on the other hand, "warm" exergy goes out from the boundary air layer and comes into the room air. In other words, the direction of the arrows shown in the drawing attached to the upper graph becomes opposite. In these cases, the inflow and outflow of exergy in Eq. (7.48) change their positions to outflow and inflow, respectively, as follows.

$$\left\{ -\left(1 - \frac{T_o}{T_1}\right) q_{cv} \right\} - x_{c\_cv} = \left\{ -\left(1 - \frac{T_o}{T_{ia}}\right) q_{cv\_ia} \right\}. \qquad (7.53)$$

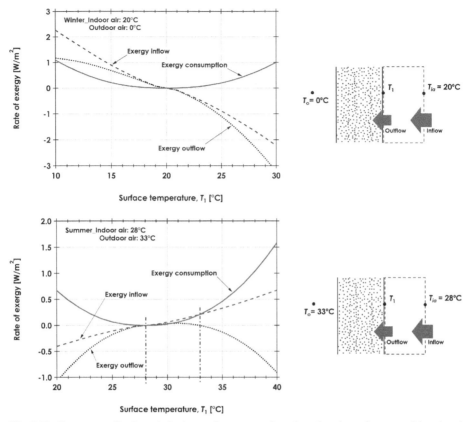

Fig. 7.22: Exergy transfer through the layer where convection takes place in a winter case (above) and in a summer case (below).

The first term of left-hand side of Eq. (7.53) represents the "warm" exergy flowing into the boundary air layer and the right-hand side represents the "warm" exergy flowing into the room air. Note that both terms have positive values. Since $T_o < T_1$ in winter, $\left(1 - \dfrac{T_o}{T_1}\right) > 0$ and in the case of $T_{ia} < T_1$, $q_{cv} < 0$. Therefore, the first term of Eq. (7.53) becomes positive. The same procedure confirms the last term in the right-hand side of Eq. (7.53) to be positive.

Let us confirm what Eq. (7.53) implies by a numerical example: in the case of the surface temperature at 25°C, "warm" exergy coming out from the wall surface at the rate of 1.1 W/m² is consumed at the rate of 0.2 W/m² and the resulting rate of exergy at 0.9 W/m² comes into the room air. This corresponds to such a case of radiant panel heating, with which "warm" exergy transfer by convection toward the room space emerges.

For summer cases, the exergy consumption rate varies quadratically as the surface temperature either increases or decreases; this is the same as in winter cases. In the cases of exergy outflow where the surface temperature is lower than 28°C, "cool" exergy flows out from the wall surface and into the boundary air layer and then, as the result of exergy consumption, "cool" exergy comes into the room air. In the cases of the surface temperature between 28°C and 33°C, "cool" exergy comes into the boundary air layer from the room air and then smaller "cool" exergy goes into the wall surface.

In the case of the surface temperature higher than 33°C, "warm" exergy comes out from the wall surface and goes into the boundary air layer, while at the same time, "cool" exergy also comes into the boundary air layer from the room air. Therefore, "cool" and "warm" exergies meet with each other within the boundary air layer and they are totally consumed into nothing. This can be expressed as follows, again by transforming the expression of Eq. (7.48). That is

$$\left(1-\frac{T_o}{T_{ia}}\right)q_{cv\_ia} + \left\{-\left(1-\frac{T_o}{T_1}\right)q_{cv}\right\} - x_{c\_cv} = 0. \qquad (7.54)$$

Equation (7.54) indicates that the two exergy inputs, one being "warm" exergy and the other "cool" exergy, are totally consumed and turn into nothing, that is, no output. "Warm" exergy meeting with "cool" exergy tends to make the exergy consumption rate larger. Such tendency grows as the surface temperature increases as can be seen in Fig. 7.22.

This suggests that poor insulation and solar control over the building envelopes results in a large exergy consumption, which inevitably brings about further large exergy consumption for an active cooling system. As a whole, either in winter or in summer, we confirm that making the difference in temperature between the interior surface of the walls and room air as small as possible is important in reducing the rate of exergy consumption due to convection.

7.6.3 Conduction

What has been described so far with respect to convection can be extended to exergy transfer by conduction. For the cases of "steady-state" conduction, in which the effects of heat capacity are neglected or negligible, provided that the heat transfer coefficient, h_{cv}, is simply replaced by the thermal conductance, C [W/(m²K)], for solid materials, then the above discussion on convection applies exactly to conduction (Shukuya 2013). For the cases of "unsteady-state" conduction, it becomes necessary to consider the time-wise variation of "warm" and "cool" exergy in addition to what we discussed in 7.6.2 (Shukuya 2013).

Since outdoor temperature has both daily and seasonal variations, it is worth developing such consideration further not only for pure academic interest alone but also for designing rational passive system having reasonably heavy building envelopes with appropriate thermal insulation. This should lead to a rational development of active systems for heating and cooling: for example, what are the keys for smart use of "warm" or "cool" exergy storage effects within building envelopes, as was discussed with some examples shown in 6.5.2. There is a lot to be investigated in relation to such characteristics (Li et al. 2014, Kazanci et al. 2016a and 2016b, Menberg et al. 2017, Choi et al. 2018). More on thermal conduction will be discussed in 9.4, Chapter 9.

References

Cardwell D. S. L. 1971. From Watt to Clausius—the rise of thermodynamics in the early industrial age. Heinemann Educational Books. London.

Carnot S. 1824. Reflections on the Motive Power of Heat (R. H. Thurston's translation and edition in 1987).

Choi W., Ooka R. and Shukuya M. 2018. Exergy analysis for unsteady-state heat conduction. International Journal of Heat and Mass Transfer 116: 1124–1142.

Harman P. M. 1982. Energy, Force, and Matter—The Conceptual Development of Nineteenth-century Physics. Cambridge University Press.

Kazanci O. B., Shukuya M. and Olesen B. W. 2016a. Exergy performance of different space heating systems: A theoretical study. Building and Environment 99: 119–129.

Kazanci O. B., Shukuya M. and Olesen B. W. 2016b. Theoretical analysis of the performance of different cooling strategies with the concept of cool exergy. Building and Environment 100: 102–113.

Lewis G. N. and Randall M. 1961. Thermodynamics 2nd ed. McGraw-Hill.

Li R., Ooka R. and Shukuya M. 2014. Theoretical analysis on ground source heat pump and air source heat pump systems by the concepts of cool and warm exergy. Energy and Buildings 447–455.

Menberg K., Heo Y., Choi W., Ooka R., Choudhary R. and Shukuya M. 2017. Exergy analysis of a hybrid ground-source heat pump system. Applied Energy 204: 31–46.

Shukuya M. 2013. Exergy—Theory and applications in the built environment. Springer-Verlag London.

Stirling engine. 2016. https://en.wikipedia.org/wiki/Stirling engine.

Tsuchida A. 1992. Thermal Science—on Open Systems Including Life and Environment. Asakura-Shoten Publishers (in Japanese).

Yamamoto Y. 2008. Historical Development of Thoughts on Thermo-Physical Phenomena—Heat and Entropy. Chikuma-Shobo Publishers (in Japanese).

Chapter 8

Air and Moisture

8.1 Atmosphere and vacuum space

Air, as the primary constituent of atmosphere, surrounds us everywhere. We humans tirelessly breathe in and out its tiny portion all the time as long as we live. All of us are aware of this, but almost unconscious that we live under the sea of atmospheric air near its bottom as abyssal fish lives deep under the sea water, since we do not realise how deep the atmosphere is.

Although we discussed the heat transfer by collective air movement, convection, in Chapter 6 and also discussed the role of air as a working fluid for a Carnot heat engine in Chapter 7, we have not yet discussed the role of atmospheric air with moisture that surrounds us indoors and outdoors. Therefore, to start our discussion here in this chapter, let us first describe how we can recognize the presence of atmosphere and also vacuum.

The depth of the atmosphere was not known until the mid-seventeenth century, but there was an empirical knowledge that was shared until then, especially by those people involved in digging deep wells for fresh water, coals and other materials. It was the fact that they could not pump up water existing under the ground lower than ten meters.

In ancient time, there was a philosophical thought that the nature, as its intrinsic characteristic, does not allow the vacuum space to exist; this meant that the space was filled with some continuous matter. On the other hand, there was another thought that the matter is made of invisible tiny particles and there should be vacuum space here and there between them. These two thoughts were put on focus first for experimental examination in medieval centuries; those famous for such experimentation were E. Torricelli (1608–1647), B. Pascal (1623–1662) and O. von Guericke (1602–1686). From the time of their investigations on to the early twentieth century, the associated scientific questions were kept being asked and answered until the atomic theory was confirmed valid by Planck (1858–1947), Perrin (1870–1842), Einstein (1879–1953), and others (Itakura 2004).

Let us first imagine the essence of the empirical knowledge owned by medieval engineers involved in digging deep wells by imagining a case that we drink juice with a very long straw as shown in Fig. 8.1. In order to drink the juice with such a long straw, you need to have a certain muscular power to keep closing your upper and lower lips tight so that the whole of lips, tongue and cheeks works as a pump

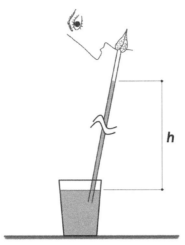

Fig. 8.1: Sucking the air inside a straw tube allowing us to drink juice, but there is a limit height, with which the juice can never climb up.

to let the juice climb up into your mouth. This is exactly to suck the air filling the space inside the straw between your lips and the upper surface of the juice so that the juice comes up towards your mouth. The old belief that vacuum space was not possible must have come from the fact that sucking air by a pump seemingly caused the tendency of water to fill the space where air originally existed.

Torricelli found that it was not true by performing an experiment as shown in Fig. 8.2. First we prepare an amount of mercury, a liquid metal, whose density is 13.6 g/cm³ (= 13.6 Mg/m³), which is 13.6 times larger than that of water, 1 g/cm³ (= 1000 kg/m³). We fill a shallow pool with mercury and then bathe a long glass tube having about one meter or so, which is also filled with mercury as shown in Fig. 8.2. By turning this long tube with the closed end upright, we will find that the mercury in the tube drops and stops at the height of about 760 mm. There emerges a seemingly hollow space above the upper mercury surface in the tube. Whether this is really hollow or not has to be confirmed by pouring the corresponding amount of water

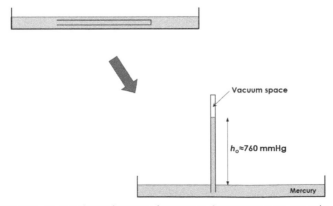

Fig. 8.2: An experimental setup to demonstrate that vacuum space can exist.

by a syringe from the open bottom end of the glass tube standing upright. Because of density of water being less than mercury as mentioned above, all the liquid water climbs inside the glass tube and reaches above the upper surface of mercury. We can fill the whole space with water; this implies that the space above mercury was originally vacuum before filling with water.

Supposing that the sectional area of the glass tube is denoted by A_S [m²], then the total mass of mercury with the height of 760 mm is expressed as $(0.76 \text{ m} \times A_S) \times 13.6 \times 10^3 \text{ kg/m}^3 = 10336 \, A_S$ [kg]. If we replace mercury with water, whose height is h_w [m], then the mass should be expressed as $(h_w \times A_S) \times 1000 \text{ kg/m}^3 = 1000 \, h_w A_S$ [kg]. We may take this to be equal to the mass of mercury, $10336 \, A_S$, since the mass of water that would result in making vacuum space occur above the upper surface of water should be the same as the mass of mercury. Then we come to know that h_w is 10.336 m. This is exactly what was known as an empirical knowledge in the medieval era.

Figure 8.3 shows another experiment that examines how vacuum space emerges above mercury. Suppose that a shallow pool of mercury with a glass tube at upright position shown in Fig. 8.2 is placed in a glass house. The left plate represents the same condition as Fig. 8.2 since a window of the glass house is open. On the other hand, the right plate represents a condition in which the window is closed and firmly sealed, while at the same time the air is sucked by a fan as we sucked the air in the straw above the juice shown in Fig. 8.1.

What we can see in this course of sucking the air is that mercury gradually descends from the original position. If you open the window again, then the height of mercury returns to 760 mm. This experiment lets us notice that the mercury surrounding the glass tube is kept being pushed by the air above the mercury surface. Pascal performed an experiment equivalent to this experiment for the first time by bringing the set of mercury-filled pool and glass tube standing upright to mountain areas, where the elevation is higher than that of plain area. His finding was that the higher the elevation, the lower the mercury height. It suggested that there would be the finite height of atmosphere at which the mercury height inside the tube becomes exactly the same as that outside. This implies that the outside space is vacuum; it must have been a surprising result that there was vacuum space above the atmosphere, far high away above our heads. What Pascal et al. did some four hundred years ago was an important epoch leading to a big change in human recognition of the environmental space, that is, we humans live under the atmospheric sea near its bottom, as mentioned in the beginning of this section.

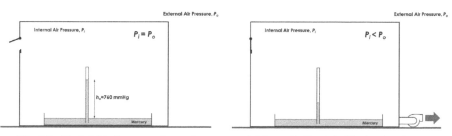

Fig. 8.3: The height of mercury descends as a powerful fan sucks the air inside the glass house.

The concept of pressure, which is of primal importance parallel to the concept of temperature, is in fact quantified in relation to what has been described above. Pressure is defined to be the force, the product of mass and gravitational acceleration, per unit area. Using the mass of mercury, 10336 A_s mentioned above, the force can be expressed as 10336 $A_s \times$ 9.8 m/s² = 101292.8 A_s [N(= kg·m/s²)]. Therefore, dividing this by the surface area becomes the atmospheric pressure, 101293 Pa (= N/m²). "Pa" is the international unit of pressure, which comes from Pascal. Pressure was conventionally expressed by the height of mercury or water as demonstrated in Fig. 8.2. With mercury, a certain atmospheric pressure was expressed as 760 mmHg or 760 Torr, where "Hg" is the chemical symbol of mercury and "Torr" comes from Torricelli, who was one of the first experimenters investigating what pressure is. With water, it was expressed as 10.3 mAq, where "Aq" comes from aqua.

The fact that pressure can be expressed by the height of mercury or water implies that pressure is an intensive quantity like temperature. This characteristic is associated with the fact that the levels of water surface become the same regardless of the size and the shape of the openings in a water-filled communicating vessel, as shown in Fig. 8.4, which can be applied to the determination of truly horizontal surface.

The atmospheric pressure is not always constant; the ceaseless change of weather here and there over the Earth surface is in association with the variation of atmospheric pressure. Figure 8.5 shows an example of the variation of atmospheric air pressure in Yokohama, Japan (JMA 2017). The day-to-day fluctuation is caused by the alteration of high-pressure and low-pressure patterns moving from west to east or from south to north over the whole of Japan. We can see that there is an overall gentle variation of atmospheric air pressure between 1000 and 1030 hPa; it is generally a little higher in winter and a little lower in summer in Yokohama area. Annual average of the atmospheric pressure is 1014 hPa (= 101.4 kPa). The atmospheric pressure used as the norm in a variety of calculation in physics and chemistry is 101.325 kPa.

As a Stirling engine shown in Fig. 7.2 demonstrated, air as a gaseous matter can be used as the working fluid. This is because air can easily be expanded and contracted depending on the pressure and temperature. The quantitative relationship

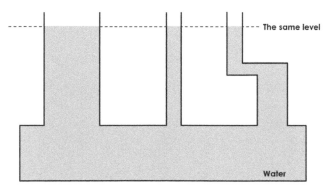

Fig. 8.4: A communicating vessel demonstrating the levels of water surface of three openings being identical.

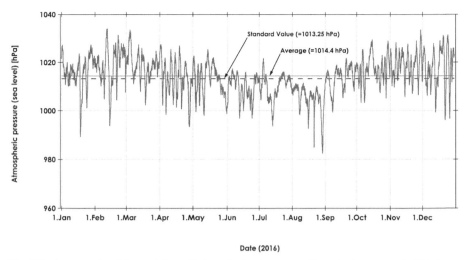

Fig. 8.5: An example of the variation of atmospheric air pressure throughout one year in Yokohama (data quoted from JMA (2017)).

between pressure, volume, and temperature given as the characteristic equation of ideal gas was gradually formulated after the aforementioned experiments were performed by Torricelli, Pascal and others.

The pressure and volume relationship was first investigated by H. Towneley (1629–1707) with respect to expansiveness of air using an experimental setup as shown in Fig. 8.6a. This experiment was extended so that the contractiveness of air was also examined by R. Boyle (1627–1691) and R. Hooke (1635–1703).

Figure 8.6b depicts what was found by Boyle. In their days, there was no custom of graphical presentation and their results were tabulated as a series of natural numbers together with fractions. The plots given in Fig. 8.6b are based on the values shown in Boyle's then treatise (West 2005, Yamamoto 2008). The horizontal axis represents the height of air space, that is, L in Fig. 8.6a, while on the other hand, the vertical axis indicates the difference between the maximum mercury height to be expected as $h_o = 76$ cm and the measured mercury height, that is, h in Fig. 8.6a. As can be seen, $(h_o - h)$ looks inversely proportional to L. This relationship turns out that the product, P, and volume, V, becomes constant as long as the air temperature is held constant.

In order to make this relationship inclusive of temperature, another experiment as shown in Fig. 8.7a, which was to measure how the air volume increases as the air temperature rises, was performed by Mariotte (1620–1684), Charles (1746–1823) and Gay-Lusaac (1778–1850). The result was that there was a linearity in the relationship between the volume and temperature as shown in Fig. 8.7b. The extrapolation of a fitted straight line giving the imaginary condition that the volume turns into nothing corresponds to –273.15°C, which is, as we already discussed in Chapter 7, the degree of absolute zero. Combining the two results, one shown in Fig. 8.6b and the other in Fig. 8.7b, provides us with the characteristic equation of atmospheric air including ordinary dilute gas expressed in Eq. (7.17).

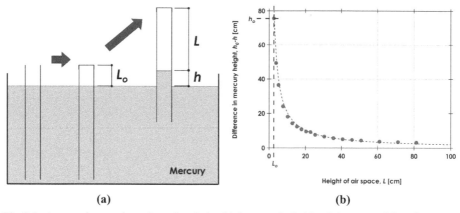

(a)   **(b)**

Fig. 8.6: An experiment to investigate the relationship between the height of air space and that of mercury, which was made by Towneley, Boyle and Hooke.

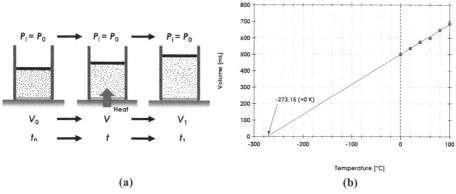

(a)   **(b)**

Fig. 8.7: An experiment to investigate the relationship between an increase of air temperature and the corresponding volume, which was made by Mariotte, Charles and Gay-Lusaac.

Before moving on to the next section, let us briefly discuss the major constituents of atmospheric air. As described in Chapter 3, the constituents of the atmosphere are considered to have changed gradually in the time frame of billion years in which geological and biological evolution took place, but for a much shorter period of time, the major constituents can be considered to be constant. Table 8.1 shows their present-day mole fractions in percentage. Mole fraction is the ratio of molar number of a constituent to the total of molar number of air, which is known to be equal to the partial pressure of constituent in dilute gas such as atmospheric air as clarified by Dalton (1766–1844). There are two columns of the mole fractions: the left one is for the air without water vapour and the right one for the air including water vapour. As can be seen, nitrogen comprises most, at about 78% and oxygen the second most, at 21%. The rest, about 1%, comprises mostly of argon and carbon dioxide if water vapour is excluded. If water vapour is included, argon and water vapour are comparable and they occupy about 2% together. The fraction of carbon dioxide is in the order of 0.04%, which is 400 ppm.

Table 8.1: The constituents of (dust free) atmospheric air.

Constituent	Chemical symbol	Mole fraction [%] (Without Water)	Mole fraction [%] (With Water)
Nitrogen	N_2	78.08400	77.43368
Oxygen	O_2	20.94760	20.77314
Argon	Ar	0.93400	0.92622
Carbon Dioxide	CO_2	0.03900	0.03868
Neon	Ne	0.00182	0.00180
Helium	He	0.00052	0.00052
Methane	CH_4	0.00018	0.00018
Krypton	Kr	0.00011	0.00011
Water	H_2O	–	0.82567*

* Mole fraction of water is for the condition of 15°C; 50%rh.

8.2 Saturated water vapour

The atmospheric air is more or less humid as we all know by using two words, wet and dry. These two words must have emerged as the reflection of moisture content in our surrounding air, but we cannot know how much or how little of air is humid with these words alone. Therefore, we need to know how to quantify that how much of moisture is contained by the atmospheric air. To do so, first let us consider an experiment again with a shallow pool and a glass tube filled with mercury as shown in Fig. 8.8a. We pour water again, but this time, only a tiny amount. Sooner or later, we see this tiny amount of liquid water disappear since it evaporates. In the course of evaporation, the temperature of liquid water and water vapour decreases a little and thereby some heat must flow into them from the surrounding space, whose temperature is assumed to be constant at 20°C, through the glass. Then again we pour a tiny amount of liquid water. We keep pouring water and seeing whether it evaporates or not until the last tiny amount of liquid water does not evaporate any more. This final condition is depicted in Fig. 8.8b.

With the condition of temperature at 20°C, the mercury height descends 18 mm at the maximum. If we do the same experiment at 10°C, then the mercury height descends 9 mm; if at 40°C, then 55 mm and so on. If it was done at 100°C, then the mercury height would descend 760 mm. At the temperature of 100°C, the water molecules as vapour are so vigorously bouncing around each other that they can push back collectively the molecules of mercury in the glass tube; this is exactly the condition at which water boils under ordinary atmospheric condition.

You might wonder whether the last tiny amount of liquid water poured, which remained as liquid, could also push down the mercury. But the whole weight of remaining liquid water is negligibly small so that the mercury height descending

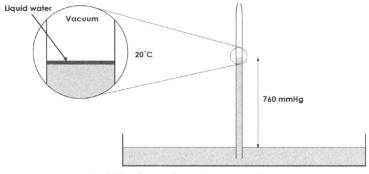

(a) Right after pouring a tiny amount of water

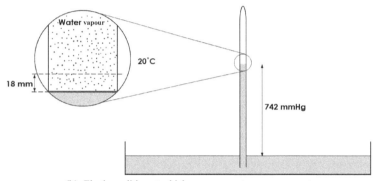

(b) Final condition at which no more water can evaporate

Fig. 8.8: Water vapour inside the space of a glass tube pushing down the mercury.

can be concluded due to the water vapour alone. Namely, the descending height of 18 mm is considered to be realized by the work exerted by the pressure made by all water vapour molecules bouncing around inside the glass tube.

The final state, in which the liquid water cannot evaporate any more, is called "saturation" and the pressure under this condition is "saturated water-vapour pressure", which is determined solely by temperature.

In what follows, let us discuss how the saturated water-vapour pressure as the function of temperature is determined on the basis of thermodynamic principle. For this purpose, as shown in Fig. 8.9, we focus on a small space in the vicinity of the boundary of water vapour and liquid water, where the upper part of space is filled with water vapour and the lower part with liquid water. We assume two tiny systems sharing the boundary surface with each other.

The fact that the water vapour behaves in a manner pushing down the whole of mercury in the glass tube, as demonstrated in Fig. 8.8, implies that the boundary surface shared by these two systems is open to energy transfer by work. During the course of work performed, the energy transfer by heat also takes place, since the water molecules carry away a lot of energy by evaporation and thereby the temperature of liquid water decreases.

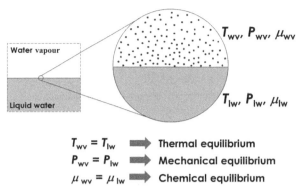

Fig. 8.9: Two tiny systems sharing the boundary, through which water molecules can cross and carry energy in or out, in addition to energy transfer by heat and work.

As pointed out in Chapter 2, a system, whose boundary surface allows heat and work, but no matter, to flow in or out is called a "closed" system, while on the hand, a system, whose boundary surface allows matter to flow in or out in addition to heat and work, is called an "open" system. The two systems here on focus are open systems, since the transportation of water molecules from liquid phase to vapour phase or vice versa are involved.

Both systems are tiny, but they can still be sufficiently large for thermodynamic thinking, due to the following reasoning: if each of the tiny systems is assumed to be a cube whose edge is 0.1 mm, then it contains about 34×10^{15} molecules for the liquid-water system and 6×10^{11} molecules for the water-vapour system. These numbers are large enough for the tiny systems to be regarded as thermodynamic systems. The two systems are considered to be in thermal, mechanical, and chemical equilibria.

Suppose that there is an infinitesimally small change in internal energy, dU, held by a system either of water-vapour system or of liquid water system. This change is brought by either an infinitesimally small flow of energy as heat, δQ, or work, δW, or by an infinitesimally small collective flow of water molecules, δY. We may write the relationship between all of three inflows of energy and the energy change in the system as follows, according to energy conservation.

$$\delta Q + \delta W + \delta Y = dU \qquad (8.1)$$

Note that two types of symbols representing infinitesimal smallness, "δ" and "d", are used again as used in Chapter 7; the former is for quantity of "flow" and the latter for quantity of "state".

We call two systems in contact with each other having the same temperatures to be in thermal equilibrium as discussed in Chapter 6. Similarly, we call two systems having the same pressures as being in mechanical equilibrium as we have come to know through aforementioned discussion in the present chapter. Taking a look at three terms in the left-hand side of Eq. (8.1), temperature is in association with the first term and pressure with the second term.

Then, a question arises: what kind of variable, having the characteristic of intensive quantity like temperature and pressure, can be considered in association with the third term? The answer is "chemical potential", which was first conceived by J. W. Gibbs (1839–1903), who was the foremost founder of physical chemistry or chemical thermodynamics that was developed so as to be able to explain how chemical reactions proceed.

The positions of three intensive quantities: temperature, pressure, and chemical potential, are equivalent to each other in the sense that they are all associated with characterising three types of equilibrium: thermal equilibrium with temperature, mechanical equilibrium with pressure and chemical equilibrium with chemical potential.

The infinitesimal amount of heat, δQ, can be expressed as the product of absolute temperature and an infinitesimal change of entropy as described in Chapter 7 with Eq. (7.30), that is, $\delta Q = TdS$. In a similar manner, the infinitesimal work, δW, can be expressed as the product of pressure, P, and an infinitesimally shrunken volume, $-dV \, (= A(-dx))$, that is, $\delta W = -PdV$. Note that either heat or work is expressed as the product of an intensive quantity such as T and P and an infinitesimally small change in extensive quantity such as dS and dV. With this nature in mind, what should be multiplied by the chemical potential as intensive quantity is an infinitesimal change in molar number. Denoting the chemical potential by μ and the molar number of the system by N, an infinitesimally small collective flow of molecules, δY, has to be expressed as $\delta Y = \mu dN$.

With all three infinitesimal inflow of energy by heat, work and mass transfer as described above, we may rewrite Eq. (8.1) for two systems as follows, denoting water vapour with the subscript "wv", and liquid water with "lw",

$$T_{wv}dS_{wv} - P_{wv}dV_{wv} + \mu_{wv}dN_{wv} = dU_{wv}, \tag{8.2}$$

$$T_{lw}dS_{lw} - P_{lw}dV_{lw} + \mu_{lw}dN_{lw} = dU_{lw}. \tag{8.3}$$

Since the right-hand sides of Eqs. (8.2) and (8.3) are infinitesimally small changes of extensive variables, internal energy of water vapour and liquid, U_{wv} and U_{lw}, respectively, the whole of left hand sides of these equations are the weighted sum of infinitesimally small change of extensive variables: S_{wv}, S_{lw}, V_{wv}, V_{lw}, N_{wv} and N_{lw}. Therefore, we can integrate the above two equations, respectively, as follows. By taking the temperature, volume, and chemical potential as constant, and thereby integrating the infinitesimal changes of entropy, volume, molar number, and internal energy respectively,

$$T_{wv}S_{wv} - P_{wv}V_{wv} + \mu_{wv}N_{wv} = U_{wv}, \tag{8.4}$$

$$T_{lw}S_{lw} - P_{lw}V_{lw} + \mu_{lw}N_{lw} = U_{lw}. \tag{8.5}$$

These equations can be further rewritten as follows,

$$G_{wv} = \mu_{wv}N_{wv} = U_{wv} + P_{wv}V_{wv} - T_{wv}S_{wv}, \tag{8.6}$$

$$G_{lw} = \mu_{lw}N_{lw} = U_{lw} + P_{lw}V_{lw} - T_{lw}S_{lw}. \tag{8.7}$$

These quantities, G_{wv} and G_{lw}, were first conceived by Gibbs. Due to its importance, they are called "Gibbs free energy", which has extensive characteristic as can be seen in these equations. Dividing these two Gibbs free energies by the respective molar numbers gives the definition of chemical potential as an intensive quantity. Namely,

$$\mu_{wv} = \frac{G_{wv}}{N_{wv}} = u_{wv} + P_{wv}v_{wv} - T_{wv}s_{wv}, \tag{8.8}$$

$$\mu_{lw} = \frac{G_{lw}}{N_{lw}} = u_{lw} + P_{lw}v_{lw} - T_{lw}s_{lw}. \tag{8.9}$$

where u_{wv}, v_{wv}, s_{wv}, u_{lw}, v_{lw}, s_{lw} are internal energy, volume and entropy of water vapour per molar number of water vapour or liquid water system, that is, specific internal energy, specific volume, and specific entropy.

An infinitesimal change in Gibbs free energy, either of water-vapour system or liquid water system, is expressed as follows by applying the product rule of differentiation (upon necessity, see Fig. 9.14 in Chapter 9 to confirm this rule).

$$dG_{wv} = d\mu_{wv}N_{wv} + \mu_{wv}dN_{wv}, \tag{8.10}$$

$$dG_{lw} = d\mu_{lw}N_{lw} + \mu_{lw}dN_{lw}. \tag{8.11}$$

Note that the second terms in the right-hand sides of Eqs. (8.10) and (8.11) are exactly the same as what appeared in the third terms of the left-hand sides of Eqs. (8.2) and (8.3); this implies that these terms appear uniquely in open systems and the first terms of Eqs. (8.10) and (8.11) appear commonly, either for closed or for open system.

An infinitesimal change in chemical potential, $d\mu_{wv}$ or $d\mu_{lw}$, can be expressed as follows, again applying the product rule of differentiation to Eqs. (8.8) and (8.9) and also substituting the infinitesimal change in specific internal energy, du, to be expressed in general as $Tds - Pdv$ in the case of a closed system into the resulting equations. The expression as $Tds - Pdv$ can be derived from Eqs. (8.2) or (8.3) assuming no transfer of matter, that is, $dN_{wv} = dN_{lw} = 0$ and dividing the whole of equation by N_{wv} and N_{lw}, respectively. The results are

$$d\mu_{wv} = v_{wv}dP_{wv} - s_{wv}dT_{wv}, \tag{8.12}$$

$$d\mu_{lw} = v_{lw}dP_{lw} - s_{lw}dT_{lw}. \tag{8.13}$$

In order to make it easier to grasp the relationship among the quantities appearing in these equations and thereby clarifying the implication of chemical potential, let us draw a three dimensional coordinate space for specific internal energy, u, specific entropy, s, and specific volume, v, as shown in Fig. 8.10a. The respective gradients of two ascending lines, one parallel to $u - s$ plane and the other parallel to $u - v$ plane, are

$$T = \left(\frac{\partial u}{\partial s}\right)_v, \qquad -P = \left(\frac{\partial u}{\partial v}\right)_s. \tag{8.14}$$

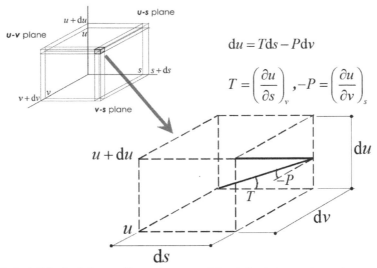

$$du = Tds - Pdv$$

$$T = \left(\frac{\partial u}{\partial s}\right)_v , -P = \left(\frac{\partial u}{\partial v}\right)_s$$

(a) An infinitesimal change of internal energy and its relation to temperature and pressure

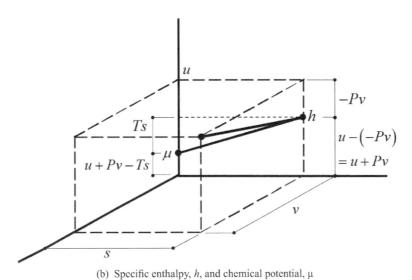

(b) Specific enthalpy, *h*, and chemical potential, μ

Fig. 8.10: Geometrical presentation of the relationship between specific internal energy, volume, and entropy for clarifying the meaning of chemical potential.

In these two equations, the lower-right subscripts denote that they are held constant.

A vertical line element given as the product of a length representing the specific volume *v* and the slope as *−P*, namely *−Pv*, subtracted from specific internal energy, *u*, is called "specific enthalpy" as shown in Fig. 8.10b.

$$h = u - (-Pv) = u + Pv. \tag{8.15}$$

Exactly in the same manner as we have had specific enthalpy, subtracting a vertical line element given as the product of a length representing specific entropy s and the slope as T, namely Ts, from specific enthalpy, h, provides us with chemical potential, $\mu = u + Pv - Ts$, again as shown in Fig. 8.10b. The chemical potential, geometrically expressed in Fig. 8.10b, is exactly what Eq. (8.8) shows for water vapour and Eq. (8.9) for liquid water.

The specific internal energy, volume and entropy of water vapour are different from those of liquid water, since their phases are different; for example, at 20°C, for which the saturated water-vapour pressure is 2.339 kPa, the specific internal energy, volume and entropy of water vapour are 2402.3 kJ/kg, 57.76 m³/kg, and 8.666 kJ/(kg·K), while on the other hand, those of liquid water are 83.9 kJ/kg, 0.001002 m³/kg, and 0.2965 kJ/(kg·K). But, their chemical potentials, μ_{wv} and μ_{lw}, are equal to each other as we have come to know that the water vapour and liquid water are in equilibrium as demonstrated in Figs. 8.8b and 8.9.

Figure 8.11 shows the same three-dimensional coordinate space as the one presented in Fig. 8.10, but from a different viewpoint. Water vapour and liquid water are represented by the corresponding two points, respectively: (s_{wv}, v_{wv}, u_{wv}) for water vapour and (s_{lw}, v_{lw}, u_{lw}) for liquid water, at which the water vapour and liquid water reach the condition shown in Fig. 8.8b. This state of equilibria is the condition that $T_{wv} = T_{lw}$ and $P_{wv} = P_{lw}$ are realized as shown in Fig. 8.9. On the three dimensional coordinate space shown in Fig. 8.11, it is equivalent to two flat planes defined by respective two gradients, T and $-P$, one for water vapour and the other for liquid water, being completely overlapped. Therefore, their chemical potentials become exactly the same as each other, that is, $\mu_{wv} = \mu_{lw}$ as indicated in Fig. 8.11.

Let us suppose that we turn the two flat planes, one for water vapour and the other for liquid water along the line element that connects the two points, u_{wv} and u_{lw}, as the axis of rotation so as to make an infinitesimal change in temperature and pressure occur by dT and dP. The resulting flat planes for water vapour and liquid water which overlap each other are expressed by dashed lines in Fig. 8.11. This infinitesimal shifting is exactly what $d\mu_{wv}$ and $d\mu_{lw}$, expressed by Eqs. (8.12) and (8.13), imply. Referring to the coordinate space shown in Fig. 8.11, $-P$, $-P + dP$, dP, T, $T - dT$ and dT can be expressed as follows.

$$-P = \frac{u_{wv} - z}{v_{wv} - v_{lw}}, \quad -P + dP = \frac{u_{wv} - z^*}{v_{wv} - v_{lw}}, \quad dP = \frac{z - z^*}{v_{wv} - v_{lw}}, \tag{8.16}$$

$$T = \frac{z - u_{lw}}{s_{wv} - s_{lw}}, \quad T - dT = \frac{z^* - u_{lw}}{s_{wv} - s_{lw}}, \quad dT = \frac{z - z^*}{s_{wv} - s_{lw}}. \tag{8.17}$$

With these geometrical relationships in terms of pressure and absolute temperature, we can derive the following equation that represents the ratio of infinitesimal change in pressure, dP, to infinitesimal change in temperature, dT.

$$\frac{dP}{dT} \left(= \frac{dP_{wv}}{dT_{wv}} = \frac{dP_{lw}}{dT_{lw}} \right) = \frac{s_{wv} - s_{lw}}{v_{wv} - v_{lw}}. \tag{8.18}$$

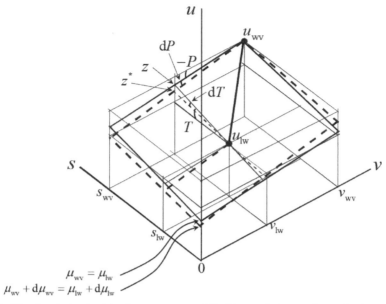

Fig. 8.11: The chemical potentials of water vapour and liquid water being equal to each other and their infinitesimally small change.

The difference in entropy between vapour phase and liquid phase that appeared in the numerator in Eq. (8.18) can be expressed as the product of dispersality, $1/T$, that was introduced in Chapter 7 and the specific value of latent heat, L, that is thermal energy to be carried away by a unit mass of water-vapour molecules in the course of evaporation. The denominator of Eq. (8.18), $v_{wv} - v_{lw}$, may be reduced to v_{wv} alone, since the specific volume v_{wv} is much greater than v_{lw}, as already shown for example, $v_{wv} = 57.84$ m³/kg and $v_{lw} = 0.001$ m³/kg at 20°C. The characteristic equation of water vapour expressed as $v_{wv} = (R/\mathfrak{M}_w)T/P_{vS}$ can be substituted into Eq. (8.18), where R is gas constant ($= 8.314$ J/(mol·K)), and $\mathfrak{M}_w$ is molar mass of water ($= 18.015 \times 10^{-3}$ kg/mol).

Assuming that the specific latent heat value of water vapour can be taken as a constant being equal to 2450 kJ/kg for the range of 0 to 40°C, though this value is 2.0% smaller than the precisely measured values of latent heat at 0°C and 1.8% larger than that at 40°C, the integral operation of Eq. (8.18) brings about the saturated water-vapour pressure, p_{vS} [Pa], expressed as Napier number ($e = 2.71828\cdots$) powered by a function of absolute temperature, T [K] as follows.

$$p_{vS} = e^{(25.89 - \frac{5319}{T})}. \tag{8.19}$$

There are various empirical formulae in similar forms to Eq. (8.19) so that it is not necessarily required to develop a new formula here, but the purpose of discussion above was to confirm how the thermodynamic principle powerfully allows us to deduce the relationship between saturated water-vapour pressure and temperature.

8.3 Hygrometry

Based on the fundamental characteristics we discussed in previous two sections, we now turn our focus on how to quantify the moisture in the atmospheric air. We consider a volume of humid air as the mixture of two constituents: one is "dry air" and the other water vapour. Although there is not a single element called "dry air" since the air is the mixture of nitrogen, oxygen, argon, carbon dioxide and others as shown in Table 8.1, we assume an imaginary matter called "dry air" to be mutually mixed with water vapour.

The saturated water vapour as a function of absolute temperature obtained from our discussion in the previous section was based on the case that a given space is occupied not by the mixture of water vapour and dry air, but by water vapour alone. Therefore, we need to elaborate how we can express the saturated water vapour pressure under the condition that water vapour exists together with dry air. Fortunately, what we have come to know in the previous section is exactly valid in the case of dry air existing in the space instead of vacuum. The reason for this is that the density of dry-air molecules in our surrounding space is sufficiently small. In other words, the volumetric size allowing one molecule of air is roughly 1000 times larger than liquid water so that whether there are dry air molecules or not does not matter at all with respect to the saturation of water vapour; one molecule of air occupies on average in a cube with the side length of 3.5 nm, while on the other hand, one molecule of liquid water occupies in a cube of 0.32 nm.

The most common expression of humidity is relative humidity, which is defined to be the ratio of water-vapour pressure within a volume of humid air, p_v [Pa], to the saturated water vapour pressure, $p_{vS}(T)$ [Pa], as a function of temperature, T [K]. Namely, the relative humidity, φ_a [%], is expressed as

$$\varphi_a = 100 \times \frac{p_v}{p_{vS}(T)}. \tag{8.20}$$

Figure 8.12 demonstrates the relationship between the water vapour pressure, relative humidity, and temperature in the range of 0 to 40°C. As pointed out in the first section in this chapter, the pressure can be expressed by the height of mercury (or liquid water) so that the mercury height is also shown together with the water vapour pressure in the unit of Pascal in Fig. 8.12. The curved line corresponding to the relative humidity of 100% represents Eq. (8.19) itself and other lines represent the values of relative humidity divided by 100 and then multiplied by saturated water vapour pressure. We can see that the water vapour pressure under ordinary room air conditions range from 1000 to 2000 Pa.

In addition to relative humidity, there are some other common ways of expression for humidity. Here two of them are introduced: one is the concentration of water vapour, c_{wv} [g/m³], and the other the mixing ratio, χ_m [g/kg]. They are expressed as follows, respectively, in relation to water vapour pressure, temperature and other parameters.

$$c_{wv} = 1000 \times \left(\frac{\mathfrak{M}_w}{R} \right) \frac{p_v}{T}, \tag{8.21}$$

$$\chi_m = 1000 \times \left(\frac{\mathfrak{M}_w}{\mathfrak{M}_a} \right) \frac{p_v}{P - p_v}. \qquad (8.22)$$

where R and $\mathfrak{M}_w$ are gas constant and molar mass of water as already introduced in the previous section, $\mathfrak{M}_a$ is molar mass of dry air (= 28.96 × 10⁻³ kg/mol) and P is the total pressure of dry air and water vapour, which may be taken to be constant at 101325 Pa as the norm. The values of c_{wv} are about 1.2 times larger than those of χ_m in ordinary range of indoor or outdoor air temperature. VBA codes for the calculation of water-vapour concentration and other related quantities are listed in Appendix 8.A.

Figure 8.13 shows the relationship between the concentration of water vapour, relative humidity and temperature in the range of 0 to 40°C. The concentration of water vapour under ordinary room air conditions ranges from 7 to 18 g/m³. More comprehensive and complex graphical presentation than either Fig. 8.12 or Fig. 8.13 called "psychrometric charts", which include the lines representing specific enthalpy, wet-bulb temperature and others, can be found in various sources of reference (ASHRAE, CIBSE, SHASEJ and others).

Figure 8.14 demonstrates the hour-by-hour variation of water-vapour concentration outdoors throughout one year together with the corresponding outdoor air temperature in Yokohama, 2016 (JMA 2017). The variation of water vapour concentration looks quite similar to that of air temperature. The weather profile in Yokohama is, in general, mildly cold and very dry in winter, while on the other hand very hot and humid in summer. These qualitative features are represented by the set of air temperature and water vapour concentration as 2 to 5°C and 5 to 8 g/m³ in winter and 27 to 32°C and 18 to 23 g/m³ in summer.

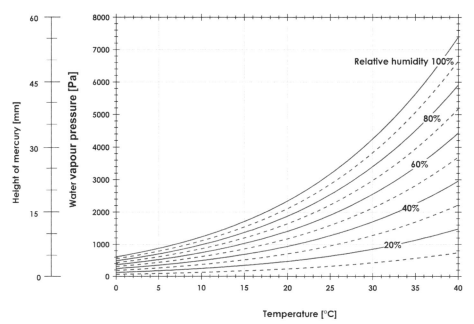

Fig. 8.12: The relationship between relative humidity, water vapour pressure and temperature in the range of 0 to 40°C.

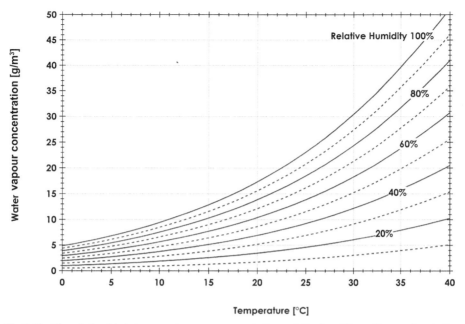

Fig. 8.13: The relationship between water vapour concentration, relative humidity and temperature in the range of 0 to 40°C.

8.4 Wet and dry exergies

Suppose that there is a room with an air-conditioning unit running under hot and humid outdoor air condition in summer. From the outlet of the air-conditioning unit, cold and less humid air comes out and sooner or later it disperses into the room air so as to let the room air temperature and humidity decrease towards their desired levels. People occupying this room inhale a portion of room air and also exhale very humid and warm air, moistened and heated in their lungs into the room space. Their skin surfaces are more or less wet so that the evaporation of water necessarily takes place. The air-conditioning unit swallowing a portion of room air lets the water vapour condense and thereby make that air less humid. This is a separation process, the opposite of dispersion, so that it cannot occur spontaneously. Why is it possible then? This is because more dispersion due to heat transfer by convection and conduction takes place inside the pipe walls and also between the pipe surface and chilled water or refrigerant flowing inside. Humid outdoor air comes into the room space more or less through the small openings along the frames of windows and then mixes with the less humid room air. This is of course the mutual dispersion of more humid air and less humid air.

On the other hand, in winter, suppose that the same air-conditioning unit as in summer is heating a portion of room air so that room air temperature increases towards the desired level, but is not humidifying. This results in the room air becoming relatively less humid, even if the water vapour concentration remains unchanged. In general, infiltration is more likely to take place in winter than in summer since the difference in indoor and outdoor temperature is much larger in winter than in

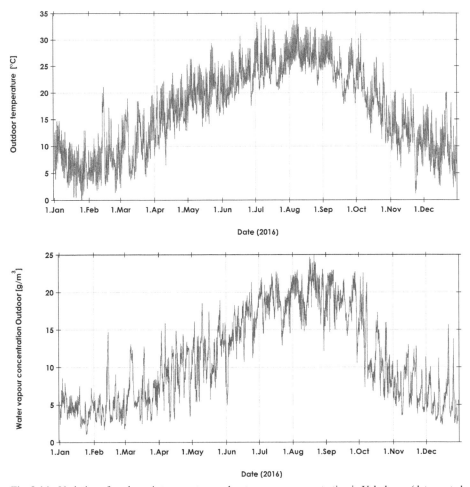

Fig. 8.14: Variation of outdoor air temperature and water vapour concentration in Yokohama (data quoted from JMA (2017)).

summer. Cold and dry air comes into the room space more or less from some small openings along the frames of windows and doors and then disperses into the room space so as to let the room air temperature and humidity decrease. Exhaled humid air from our mouths and noses also disperse with less humid air in the room space.

All of these examples in summer and in winter are typical mutually dispersing process of more humid air and less humid air. In order to know how much of dispersion takes place, the concept of exergy can be applied as warm and cool exergies are useful for a better understanding of heat transfer. Let us imagine that there is a vessel containing an amount of humid air whose relative humidity is much higher than the room air that surrounds the vessel as shown in Fig. 8.15. Both the vessel and the room space are surrounded by outdoor humid air, whose relative humidity is higher than that of the room air but lower than that inside the vessel.

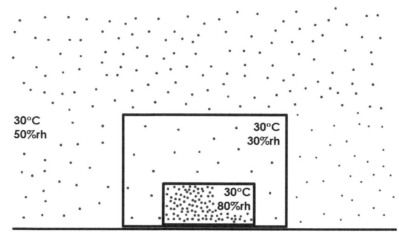

Fig. 8.15: A vessel containing humid air, which is surrounded by a room space also containing humid air. Both are surrounded by humid air outdoors.

In this example, there is no thermal exergy to be calculated from the right-hand side of Eq. (7.33) given in Chapter 7, since the indoor air temperature and the air temperature inside the vessel are assumed to be equal to outdoor air temperature. The whole pressure, the sum of dry-air and water-vapour pressures of room air and the air inside the vessel are also assumed to be equal to the whole outdoor pressure. Therefore, the air inside the vessel, the room air and outdoor air are in thermal and mechanical equilibria. But, they are not in chemical equilibrium. The concentrations of water vapour in the vessel and in the room space are 24 and 9 g/m³, respectively, while on the other hand, that of the outdoor space is 15 g/m³, as all three of them can be read from Fig. 8.13. Because of these different values of water-vapour concentration, mass diffusion occurs once either the vessel or the room space is opened.

We regard the vessel under such a condition to contain "wet" exergy since the water vapour can disperse either into the room space or into the outdoor environmental space. Similarly, we regard the room air to contain "dry" exergy since the water vapour in the outdoor environmental space may disperse into the room space. The concept of "wet" and "dry" exergies is analogical to that of "warm" and "cool" exergies which exist depending on the relationship in temperature between the system and the environment. "Wet" and "dry" exergies are one kind of chemical exergy to be determined by the relationship between chemical potential of water vapour and dry air in the system and the environment (Shukuya 2013).

In general, the exergy contained by humid air is expressed as the sum of thermal and chemical exergies; the former exists due to a difference in temperature between the system and its environment as described in Chapter 7, and the latter due to a difference in chemical potential of water vapour and dry air between the system and its environment to be described in what follows in the next subsection.

In addition to thermal and chemical exergies, there is also one more type of exergy, "high-pressure" and "low-pressure" exergies to be determined in relation to mechanically non-equilibrium condition and to be regarded as the actual potential that drives ventilation either by passive means or active means; this last type of exergy will be discussed in Chapter 11.

8.4.1 Humid air

In order to quantify "wet" and "dry" exergies of humid air, what we first need to know is the amount of entropy to be increased by free expansion. This is because these exergies are the function of chemical potentials of water vapour and dry air, which are determined by entropy changes in relation to mass diffusion. Figure 8.16 shows the process of a thought experiment for the determination of entropy increase by free expansion. In the beginning, dry-air molecules are confined to a compartment having the volume of V_0, which is ξ times smaller than the whole volume of container, V_1, and the space above the compartment, whose volume is $V_1 - V_0$, is vacuum. If the middle lid is slid, the dry-air molecules disperse freely and sooner or later reach the state of equilibrium with the volume of V_1. The dry-air molecules are assumed to be at the atmospheric pressure, P, in the beginning and to be at a lower pressure, p_a, after free expansion.

Dry air that we are focusing on here is a kind of dilute gas which never decreases its temperature as expanding freely, since there is no attractive force to be exerted between the molecules due to their sparseness though they collide with each other. Therefore, the temperature of dry air stays the same at T before and after free expansion.

In order to know how much of increase in entropy emerges, let us think about returning the gas with the volume of V_1 again to the original volume of V_0. To do so, we need to let a weight work isothermally, that is, keeping the temperature of dry air unchanged. This process is called isothermal compression and it requires the disposal of heat whose amount is exactly equal to the work performed. Since the last state of dry air is exactly the same as the original state, the isothermal compression squeezes the entropy increased in the course of free expansion. In other words, dry air had

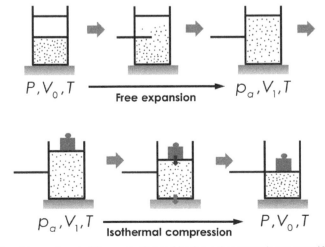

Fig. 8.16: A thought experiment of free expansion and isothermal compression to quantify the amount of entropy held by dilute gas.

this amount of entropy after it was freely expanded. This can be calculated from the following equation (Shukuya 2013).

$$\Delta S_a = \left(\frac{1}{T}\right) Q = R n_a \ln\left(\frac{V_1}{V_0}\right) = R n_a \ln\left(\frac{P}{p_a}\right). \tag{8.23}$$

where ΔS_a is the amount of entropy contained by dry air at the pressure of p_a relative to the pressure P and n_a is molar number of dry air. We apply the same discussion to water vapour and express its entropy by the following equation

$$\Delta S_{wv} = R n_{wv} \ln\left(\frac{P}{p_v}\right). \tag{8.24}$$

where ΔS_{wv} is the amount of entropy contained by water vapour at the pressure of p_v relative to the pressure P and n_{wv} is molar number of water vapour.

The entropy held by the mixture of dry air and water vapour, ΔS, can simply be expressed as the sum of respective entropy expressed by Eqs. (8.23) and (8.24) according to Gibbs theorem, with which the summation is proved to be valid for the mixture of dilute gases (Zemansky and Dittman 1981, Shukuya 2013). That is,

$$\Delta S = \Delta S_{wv} + \Delta S_a = R n_{wv} \ln\left(\frac{P}{p_v}\right) + R n_a \ln\left(\frac{P}{p_a}\right). \tag{8.25}$$

The next step that we must consider is to derive the equation of entropy increased by mutual dispersion of a system of humid air and another system of humid air whose water-vapour pressure is higher or lower than the former. Taking a look at Fig. 8.17, we suppose that there is an aggregate system consisting of four subsystems, each of which contains either of water-vapour molecules or of dry-air molecules alone. We denote this aggregate system as "A". We also suppose that there is another aggregate system consisting of two systems, each of which contains again either of water-vapour molecules or of dry-air molecules alone. The amount of entropy held by the aggregate system "A" is exactly the same as that by "A*". This is because the subsystem consisting of water vapour alone in the aggregate system "A*" is exactly the two subsystems of water vapour combined in the aggregate system "A" and the same applies to the subsystem of dry air in "A*".

The amount of entropy increased by mutual dispersion from "A" to "B" and also that from "A*" to "C" can be expressed by applying Eq. (8.25). Taking their difference yields the following equation that is exactly the difference in entropy between the aggregated system "B" and the system "C".

$$\Delta S_{\text{B-C}} = R n_{wv} \ln\left(\frac{p_v}{\hat{p}_{v0}}\right) + R n_{wv0} \ln\left(\frac{p_{v0}}{\hat{p}_{v0}}\right)$$
$$+ R n_a \ln\left(\frac{p_a}{\hat{p}_{a0}}\right) + R n_{a0} \ln\left(\frac{p_{a0}}{\hat{p}_{a0}}\right), \tag{8.26}$$

where n_{wv0} and n_{a0} denote the molar number of water vapour and dry air molecules contained by the subsystems having the volumes of V_{v0} and V_{a0}, respectively, within the aggregate system "A".

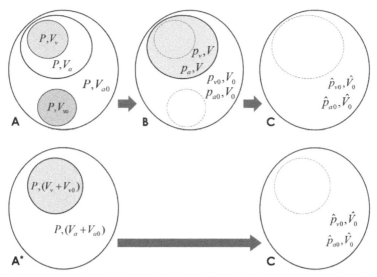

Fig. 8.17: Entropy increase brought by a series of mutual dispersion of water vapour and dry air molecules.

If we let the size of the outer subsystem having the volume of V_0 surrounding the inner subsystem inside the aggregated system "B" be infinitely large and thereby replace the water-vapour pressure, p_{v0}, and dry-air pressure, p_{a0}, in the outer subsystem with $\hat{p}_{v0}$ and $\hat{p}_{a0}$, then the inner subsystem, in which the water-vapour pressure and dry-air pressure are p_v and p_a, may be regarded as the humid air system surrounded by the environment in which the water-vapour pressure and dry-air pressure are p_{v0} and p_{a0}.

In this particular case, the whole change in the entropy contained by the system and the environment between before and after the humid-air system dispersing into the environment, ΔS, can be reduced from Eq. (8.26) to the following equation with a series of mathematical limiting operation in terms of the whole molar number of molecules existing in the outer subsystem of the aggregated system "B" shown in Fig. 8.17.

$$\Delta S = Rn \left\{ \frac{p_v}{P} \ln \left(\frac{p_v}{p_{v0}} \right) + \frac{P - p_v}{P} \ln \left(\frac{P - p_v}{P - p_{v0}} \right) \right\}, \tag{8.27}$$

where n is the sum of molar numbers of water-vapour and dry-air molecules in the humid-air system. Symbol "ln" is natural logarithm and its explanation is given in Chapter 9. An important feature of this equation is that ΔS is always greater than null for both cases of $p_{v0} < p_v$ and $p_v < p_{v0}$ (Shukuya 2013). According to the "exergy-consumption" theorem described in Chapter 7, we can express the general form of exergy balance equation for the system as follows.

$$X_m - S_g T_o = 0. \tag{8.28}$$

Let us suppose that X_m is an amount of "wet" or "dry" exergy contained by a humid-air system in which the partial pressures of water vapour and dry air are p_v and $P - p_v$, respectively, surrounded by the environment, whose partial pressures are p_{vo} and $P - p_{vo}$, respectively. The entropy generation, S_g, is nothing other than the entropy increase expressed by Eq. (8.27). Therefore, the product of Eq. (8.27) and the environmental temperature, T_o, turns out to be the exergy consumption, the second term in the left-hand side of Eq. (8.28), and hence the exergy contained by the moist-air system can be expressed as follows.

$$X_m = Rn\left\{\frac{p_v}{P}\ln\left(\frac{p_v}{p_{vo}}\right) + \frac{P - p_v}{P}\ln\left(\frac{P - p_v}{P - p_{vo}}\right)\right\}T_o. \tag{8.29}$$

It is convenient to express "wet" or "dry" exergy as a volumetric intensive quantity, x_m, using the characteristic equation of ideal gas, $PV = R(nT)$, for the whole of humid air. That is,

$$x_m = \left\{p_v\ln\left(\frac{p_v}{p_{vo}}\right) + (P - p_v)\ln\left(\frac{P - p_v}{P - p_{vo}}\right)\right\}\frac{T_o}{T}. \tag{8.30}$$

The volumetric thermal exergy contained by a humid air system, x_{th}, is expressed as follows referring to Eq. (7.33) in Chapter 7.

$$x_{th} = (c_{p\_wv}\rho_{wv} + c_{p\_a}\rho_a)\left\{(T - T_o) - T_o\ln\left(\frac{T}{T_o}\right)\right\}, \tag{8.31}$$

where c_{p_wv} and ρ_{wv} are the specific heat capacity and the density of water vapour, respectively, and c_{p_a} and ρ_a are those of dry air. The heat capacity of water vapour and dry air can be assumed to be constant at the values of 1846 J/(kg·K) and 1005 J/(kg·K), respectively.

The density of water vapour, ρ_{wv}, ranging approximately from 0.005 to 0.025 kg/m³, can be determined from Eq. (8.21) for the concentration of water vapour, c_{wv}, since $\rho_{wv} = 0.001 \times c_{wv}$, provided that water vapour pressure, p_v, is given. The density of dry air, ρ_a, ranging from 1.15 to 1.25 kg/m³, to be consistent with the value of ρ_{wv}, can be calculated from the characteristic equation $(P - p_v)V = R(n_a T)$. Denoting the mass of dry air in the volume V as M_a and using the molar mass of dry air, $\mathfrak{M}_a (= 28.96 \times 10^{-3}$ kg/mol), the molar number of dry air, n_a, is expressed as $n_a = M_a\big/\mathfrak{M}_a$. Therefore, the value of ρ_a can be calculated from

$$\rho_a = \frac{M_a}{V} = \frac{(P - p_v)\mathfrak{M}_a}{RT}. \tag{8.32}$$

Figure 8.18 shows two numerical examples of "wet" or "dry" exergy as a function of temperature calculated from Eq. (8.30), together with "warm" or "cool" exergy calculated from Eq. (8.31) for winter and summer conditions. The environmental temperature is assumed to be 5°C for winter and 32°C for summer. The relative-humidity is assumed to be constant both in the system and in the environment at 60%. In winter, a unit volume of room air conditioned at the temperature of 20°C and the relative humidity of 60%, for example, contains 500 J/m³ of "wet" exergy, which is almost the same order of "warm" exergy. This confirms the importance of air tightness for building envelopes. Making a building air-tight is to minimize not only the consumption of "warm" exergy but also that of "wet" exergy. Some

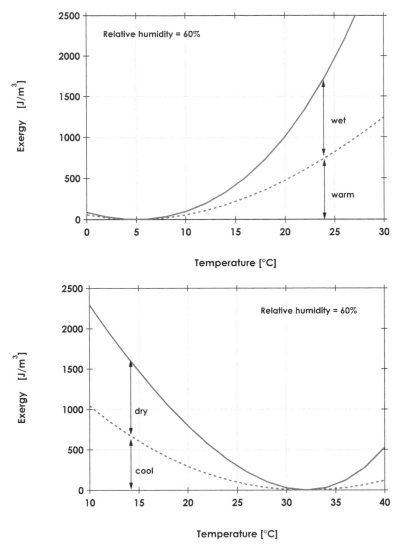

Fig. 8.18: The relationship between "wet" or "dry" exergy together with warm or cool exergy as a function of temperature with the environmental temperature at 5°C in winter and 32°C in summer. Relative humidity is assumed to be constant at 60% both for the system and the environment.

materials, which absorb and desorb water vapour depending on the surrounding conditions, may perform to reduce the necessity of supplying excess "wet" exergy to the room space. The moisture mass of such materials may act to make the indoor water vapour concentration remain above a certain critical value in a similar manner to the thermal mass, which is effective in abating the fluctuation of temperature. This must be interesting to take a look at for a further development of low-exergy systems for humidity control.

In summer, a unit volume of room air conditioned at the temperature of 26°C and the relative humidity of 60% contains about 150 J/m³ of "dry" exergy, which is more than twice larger than "cool" exergy of about 70 J/m³. In order to realize such a condition of room air, it is necessary to have an amount of conditioned air coming out from the outlet of the air-conditioning unit and let it circulate to disperse in the room space. For example, if the outlet air temperature and the relative humidity are assumed to be 16°C and 90%, respectively, its corresponding "dry" and "cool" exergies are 369 J/m³ and 534 J/m³, respectively. These values are much larger than those values held by the room air. In particular, cool exergy held by the outlet air is almost eight times larger. This confirms that ordinary dehumidification process by lowering temperature is, exergy-wise, very intensive. Exergy balance in relation to the dehumidifying process will be further demonstrated later in this section.

Appendix 8.A lists the VBA codes for the calculation of warm/cool exergy and wet/dry exergy of humid air.

8.4.2 Liquid water

In the case of liquid water as an open system surrounded by humid air, we all know that a portion of liquid water necessarily evaporates. This implies that an amount of liquid water contains "wet" exergy. In general, the exergy contained by liquid water is expressed as the sum of thermal exergy, which is "warm" or "cool", and "wet" exergy. What follows describes how to derive the formula of "wet" exergy for liquid water.

Suppose that there is a closed system containing water-vapour molecules alone with the molar number of n_{wv}, whose corresponding pressure is p_{vo}, as shown in the upper-left drawing in Fig. 8.19. Starting with this condition, we may consider a series of thought experiments. We first compress this system of water vapour until it is fully saturated and then condense it into liquid water. The whole of these processes are assumed to be done isothermally, that is, the temperature of the system remains unchanged at T_o.

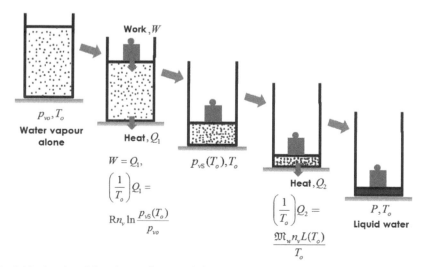

Fig. 8.19: A series of thought experiments to bring a water-vapour system into a liquid-water system.

The change in enthalpy, ΔH, and entropy, ΔS, of the closed system in the course of isothermal compression and liquefaction are expressed as follows

$$\Delta H = -\mathfrak{M}_w n_{wv} L(T_o), \tag{8.33}$$

$$\Delta S = -Rn_{wv} \ln \frac{p_{vS}(T_o)}{p_{vo}} - \mathfrak{M}_w n_{wv} \frac{L(T_o)}{T_o}. \tag{8.34}$$

where $\mathfrak{M}_w$ is the molar mass of water molecules (18.015×10^{-3} kg/mol), $L(T_o)$ is specific latent-heat value at the environmental temperature, T_o, [J/kg] and $p_{vS}(T_o)$ is saturated water-vapour pressure also at the environmental temperature, T_o, to be calculated from Eq. (8.19) [Pa]. Since $\mathfrak{M}_w > 0$, $n_{wv} > 0$, and $L(T_o) > 0$, $\Delta H < 0$; this indicates that the amount of heat expressed by Eq. (8.33) is squeezed out from the water vapour. In parallel to the change of enthalpy, the amount of entropy given off is expressed as the sum of entropy thrown away in both processes by isothermal compression and by liquefaction; the former is the first term of the right-hand side of Eq. (8.34) and the latter the second term.

Substitution of Eqs. (8.33) and (8.34) into the general form of exergy formula yields the following equation of "wet" exergy, X_{m_lw}.

$$X_{m\_lw} = \Delta H - T_o \cdot \Delta S = Rn_{wv} \ln \frac{p_{vS}(T_o)}{p_{vo}}. \tag{8.35}$$

Since $\mathfrak{M}_w n_{wv}$ is equal to $\rho_w V$, where ρ_w is the density of liquid water, 1000 kg/m³, and V its volume, and the relative humidity, φ_o, in percentage value is equal to the ratio of p_{vo} to $p_{vS}(T_o)$ multiplied by 100, the volumetric value of "wet" exergy of liquid water can be calculated from

$$x_{m\_lw} = \frac{R}{\mathfrak{M}_w} \rho_w T_o \ln \frac{100}{\varphi_o}. \tag{8.36}$$

Figure 8.20 shows "wet" exergy contained by a unit volume of liquid water calculated from Eq. (8.36) under the condition of the environmental temperature at 32°C and relative humidity at 60%. Shown together is thermal exergy contained by a unit volume of liquid water. "Wet" exergy is constant as far as the environmental temperature and relative humidity are constant as can be expected from Eq. (8.36). It is much larger than "warm" or "cool" exergy. The "wet" exergy of liquid water is in the order of one-hundred-thousand to one-million times larger than the "wet" or "dry" exergy of humid air, 50 to 200 J/m³, to be calculated from Eq. (8.30).

Appendix 8.A lists the VBA codes for the calculation of warm/cool exergy and wet exergy of liquid water.

8.4.3 Dehumidifying and mixing processes

As briefly introduced in the beginning of present section, in the process of air conditioning, the following four phenomena take place: humidification, dehumidification, condensation, and evaporation. As a volume of humid air at a

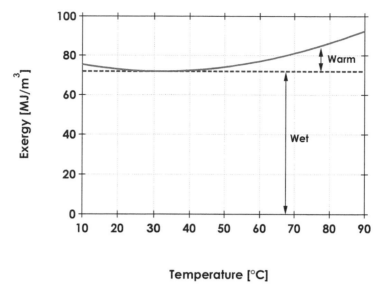

Fig. 8.20: "Wet" exergy and "warm" or "cool" exergy as a function of temperature under the condition of environmental temperature at 32°C and relative humidity at 60%.

certain temperature and humidity is mixed with the other volume of humid air at a different temperature and humidity, respective volumes of humid air become more humid air or less humid air depending on their primary conditions before being mixed. In this process, total mass and energy of volumes of humid air are necessarily conserved and the resulting conditions of temperature and humidity are determined so that the laws of conservation are satisfied. But this calculation alone cannot let us know how much of dispersion takes place.

The same applies to dehumidification and condensation. In order to dehumidify a volume of humid air, it is necessary to squeeze and remove a required amount of water vapour from the humid air and thereby make the humid air be less humid. This is the process of condensation. In this process too, the amount of condensed water is determined so that the total mass and energy involved are conserved. In addition to determining the temperature and humidity of outflowing volumes of humid air and also liquid water, how much of exergy is consumed has to be determined using the equations introduced in the previous subsections.

Figure 8.21 shows a dehumidifying process in an air-conditioning unit under the outdoor air condition of air temperature at 32°C and relative humidity at 60%. In this example, chilled water at 7.5°C is assumed to flow into the air-conditioning unit and flows out at 12.5°C. The incoming humid air at the rate of 0.21 m³/s (= 756 m³/h) is at 30°C and 50%rh and the outgoing humid air is at 15.48°C and 100%rh. A portion of water vapour carried by the incoming volume of humid air at 15 g/m³, which corresponds to 30°C and 50%rh, is condensed at the rate of 0.555 g/s ($\approx$ 2 kg/h), and as a result the outgoing volume of humid air becomes less humid at 13.1 g/m³ corresponding to 15.48°C and 100%rh. The numbers in the rectangles indicate the rate of exergy consumption.

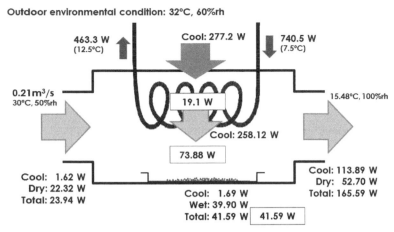

Fig. 8.21: Exergy inflow, consumption and outflow in the process of dehumidification process.

Liquid water is produced from the humid air at the rate of "cool" exergy of 1.69 W and "wet" exergy of 39.9 W; the latter is 23.6 times larger than the former. These numbers were determined from the aforementioned formulae described in 8.4.1 and 8.4.2. The sum of "wet" and "cool" exergies of liquid water generated is 41.59 W and it is totally consumed in the course of being dumped as a waste. The rate of exergy amounting to 41.59 W is about 56% of exergy consumption rate for dehumidifying process, 73.88 W, and about 30% of the difference in the total of "cool" and "dry" exergies given to the conditioned air, 141.65 (= 165.59–23.94) W. With the concepts of mass and energy alone, liquid water generated in the process of air conditioning tends to be out of focus because of being regarded as waste. But, with the concept of exergy, what we can see becomes different. Generated liquid water is not nothing but has quite a large potential.

In this whole process, 277.2 W of "cool" exergy is supplied towards the pipe surface where condensation takes place. Within the pipe wall, 19.1 W of "cool" exergy is consumed and as its result 258.12 W is supplied to the air to be conditioned. The incoming humid air contains "cool" exergy of 1.62 W and "dry" exergy of 22.32 W; this means that the incoming humid air is quite dry. The outgoing humid air contains a rate of "cool" exergy at 113.89 W, which is more than 70 times larger than the incoming humid air and also "dry" exergy at 52.7 W, 2.4 times larger than the incoming humid air.

The rate of thermal energy extracted by the process of cooling together with dehumidifying shown in Fig. 8.21 is 3562 W, which is much larger, at least twenty times larger or so, than the fan power for the circulation of humid air and the pump power for that of chilled water. Therefore, it may not look worth taking a careful consideration on fans and pumps, but, what is to be compared is in fact the "cool" and "dry" exergies involved and they are quite comparable to the fan and the pump powers. If this is taken into consideration, designing the whole of air-conditioning process may be worth re-consideration for a better process, which requires much smaller and rational exergy consumption. More on the related topics will be discussed in Chapter 11.

Figure 8.22 shows another example of air-conditioning process: the mixing of two volumes of humid air, one is less humid and the other more humid. Again, also in mixing process, the total mass and energy involved are conserved. In this example, the less humid air with the flow rate of 0.628 m³/s (= 2261 m³/h) carries thermal energy at the rate of –4480 W, which implies that thermal energy contained by this aggregate of less humid air is smaller than that contained by the same aggregate of humid air at the environmental condition. The humid air to be mixed with this less humid air is at the same temperature and humidity as the environment so that its relative content of energy is null.

The humid air flowing out after mixing process has exactly the same rate of thermal energy according to the principle of conservation. Therefore, with the concept of energy alone, mixing process may look all right. But with the concept of exergy, it becomes clearer what really happens in the process of mixing. In this example, "cool" exergy of 11.22 W and "dry" exergy of 52.32 W is consumed. The latter is 4.6 times larger than the former. This is because the less humid air before mixing has a large rate of "dry" exergy, 185.9 W, about four times larger than "cool" exergy. The total rate of exergy consumption is 27.6% of total exergy supplied before mixing. Redesigning such mixing process towards the one requiring less exergy consumption must be an interesting topic for a low-exergy air conditioning system.

Outdoor environmental condition: 32°C, 60%rh

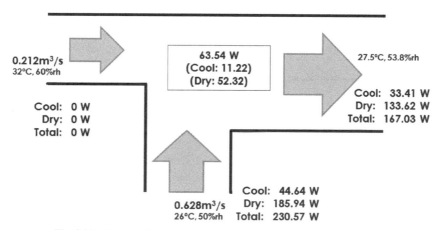

Fig. 8.22: Exergy inflow, consumption and outflow in the process of mixing.

8.5 Evaporative cooling

The fact that liquid water is very rich in "wet" exergy implies that its consumption may become useful either for decreasing unwanted "warm" exergy or for generating "cool" exergy depending on the surrounding condition. This process is very important among a variety of thermal phenomena that we encounter within the built environment such as the human-body, plant leaves, and evaporative cooling systems. Evaporation is one of the four heat-transfer paths introduced in the beginning of

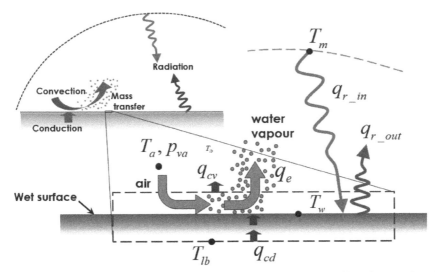

Fig. 8.23: A wet surface where the fourth path of heat transfer, evaporation, is taking place together with other three paths, radiation, convection, and conduction.

Chapter 6, in which the fundamentals of three out of four, radiation, convection, and conduction, were discussed. Here in this section heat transfer involving evaporation of liquid water is going to be discussed from the viewpoint of exergy.

Let us suppose a wet surface, as shown in Fig. 8.23, whose temperature, T_w, is surrounded by humid air above having the temperature of T_a and the water-vapour pressure of p_{va} and is also surrounded by a matter below, whose boundary temperature is T_{lb}. The matter below may be either of liquid water, the skin tissue of human body, the skin tissue of a plant leaf, or a building envelope material.

Energy balance equation at a very thin system including the wet surface may be expressed as [Energy input] = [Energy output]. The energy input is the sum of two components: one is the incoming radiant energy from the adjacent upper surface at the rate of q_{r_in} and the other the incoming thermal energy by conduction from the liquid water just below the wet surface at the rate of q_{cd}. The energy output is the sum of three components: one is the outgoing radiant energy from the wet surface at the rate of q_{r_out}, another is the outgoing thermal energy by convection at q_{cv}, and the last is the thermal energy carried away by the evaporation of water at q_e. That is

$$q_{r\_in} + q_{cd} = q_{r\_out} + q_{cv} + q_e. \tag{8.37}$$

The energy carried away by the evaporation of water may be expressed as the difference in enthalpy between liquid water and water vapour. Assuming that an amount of liquid water, whose enthalpy is ΔH_{lw}, which is the quantity of state measured as a difference from the enthalpy at a reference condition, is evaporated into water vapour having the enthalpy of ΔH_{wv}, which is also measured as a difference from the enthalpy at the same reference condition. Using the rate of energy carried away by the evaporation of liquid water denoted by q_e, ΔH_{lw} and ΔH_{wv} can be related to each other as follows based on the law of energy conservation.

$$\Delta H_{lw} + q_e = \Delta H_{wv}. \tag{8.38}$$

The energy carried away by evaporation, q_e, may be expressed as

$$q_e = wh_e \{p_{vS}(T_w) - p_{va}\}. \tag{8.39}$$

where w is wetness factor which should be at some value between 0 for totally dry and 1 for fully wet as liquid-water surface, h_e is thermal-energy transfer coefficient due to the evaporation of water. As we all know by our experience, for example, as soon as we start feeling hot, we usually start fanning by hand or a fan. This is the behaviour for the enhancement of convection and if the skin surface is wet, the heat transfer is further enhanced because of evaporation together with convection. Namely, the convection of air relates strongly to the evaporation of water. We can determine the value of h_e in relation to the value of convective heat-transfer coefficient, h_c, with the proportional constant, l_r, which is equal to 16.5×10^{-3} K/Pa (Kerslake 1972, ASHRAE 2005), that is, $h_e = l_r h_c$ in the unit of W/(m²Pa).

The specific latent heat, $L(T)$, at an arbitrary value of absolute temperature, T, can be approximated very well by the following equation (Shukuya 2013),

$$L(T) = L(273.15) + (c_{p\_wv} - c_{p\_lw})(T - 273.15), \tag{8.40}$$

where $L(273.15)$ is the specific latent heat of evaporation at 0°C (= 273.15 K) being equal to 2500 kJ/kg and c_{p_wv} and c_{p_lw} are specific heat capacity of water vapour (= 1.846 J/(g·K)) and liquid water (= 4.186 J/(g·K)), respectively. For the environmental temperature, T_o, the latent heat is expressed as follows.

$$L(T_o) = L(273.15) + (c_{p\_wv} - c_{p\_lw})(T_o - 273.15). \tag{8.41}$$

Rewriting each term of Eq. (8.37) with the respective heat-transfer coefficient and the corresponding difference in temperature together with the relationship expressed in Eq. (8.39), the resultant equation can be solved for the wet-surface temperature, T_w (Shukuya 2013). Using the result obtained, the value of energy carried away by evaporation, q_e, can be determined by Eq. (8.39). Denoting the rate of water molecules to be evaporated as n_{wv} [mol/s], and the molar mass as $\mathfrak{M}_w$ (= 18.015×10^{-3} kg/mol), the rate of energy carried away by evaporation, q_e, turns out to be $\mathfrak{M}_w n_{wv} L(T_w)$. Therefore, the rate of water molecules turning into the state of vapour from the state of liquid can be determined from the following equation.

$$n_{wv} = \frac{q_e}{\mathfrak{M}_w L(T_w)}. \tag{8.42}$$

Subtraction of Eq. (8.41) from Eq. (8.40) with the wet surface temperature at T_w and multiplication of $\mathfrak{M}_w n_{wv}$ over the whole resulting equation yields,

$$\begin{aligned} \mathfrak{M}_w n_{wv} \{-L(T_o) + c_{p\_lw}(T_w - T_o)\} \\ + \mathfrak{M}_w n_{wv} L(T_w) = \mathfrak{M}_w n_{wv} c_{p\_wv}(T_w - T_o). \end{aligned} \tag{8.43}$$

The first term on the left-hand side of Eq. (8.43) corresponds to ΔH_{lw} in Eq. (8.38) and the right-hand side to ΔH_{wv}.

The equation with respect to entropy, which is parallel to Eq. (8.43) for enthalpy, can be derived by integrating Eq. (8.18) for the range of temperature from T_o to T_w and that of saturated water-vapour pressure from $p_{vS}(T_o)$ to $p_{vS}(T_w)$. In the integral operation of Eq. (8.18), the relationship of Eq. (8.40) for the wet-surface temperature, T_w, is used instead of a constant value used for deriving Eq. (8.19). The result is as follows,

$$\Delta S_{lw} + n_{wv} \left\{ \mathfrak{M}_w \frac{L(T_w)}{T_w} + R \ln \frac{p_{vS}(T_w)}{p_{va}} \right\} = \Delta S_{wv}, \qquad (8.44)$$

where

$$\Delta S_{lw} = -R n_{wv} \ln \left(\frac{p_{vS}(T_o)}{p_{vo}} \right) - \mathfrak{M}_w n_{wv} \left\{ \frac{L(T_o)}{T_o} - c_{p\_lw} \ln \left(\frac{T_w}{T_o} \right) \right\}, \qquad (8.45)$$

$$\Delta S_{wv} = -R n_{wv} \ln \left(\frac{p_{va}}{p_{vo}} \right) + \mathfrak{M}_w n_{wv} c_{p\_wv} \ln \left(\frac{T_w}{T_o} \right). \qquad (8.46)$$

The second term of the left-hand side of Eq. (8.44) is greater than zero as long as the water-vapour pressure in the surrounding humid air is lower than the saturated water-vapour pressure at the surface temperature of T_w, since $0 < \ln \left(\frac{p_{vS}(T_w)}{p_{va}} \right)$ in addition to $0 < n_{wv}$, $0 < \mathfrak{M}_w$, $0 < L(T_w)$, $0 < T_w$, and $0 < R$. This confirms that the evaporation of water, a typical mass-diffusion phenomenon, is exactly accompanied by an amount of increased entropy quantified by the second term of Eq. (8.44).

The water vapour does not exist itself alone after its evaporation, but it disperses spontaneously with the humid air nearby. The rate of molar number of dry-air molecules to be dispersed mutually with water-vapour molecules can be expressed as follows according to the ideal-gas equation relationship both for water vapour and for dry air, assuming that the molar number of evaporated water molecules of n_{wv} does not change the water vapour pressure in the surrounding humid air, p_{va},

$$n_a = \frac{P - p_{va}}{p_{va}} n_{wv}. \qquad (8.47)$$

We add the entropy value of dry air for the number of molecules given by Eq. (8.47) to both sides of Eq. (8.44), that is, to take the sum of ΔS_{lw} given by Eq. (8.45) and the entropy value of dry air and also to take the sum of ΔS_{wv} by Eq. (8.46) and the entropy value of dry air. Namely,

$$\Delta S_{lw}{}^* = \Delta S_{lw} - \frac{P - p_{va}}{p_{va}} R n_{wv} \ln \left(\frac{P - p_{va}}{P - p_{vo}} \right) \text{ and } \Delta S_{wv}{}^* = \Delta S_{wv} - \frac{P - p_{va}}{p_{va}} R n_{wv} \ln \left(\frac{P - p_{va}}{P - p_{vo}} \right).$$

Then, using $\Delta S_{lw}{}^*$ together with the term in Eq. (8.43) corresponding to ΔH_{lw} and the environmental temperature, T_o, the molar exergy of liquid water, x_{lw}, to be evaporated can be derived as follows

$$x_{lw} = \mathfrak{M}_w c_{p\_lw} \left\{ (T_w - T_o) - T_o \ln\left(\frac{T_w}{T_o}\right) \right\}$$

$$+ RT_o \left\{ \ln\left(\frac{p_{vs}(T_o)}{p_{vo}}\right) + \frac{P - p_{va}}{p_{va}} \ln\left(\frac{P - p_{va}}{P - p_{vo}}\right) \right\}$$ (8.48)

The first term of the right-hand side of Eq. (8.48) is thermal exergy, which is either "warm" or "cool", and the second term is "wet" exergy contained by liquid water and its associated dry air to be dispersed mutually once evaporated (Shukuya 2013).

The same procedure using ΔS_{wv}^*, together with the term in Eq. (8.43) corresponding to ΔH_{wv}, and the environmental temperature, T_o, as for Eq. (8.48) brings us the following expression of molar exergy of the mixture of water vapour and dry air dispersing into the surrounding moist air.

$$x_{wv} = \mathfrak{M}_w c_{p\_wv} \left\{ (T_w - T_o) - T_o \ln\left(\frac{T_w}{T_o}\right) \right\}$$

$$+ RT_o \left\{ \ln\left(\frac{p_{va}}{p_{vo}}\right) + \frac{P - p_{va}}{p_{va}} \ln\left(\frac{P - p_{va}}{P - p_{vo}}\right) \right\}$$ (8.49)

The first term on the right-hand side of Eq. (8.49) is "warm" or "cool" exergy and the second term is "wet" or "dry" exergy of humid air to be generated in the course of the evaporation of liquid water (Shukuya 2013).

Using Eqs. (8.48) and (8.49), the exergy balance equation for the system including the wet surface shown in Fig. 8.23 is expressed as

$$x_{r\_in} + n_{wv} x_{lw} + \left(1 - \frac{T_o}{T_{lb}}\right) q_{cd} - x_c$$

$$= x_{r\_out} + \left(1 - \frac{T_o}{T_w}\right) q_{cv} + n_{wv} x_{wv}$$ (8.50)

where x_{r_in} and x_{r_out} are the incoming and outgoing radiant exergies to be calculated from Eq. (7.44) and x_c is the exergy consumption rate, which is due to three processes: the absorption of radiation at the wet surface, the conduction of thermal exergy from the lower boundary towards the wet surface, and the evaporation of liquid water.

Figure 8.24 shows three examples of exergy balance at the wet surfaces: a pond of liquid water, the naked shoulder of a sweating human-body surface, and a leaf of a broadleaved tree. In these examples, the wet-surface temperature, T_w, was determined by solving Eq. (8.37) assuming the surrounding condition of temperature and relative humidity as shown within the respective plates in Fig. 8.24 and the heat transfer characteristics shown in Table 8.2. The wetness factor is assumed to be 1.0 for the liquid-water surface of a pond, 0.12 for the sweating surface of human body, and 0.1 for the leaf surface. The last column of Table 8.2 indicates the rates of mass transfer calculated using the relationship expressed by Eq. (8.42) in the unit of mg/(m²s) for the respective three cases. These values are comparable to the global average

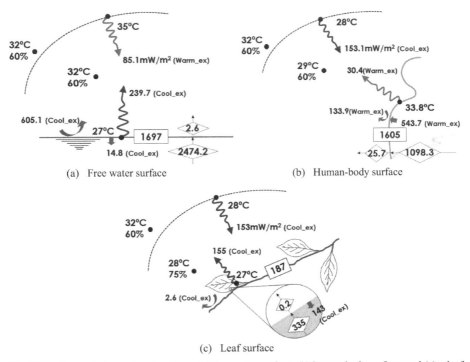

Fig. 8.24: Exergy balance at wet surfaces: (a) free water surface; (b) human-body surface; and (c) a leaf surface. The numbers in the rectangles indicate the exergy consumption rates and those in the rhombuses below and above the surfaces are the rate of wet exergy carried by liquid water and humid air, respectively.

Table 8.2: Heat-transfer characteristics assumed for the calculation of wet-surface temperature and the rate of mass transfer calculated.

case	w [-]	h_c [W/(m²K)]	$w \cdot h_e$ [W/(m²Pa)]	C [W/(m²K)]	h_r [W/(m²K)]	$n_{wv} \mathfrak{M}_w$ [mg/(m²s)]
a)	1.0	7.3	0.120	600	5.8	34.3
b)	0.12	4.8	0.0095	400	5.8	11.2
c)	0.1	7.3	0.012	400	5.8	4.6

Air velocity is assumed to be 0.8 m/s for both a) and c), and 0.4 m/s for b).
C is the conductance between the surface as the system and the lower boundary.

evaporating rate at 31.8 mg/(m²s), which is the rough estimate from the average precipitation in the global environmental system, and also consistent with 1.2 to 5 mg/(m²s) for a plant system and 5 to 40 mg/(m²s) for human body (Shukuya 2013). In Fig. 8.24, the numbers in the rectangles indicate the exergy consumption rates and those in the rhombuses below and above the surfaces are the rate of wet exergy carried by liquid water and humid air, respectively.

Case a) corresponds to such an outdoor space under hot and humid summer condition, where the intense solar radiation is well shaded by a lot of leaves above the pond surface, while on the other hand, nice breeze blows over the pond. The liquid-water surface emits "cool" radiant exergy at a large rate and also there is "cool" exergy transfer by convection towards the surrounding air. The availability of "cool" exergies by radiation together with convection must provide the persons nearby with comfortable coolness. The rate of exergy consumption indicated in the rectangle, which is almost 70% of the "wet" exergy contained by liquid water, 2474 mW/m^2, is caused by three phenomena: one is the consumption in the process of thermal exergy conduction under the liquid-water surface nearby; another is the absorption of "warm" radiant exergy at the liquid-water surface; and the last is the evaporation of liquid water. The "cool" radiant exergy emitted by the water surface is almost three times larger than the "warm" radiant exergy coming from upper surrounding surface; this is considered to be mainly provided by the evaporation of water.

At the sweating human-body surface shown in case b), there is also a large rate of exergy consumption comparable to the rate of exergy consumption at the liquid-water surface in case a). This is mainly due to "warm" exergy consumption in the course of conduction within the skin tissue and to the consumption of "wet" exergy in the course of sweat evaporation. The latter in fact plays the key role in decreasing the body-surface temperature and thereby contributes to enhancing the conduction of "warm" exergy from the human-body core towards the skin surface. This also results in reducing the outgoing rate of "warm" exergy by radiation and convection into the surrounding space. If the surrounding air temperature is lowered to secure the release of "warm" exergy from the human-body surface by convection, it could result in thermal discomfort because of the so-called cold draught. This suggests the importance of minimizing the emission of "warm" radiant exergy from the surrounding surfaces and also the importance of making available the "cool" radiant exergy emitted by the surrounding surfaces. This topic will be further discussed in Chapter 10.

In the case of a leaf, its surrounding temperature is assumed to be well controlled by the presence of other leaves. This enables the leaf to have the wetness factor comparable to the sweating human-body surface as assumed here in this example calculation. In the photosynthesis, the relative amount of liquid water to be evaporated is about one-hundred times the amount of liquid water to be fixed as a portion of glucose (Shukuya 2013). But, it does not imply that the leaf surface is as wet as liquid water surface is. The rate of exergy consumption is much smaller than the sweating human-body surface. In the case of the leaf surface, 56% of "wet" exergy contained by liquid water is consumed and thereby "cool" radiant exergy and "cool" exergy transfer by convection are provided. This confirms that plants are in general extremely good at using liquid water to grow their bodies themselves by photosynthesis while at same time creating the "cool" environmental space from which we humans can benefit for our well-being.

References

ASHRAE. 2005. ASHRAE Handbook of Fundamentals.
ASHRAE. 2017. https://www.ashrae.org/resources—publications/bookstore/psychrometrics.
CIBSE. 2017. http://www.cibse.org/Knowledge/knowledge-items/detail?id=a0q20000008I7k8.
Itakura K. 2004. The History of Atomism, Kasetsu-sha Publisher. Tokyo.

Japan Meteorological Agency (JMA). 2017. http://www.data.jma.go.jp/gmd/risk/obsdl/index.php.

Kerslake D. M. 1972. The Stress of Hot Environments. Cambridge University Press.

SHASEJ. 2016. http://www.shasej.org.

Shukuya M. 2013. Exergy—Theory and applications in the built environment. Springer-Verlag London.

West J. 2005. Robert Boyle's landmark book of 1660 with the first experiments on rarified air. Journal of Applied Physiology 98: 31–39.

Yamamoto Y. 2008. Historical Development of Thoughts on Thermo-physical Phenomena—Heat and Entropy. Chikuma-Shobo Publishers (in Japanese).

Zemansky M. W. and Dittman R. H. 1981. Heat and Thermodynamics. McGraw-Hill, Inc.

Appendix

8.A Calculation of hygrometric quantities

Tables 8.A.1a to e show the VBA function codes for the calculation of (a) saturated vapour pressure, (b) water-vapour concentration, (c) mixing ratio, (d) wet-bulb temperature, and (e) dew-point temperature, respectively.

Tables 8.A.2a to d show the VBA function codes for the calculation of (a) warm/cool exergy of humid air, (b) wet/dry exergy of humid air, (c) warm/cool exergy of liquid water, and (d) wet exergy of liquid water, respectively.

Table 8.A.1a: VBA code for the calculation of saturated water-vapour pressure.

```
Function ps(tak)
'
' Saturated water-vapour pressure [Pa]
'
'Input: taK -> absolute temperature of humid air [K]
'
ps = Exp(25.89 - 5319 / taK)
End Function
```

Table 8.A.1b: VBA code for the calculation of water-vapour concentration.

```
Function wv_c(ta, rh)
'
' The concentration of water vapour [g/m^3]
'
'Input: ta -> air temperature [degC]
'       rh -> relative humidity [%]
'
Mv = 18.015  'g/mol
R = 8.314    'J/(molK)
t = ta + 273.15
wv_c = Mv / R * (ps(t) * rh * 0.01) / t
End Function
```

Table 8.A.1c: VBA code for the calculation of mixing ratio.

```
Function h_ratio(ta, rh)
'
'The calculation of humidity ratio [g/kg]
'
'Input: ta -> air temperature [degC]
'      rh -> relative humidity [%]
'
Mv = 18.015  'g/mol
Ma = 28.96  'g/mol
Po = 101325  'Pa
R = 8.314   'J/(molK)
TK = ta + 273.15
eps = Mv / Ma
pvv = Exp(25.89 - 5319 / TK) * rh * 0.01
h_ratio = 1000 * eps * pvv / (Po - pvv)
End Function
```

Table 8.A.1d: VBA code for the calculation of wet-bulb temperature.

```
Function wb_t(ta, rh)
'
' Calculation of wet-bulb temperature [degC]
' from dry-bulb temperature [degC] and relative humidity [%]
'
' Input: ta -> air temperature [degC]
'      rh -> relative humidity [%]
'
eps = 0.01: Delta_T = 1#
alfa = 68.15
taK = ta + 273.15
tw0 = taK
Do While Delta_T > eps
y = ps(tw0) + alfa * tw0 - (alfa * taK + 0.01 * rh * ps(taK))
a = ps(tw0) * 5319 / tw0 / tw0 + alfa
tw1 = tw0 - y / a
Delta_T = Abs(tw1 - tw0)
tw0 = tw1
Loop
wb_t = tw1 - 273.15
End Function
```

Table 8.A.1e: VBA code for the calculation of dew-point temperature.

```
Function dewP_t(ta, rh)
'
' Calculation of dewpoint temperature [degC]
' from dry-bulb temperature [degC] and relative humidity [%]
'
' Input: ta -> air temperature [degC]
'      rh -> relative humidity [%]
'
t = ta + 273.15
dewP_t = 5319 / (5319 / t - Log(0.01 * rh)) - 273.15
End Function
```

Table 8.A.2a: VBA code for the calculation of warm/cool exergy of humid air.

```
Function WC_Ex_air(roh_wv, roh_air, ta, tOO)
'
' Calculation of thermal exergy contained by one cubic meters of humid air [J/m^3]
'
' Input: roh_wv -> density of water vapour(concentration of water vapour) [g/m^3]
'        roh_air-> density of dry air  [g/m^3]
'        ta -> air temperature [degC]
'        t00 -> environmental temperature [degC]
'
Cpv = 1846#: Cpa = 1005#          'J/(kgK)
taK = ta + 273.15: ToK = tOO + 273.15
CpT = roh_wv * 0.001 * Cpv + roh_air * 0.001 * Cpa
'
WC_Ex_air = CpT * ((taK - ToK) - ToK * Log(taK / ToK))
'
End Function
```

Table 8.A.2b: VBA code for the calculation of wet/dry exergy of humid air.

```
Function WD_Ex_air(ta, tOO, pv1, pvo)
'
' Calculation of wet/dry exergy contained by one cubic-meter of humid air [J/m^3]
'
'Input: ta -> air temperature [deg C]
'       t00 -> environmental temperature [deg C]
'       pv1 -> water vapor pressure [Pa]
'       pv0 -> environmental water vapor pressure [Pa]
'
POT = 101325#   'Pa
taK = ta + 273.15: ToK = tOO + 273.15
'
WD_Ex_air = (pv1 * Log(pv1 / pvo) + (POT - pv1) * Log((POT - pv1) / (POT - pvo))) * (ToK / tak)
'
End Function
```

Table 8.A.2c: VBA code for the calculation of warm/cool exergy of liquid water.

```
Function WC_Ex_1w(t, tOO)
'
' Calculation of thermal exergy contained by one cubic meter of liquid water
'
'  Input: t -> water temperature [degC]
'         t00 -> environmental temperature [degC]
'
Cpw = 4186#     'J/(kgK)
roh_w = 1000#   ' kg/m^3
TK = t + 273.15: ToK = tOO + 273.15
'
WC_Ex_lw = roh_w * Cpw * ((TK - ToK) - ToK * Log(TK / ToK))
'
End Function
```

Table 8.A.2d: VBA code for the calculation of wet exergy of liquid water.

```
Function Wet_Ex_1w(tOO, phio)
'
' Calculation of wet exergy contained by one kilogram of liquid water [J/kg]
'
' Input: tOO -> water temperature [degC]
'        phio -> environmental relative humidity [%]
'
R_gas = 8.314              'J/(molK)
M_water = 18.015 * 0.001   'kg/mol
POT = 101325#              'Pa
ToK = tOO + 273.15
'
WD_Ex_lw = R_gas / (M_water) * ToK * Log(100# / phio)
'
End Function
```

Chapter 9

Mathematical Modelling

9.1 From qualitative to quantitative thinking

The origin of thinking, uniquely realized by human nervous systems, must date as far back as more than five thousand years ago; we can say so based on the evidence of written symbols found (Lloyd 2012). Since then, the ways and the extents of thinking have developed so that a variety of cultures have flourished locally and globally as we see in contemporary societies. Along with the context here in this treatise, bio-climatology for built environment, the nature of thinking may be regarded as having two sides: one is "qualitative" and the other "quantitative".

This chapter outlines how "quantitative" thinking can be developed from its very basics to an advanced level of application such as exergy-balance analyses. For some of the readers, the title of the present section may sound that "qualitative" is placed at a lower level than "quantitative" and the latter is more advanced than the former, but this is not what the title is intended to mean. On the contrary, the discussion to be done here within this chapter is to demonstrate that "quantitative" thinking can enrich the quality of "qualitative" thinking, with which the "quantitative" thinking can also be enriched to a level with higher quality.

9.1.1 The dual characteristics of a physical entity

As described in Chapter 2, there are two distinctive aspects of physical entities, whether they are non-living ones such as buildings or living ones such as human bodies: the "structure (*Katachi*)" to see and the "function (*Kata*)" to read. Mathematics has such a rudimental characteristic allowing us to read the functional aspect (*Kata*) while seeing something having a form, the structural aspect (*Katachi*).

Let us discuss what it implies using a very simple example: a sheet of paper with the size of "A4" that we are all familiar with. First we measure the side length and find that the shorter side is about 21.0 cm and the longer side about 29.7 cm. These figures may look odd at first glance, but in fact they are not. Why aren't they odd, even though they are not exact numbers like 20 cm and 30 cm? The reason for this is the self-similarity as described below.

Two sheets of A4 paper attached to each other with their long sides become the size of A3. The shape of A3 is the same as that of A4. Similarly, a rectangle with

A5 size appears as a sheet of A4 paper is folded so that the two short sides are exactly overlapped. A5 has again the same shape as A4. Reduction from A4 to A5, to A6, and so on, can continue infinitely in theory, while on the other hand, enlargement can continue up to the size of A0, whose area is exactly 1 m².

In general, as an arbitrary rectangle is folded so that two parallel sides are completely overlapped, the resulting rectangle has a shape dissimilar to the original one, but such series of rectangles as A0, A1, A2, and so on is special because folding necessarily results in an exactly similar rectangle. Let us examine how we can read such functional aspect (*Kata*) in the whole structure (*Katachi*) of these particular rectangles by applying mathematical thinking.

Exact similarity in a series of rectangles is characterized by the constant ratio of short side to long side. Denoting the short and long sides of A4-sized rectangle to be *a* and *b*, respectively, as shown in Fig. 9.1, the geometrical similarity of A3 to A4 can be written as

$$\frac{a}{b} = \frac{b}{2a}. \tag{9.1}$$

Taking a look at A0 in Fig. 9.1, we come to know that its area being 1 m² is equal to the product of 4*a* and 4*b*. That is,

$$1 = 4a \times 4b = 2^4 (ab). \tag{9.2}$$

Substituting the relationship between *a* and *b* as $b = \sqrt{2}a$ derived from Eq. (9.1) into Eq. (9.2) yields the formula for solving the value of *a* as follows.

$$a = \frac{1}{2^2\sqrt{\sqrt{2}}}. \tag{9.3}$$

As can be seen in Eq. (9.3), *a* is expressed with natural number 2 and its squared root alone; this is exactly what is hidden as the functional aspect of rectangles having the characteristic of self-similarity. With Eq. (9.3), we now come to know that the value of *a* is 21.0 cm; with the relationship of *a* and *b* as $b = \sqrt{2}a$, we also come to know that the value of *b* is 29.7 cm. These values calculated are of course the same as what we first measured. Note that such a process comparing the quantitative observation, the measurement, with theoretical calculation is a typical method in scientific research.

Setting up a set of mathematical equations and their operation from Eq. (9.1) to Eq. (9.3), the relative size of special rectangles having the characteristic of self-similarity is found; this is exactly the process to read and extract the invisible relationship, the function (Kata), from what is visible structure (Katachi) in a set of special rectangles.

In fact, how the Carnot engine performs described in Chapter 7 was made following the same process as above, that is, by combining the characteristic equation of ideal gas shown in Eq. (7.17) and the associated energy balance equation; the former is corresponding to Eq. (9.1) and the latter to Eq. (9.2). A similar process was also taken as the equation of saturated water vapour was derived in Chapter 8. Both of them typically show the richness of mathematical approach to unveil the functional aspect of physical entities.

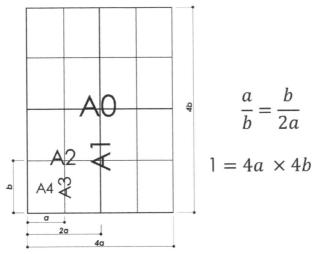

Fig. 9.1: The structure (Katachi) of self-similar rectangles and its associated functional aspect (Kata) expressed by mathematical symbols.

Similar to Eq. (9.3) with respect to a special set of rectangles, the functional aspect of circles, which are all similar to each other, can also be characterised as a combination of natural number 2. As is well known, the ratio of the circumferential length of a circle to its diameter is called pie number with the symbol, π, and it is necessarily constant as the intrinsic nature of all circles.

The formula to calculate this value can be derived according to the process shown in Fig. 9.2. We first consider a square circumscribing a circle and cut its edges to produce an octagon circumscribing the circle. Next, we do the same to produce a hexadecagon having sixteen edges. In this process, as can be seen in Fig. 9.2, we can find the relationship between the half of each side length and the circumference. Setting mathematical equations characterising this series of process and the mathematical operation provide us with the following beautiful equation, which is regarded to have been derived by François Viète (1540 ~ 1603) for the first time.

$$\pi = 2 \cdot \frac{2}{\sqrt{2}} \cdot \frac{2}{\sqrt{2+\sqrt{2}}} \cdot \frac{2}{\sqrt{2+\sqrt{2+\sqrt{2}}}} \cdot \frac{2}{\sqrt{2+\sqrt{2+\sqrt{2+\sqrt{2}}}}} \cdots\cdots \qquad (9.4)$$

We can recognize that Viète did find the relationship between the whole structure of circles having universal similarity (*Katachi*) and its associated function (*Kata*). Of course, we can reach the well-known value of $\pi = 3.1415926\cdots$ by calculating Eq. (9.4).

The process of understanding physical entities such as a square, a circle, a heat engine, an amount of humid air and so on, emerges as one of a variety of neurological processes, in which the visual information given through our eyes is connected with the auditory information given through our ears, as demonstrated with a drawing (left) in Fig. 9.3, as any language functions, and thereby it is strengthened by the output performed by the associated motor system as shown with the other drawing

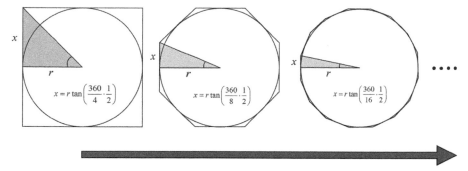

Fig. 9.2: Process of cutting the edges of a square to have octagon having eight edges, then cutting its edges to have hexadecagon having sixteen edges, and so on, finally reaching a true circle.

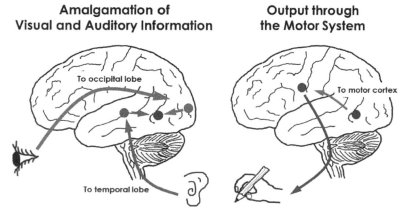

Fig. 9.3: How a brain works in relation to the structural (*Katachi*) and functional (*Kata*) aspects of physical entities.

(right) in Fig. 9.3. This is also related to structure (*Katachi*) and function (*Kata*) described in Chapter 2.

Table 9.1 lists up a variety of concepts in relation to the dual characteristics of physical entities and also to the processes shown in Fig. 9.3. The structure (Katachi) is definitely associated with "geometry", while on the other hand, function (Kata) with "algebra" used in the explanation above. As explained with Fig. 9.3, we may regard them to be associated with visual information and auditory information, respectively; the former is processed first in the "occipital lobe", while on the other hand, the latter is in "temporal lobe" together with the motor system generating the output.

Within mathematics, such "fractions" as $\frac{1}{3}$, $\frac{5}{6}$, $\frac{7}{11}$ and others, whose corresponding "recurring decimals" are 0.333⋯, 0.8333⋯, 0.636363⋯ and others, respectively, may be regarded to have the characteristic of structure (*Katachi*) with two natural numbers as the ratio, within which there is the characteristic of function (*Kata*) expressed in a series of unstoppable continuous decimals expressed by the symbol "⋯". According to quantum physics developed in the first half of twentieth century, the light that had been established as electromagnetic-wave phenomena by

Table 9.1: Dual characteristics of physical entities: structure (Katachi) and function (Kata) and their associated concepts.

Structure (Katachi)	Function (Kata)
Geometry	Algebra
Visual information	Auditory information
Occipital lobe	Temporal lobe
Fractions	Recurring decimals
Particles	Waves
Written symbols	Utterance
Musical notes	Musical sounds
Morphology	Physiology
Brain and nerves	Mind
Heart, blood vessels and blood	Flow and circulation
Air and moisture	Flow and circulation

the end of nineteenth century may be regarded to be one of the physical entities having particle-like characteristic while at the same time holding the characteristic as wave. Such dual nature is very similar to the relationship between fractions and recurring decimals.

The characters we use for writing words and sentences in various languages are nothing other than symbols to be seen by our eyes, while at the same time they are necessarily associated with utterance. This implies that the written symbols and their associated pronunciation are in the same relation to the structure (*Katachi*) and its associated function (*Kata*). Such relationship can also be seen in music: that is, musical notes and musical sounds.

In human biological science, there are two distinctive sub-fields: morphology and physiology. The former is exactly associated with structure (*Katachi*) and the latter function (*Kata*). The former identifies various structures such as brain, nerves, heart, blood vessels and others, all of which we can see because they have structure (*Katachi*). On the other hand, the latter identifies various functions such as "mind" as mental process, which is not the matter itself, and blood circulation as distributing process of nutrients and collecting process of waste, both of which are also not the matters themselves. They cannot be seen, but can be read with respect to how they function. Indoor humid air exists by being confined within the building envelope systems forming a structure and it acts the function of flow and circulation for the occupants' health and comfort. What was discussed in Chapter 8 was in fact functional (*Kata*-wise) aspect of humid air.

9.1.2 Symbols and their meanings to be assured

Mathematics may be regarded as a kind of language with which we can discern how a portion of the nature, on which we focus as a system, functions. In mathematical language, we use a variety of symbols, which necessarily carry their respective meanings; otherwise, they just display meaningless patterns. Here, before moving on to describe how mathematical modelling works, we briefly confirm two of the very basics of mathematical symbols: zero and function, both of which necessarily carry their own meanings.

In order to familiarize ourselves with the use of any language to a certain basic level of communication, it is necessary for us to know the meaning of some basic words, first through hearing the associated sounds and then through writing their associated symbols. Such process continues ceaselessly until we reach some advanced level, at which we can construct our own sentences that we need or want to express for communication. Similar features exist in mathematics. This is why we here reassure the two basic symbols, zero, "0", and function, "$f(x)$".

(a) Zero, "0"

The symbols representing numbers such as "1", "2", $\cdots$, are of course very basic and may look too obvious, but they are worth digging for reassuring the meaning of "0", since its idea is really an immense leap, which finally led to the decimal-number system that is of vital importance not only in the field of science alone but also in contemporary societies at large.

Any single object can be expressed as "1" using its symbol, two as "2" using its symbol, and so on; with these symbols, we count any number of objects. In similar manner, none is expressed as "0" as its symbol, but what it really implies becomes clearer as we express "0" together with "1", "2", $\cdots$ as a set of schemes representing their respective function (*Kata*) as demonstrated in Fig. 9.4. We can regard "0", "1", "2",$\cdots$ as containers holding respective numbers of imaginary blocks (Tooyama 1959). Zero is the container holding none.

With this special functionality, we can ascend or descend from a certain number towards a larger or smaller one as shown in Fig. 9.5. Addition of 1 to 9 is ten and it can be expressed as two containers connected in series as shown in Fig. 9.5a; one block is in the left container implying ten and no block in the right container implies zero. Thirteen, for example, can be expressed by putting three blocks in the right container as being represented by a set of symbols as "13". The same rule applies to the addition of ten to ninety as also shown in Fig. 9.5a. The important feature is that this rule can continue endlessly to the infinity. The additions of one tenth to nine tenth and also one hundredth to nine hundredth can also be done similarly as shown in Fig. 9.5b. Again, this rule can continue endlessly to the infinitesimal smallness. The whole of this decimal number system is exactly the use of self-similarity that is intrinsic in the whole of ten containers from "0" up to "9". Those who have felt some difficulty when they first learned binary system may find ease by applying the schemes of containers as demonstrated in Fig. 9.5; in the case of binary system, we imagine there are two containers, one with no block as "0" and the other with only one block allowed as "1".

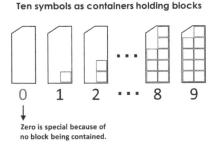

Fig. 9.4: Numbers from zero via 1, 2 ··· to 9, each of which is a kind of container holding respective number of blocks.

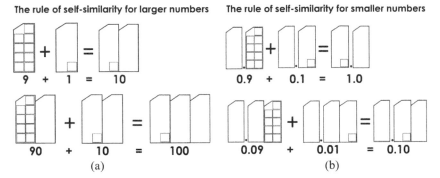

Fig. 9.5: Decimal-number system is made up of a series of containers having self-similarity.

Decimal point system that can express any numbers in between two integers was first successfully conceived and compiled by Simone Stevin (1548 ~ 1620), who was a contemporary of Francois Viète who derived the formula of pie number shown in Eq. (9.4) and also Galileo Galilei (1564 ~ 1641), who has been regarded as the precursor of Isaac Newton, the founder of classical mechanics; Stevin should be regarded as the other precursor of Newton parallel to Galilei because of the importance of decimal number system (Devreese and Berghe 2007).

(b) Function, "$f(x)$"

Let us next take a brief review on the functionality of symbol "$f(x)$". From Chapters 5 to 8, various mathematical symbols have already been used for the detailed explanation of respective topics. They are all to be recognized as the containers of either real numbers to be expressed with decimal-number system, or integers that are the whole of natural numbers with either positive or negative sign and zero. Such containers with respective symbols assigned are called variables because the numbers contained may vary from one situation to another depending on the change of other variables in association with the former.

As the simplest case, suppose that there are two variables, x and y. A change in variable x affects the corresponding change in variable y. This relationship is expressed as $y = f(x)$. This equation is read as follows: variable y is a function of

Evolution of Pictographic Symbol

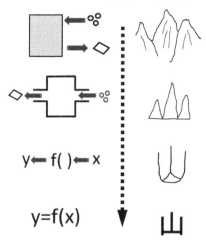

$$y \leftarrow f(\ ) \leftarrow x$$

$$y = f(x)$$

Fig. 9.6: A set of characters as mathematical function can be regarded as the abstract image of pictographic symbols, with which its meaning is easy to understand.

variable x. Such a relationship between x and y may be found here and there in our surroundings, for example, automated-teller machines, vending machines and others. Suppose such a case that you insert your coins into a machine in order to buy a train ticket as schematically shown in Fig. 9.6. The coins to be inserted corresponds to variable x, the machine to "$f(\ )$", and the ticket to y. Symbol "f" in front of the bracket "$(\ )$" comes originally from the initial of the word, function, and the bracket may be regarded as the inlet. With this image, a set of symbols $y = f(x)$ eventually becomes vivid and instructive (Tooyama 1970). In Fig. 9.6, the evolution of a Chinese character expressing mountain is demonstrated in parallel to the construction of a set of symbols $y = f(x)$. In both of them, we can see the dual characteristics in their final abstract symbols: one being structure (*Katachi*) and the other being function (*Kata*).

The notion that y is a function of x may be extended to be such expressions as "$y = y(x)$", "$z = f(x, y)$", "$\varsigma = \varsigma(T)$", "$u = u(s, v)$" and so on. With these images in mind, those who may have thought that the use of symbols in Chapters 7 and 8 were not easy to understand at first glance are recommended to try to go through those chapters again. It should be felt a bit easier than before going through what has been mentioned here in this section.

9.1.3 Proof by contradiction

Symbol "$\sqrt{\ }$", which appeared in Eqs. (9.3) and (9.4), represents the side length of a square having the area being equal to 2. The reason that $\sqrt{2}$ cannot be reduced further to be expressed by a finite or recurring decimal number is that it is an unstoppable decimal number, which is called "irrational number". Rational numbers are, on the other hand, finite decimal numbers or recurring decimal numbers such as $0.25\left(= \frac{1}{4}\right), 1.33\cdots\left(= \frac{4}{3}\right)$ and others. The name "rational" comes from that these numbers can be expressed as the ratio of two prime natural numbers. How many rational numbers are there? It is so many that we cannot count all of them directly; this implies that it is not possible for us to confirm whether $\sqrt{2}$ is really irrational or not by raising all of the rational numbers. But there is an indirect method that we can assure that $\sqrt{2}$ is irrational. Let us demonstrate this method as follows.

First, we assume that $\sqrt{2}$ is rational number; this implies that we can express $\sqrt{2} = \frac{m}{n}$, where variables m and n are arbitrary prime natural numbers, respectively. We may rewrite this relationship as $2n^2 = m^2$. This implies that m^2 is an even number and m is also even, since the product of two natural numbers can be even if one of them is even. Therefore, we can express $m = 2k$, where k is a natural number. Substituting this relationship into $2n^2 = m^2$, we find $n^2 = 2k^2$. This result indicates that n is also even for the same reason as m being even. Therefore, n can be expressed as $n = 2l$, where l is a natural number. Substitution of two relationships, $m = 2k$ and $n = 2l$, into $\sqrt{2} = \frac{m}{n}$ lets us recognize that $\frac{m}{n}$ is not the ratio of two prime natural numbers. This is contradictory to the assumption made in the beginning that $\sqrt{2}$ is rational number defined to be the ratio of two prime natural numbers. This concludes that $\sqrt{2}$ is irrational.

In Chapter 7, we discussed that a Carnot heat engine functions so as to generate the upper limit of work out of an amount of heat flowing between heat source and heat sink. We confirmed that this is true by showing that a super Carnot engine has to function in the manner that heat flows spontaneously from low temperature to high temperature. This contradicts what the nature allows. Therefore, the amount of work to be produced by Carnot engine is the maximum limit. This proving process was in fact the same as that of $\sqrt{2}$ being irrational.

9.2 Modelling the flows and changes in a system

Here in this section, keeping what was reviewed in the last section, let us demonstrate how quantitative concepts, first mass and then momentum, can be used to articulate the function (*Kata*) of some physical entities.

9.2.1 Mass balance of a water tank

Let us first suppose a water tank as shown in Fig. 9.7. There are two pipes above and below; one for inlet and the other for outlet. With respective faucets, we may control the rate of water coming in or out from the tank. If the rate of coming in, m_{in} [kg/s], is larger than that of going out, m_{out} [kg/s], then the water level in the tank increases; if it is smaller, then it decreases. As shown in Fig. 9.8, in the period of time, Δt, from t_0 to t_1, a mass of water, M_{in} [kg], flows in, while on the other hand the other mass of water, M_{out} [kg], flows out. They can be expressed as

$$M_{in} = m_{in}\Delta t, \quad M_{out} = M_{out}\Delta t. \tag{9.5}$$

In this course, the water level of the tank increases by Δh from h_0 to h_1, that is, $\Delta h = h_1 - h_0$. Note that we have used symbol "Δ" to indicate the difference of a variable. Assuming that the bottom surface area is A [m²] and the density of water is ρ_w [kg/m³], we can express the water mass in the tank at the time of t_0 as $\rho_w A h_0$ and that at the time of t_1 as $\rho_w A h_1$. Their difference, $\rho_w A \Delta h$, has to be equal to the

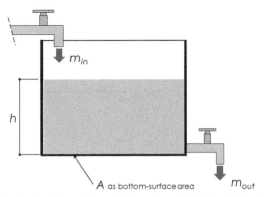

Fig. 9.7: A water tank with two pipes with faucets: one for inlet and the other for outlet.

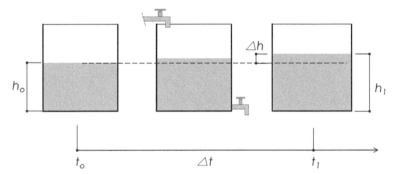

Fig. 9.8: A change of water level, Δh, in the tank within a period of time from t_0 to t_1.

difference between inflow and outflow of water, $M_{in} - M_{out}$, since no mass can emerge or disappear spontaneously by the law of mass conservation.

$$M_{in} - M_{out} = \rho_w A(h_1 - h_0) \tag{9.6}$$

We rewrite Eq. (9.6) as follows

$$M_{in} = \rho_w A\Delta h + M_{out}. \tag{9.7}$$

Equation (9.7) indicates how the water tank functions in the period of time Δt from t_0 to t_1; we can read that the mass of water flowing in has to equal the sum of the mass changed in the tank and the mass of flowing out. Note that the equal sign "=" in Eq. (9.7) represents the law of mass conservation. In setting up any balance equation in terms of other quantities having the same characteristic of conservation as mass such as momentum, energy, and electric charge, the forms of balance equation are the same. The exception are the concept of entropy and exergy, since the entropy is generated and exergy is consumed in any macroscopic process, as was demonstrated in some of previous chapters and also will be described in the next section.

We further rewrite Eq. (9.7) substituting the relationships expressed by Eq. (9.5).

$$m_{in}\Delta t = \rho_w A\Delta h + m_{out}\Delta t. \tag{9.8}$$

We may rewrite the above equation as follows.

$$m_{in} = \rho_w A \frac{\Delta h}{\Delta t} + m_{out}. \qquad (9.9)$$

Equation (9.9) shows how the mass flow rates relate to the rate of change in water level, while Eq. (9.8) represents simply the mass balance. If we let the time interval infinitesimally small to be expressed as dt, and the corresponding change in height as dh, then Eqs. (9.8) and (9.9) turn into the differential mass balance equations.

It is important to recognize the difference between "flow" and "change". In either Eq. (9.8) or Eq. (9.9), the "change" is necessarily expressed as the difference in the quantity of state in a certain period of time, while on the hand, "flow" is expressed in relation to the space-wise difference in intensive quantities. In the case of water tank in Fig. 9.7, the reason why water flows in and out is that there is a water-pressure difference between incoming side and outgoing side of each faucet.

For a certain value in the change of mass, if the bottom surface area, A, is large, then the change in height, Δh, is small. On the contrary, if A is small, Δh is large. Whether the height of water can easily change or not is very much associated with the bottom area, A. Such a role as the bottom surface area can be seen in the mass in the case of momentum balance and also in the heat capacity in the case of thermal energy balance, as will be described later in this chapter.

9.2.2 Momentum balance and atmospheric pressure

In a similar manner to setting up the mass balance equation discussed above, we can develop the momentum balance equation based on the conservation of momentum. For those who suspect this conservation law, it is recommended to take a look at an experiment with an apparatus called Newton's cradle (Wikipedia/2017). Let us first apply this law to set up the momentum balance equation to a small mass of air as a system within the atmosphere as shown in Fig. 9.9.

The concept of momentum is to quantify how much the strength of a body in motion is. It is defined as the product of the mass of a body as a system, m, and its velocity, v. Denoting the momentum by p, it is expressed as $p \equiv mv$. Note that "$\equiv$" is also a sign of equal; the reason why the symbol "$=$" is not used here is to make it clearer that "$\equiv$" indicates the definition of the concept of momentum, which is equal to the product of mass and velocity, not the expression of the relationship between flows and change.

Velocity is the rate of movement of a body from one position to another so that it is in general a three-dimensional quantity, which can be expressed by the concept of vector. Here in the present discussion, let us confine the velocity as one-dimensional in the case of a mass of air statically situated in Fig. 9.9. In such a case, the vector characteristic can be expressed either downwards or upwards alone and thereby we can assume the upward velocity to be expressed by positive values and the downward velocity by negative. In Fig. 9.9, open bold arrows indicate either inflow or outflow of momentum with respect to the system of air, while respective thin arrows within the bold arrows indicate the direction of momentum, either upward or downward.

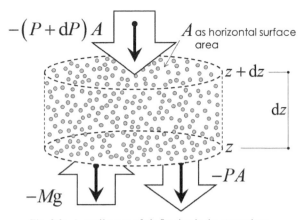

Fig. 9.9: A small mass of air floating in the atmosphere.

For the system of a small flat cylinder shown in Fig. 9.9, the momentum balance equation can be written so that the inflow of momentum is equal to the sum of the change in momentum contained by the system and the outflow of momentum. There are two inflows of momentum to the system: one is caused by pressure exerted by the atmospheric air above the system and the other by gravitation of the Earth. On the other hand, there is one outflow of momentum from the system: the one causing the pressure towards the atmospheric air under the system (Imai 2003).

The former of the two inflows can be expressed as $-(P + dP)A\,dt$, where $P + dP$ is the pressure exerted at the upper horizontal surface of the system at the height of $z + dz$ above the ground surface [m], A is the horizontal surface area of the system [m²], and dt is the infinitesimally small period of time [s]. $-(P + dP)A$ is the force acting on the upper surface and its direction is downwards so that there is negative sign. Since the force is the rate of momentum to be exerted, the product of force and time becomes the flow of momentum (Maxwell 1891, Imai 2003). Denoting the mass of the system by M [kg] and the absolute value of gravitational acceleration rate by g [m/s²], the inflow of momentum caused by gravitation can be expressed as $-Mg\,dt$. Again, negative sign represents downwards. In a similar manner to these two inflows, the outflow of momentum from the bottom surface of the system can be expressed as $-PA\,dt$, where P is the pressure at the height z. If the velocity of the system is increased from v to $v + dv$ in the period of dt, the change of momentum carried by the system is expressed as $M\,dv$.

Using the inflows, the outflow and the change of momentum described so far, the momentum balance equation can be expressed as follows:

$$\{-(P + dP)A\,dt\} + (-Mg\,dt) = M\,dv + (-PA\,dt). \tag{9.10}$$

The mass of the system is expressed as $M = \rho A\,dz$, where ρ is the density of air [kg/m³]. Assuming that the system of air is moving neither upwards nor downwards, that is, $dv = 0$, Eq. (9.10) can be reduced to the following equation.

$$dP = -\rho g\,dz. \tag{9.11}$$

Since $\rho > 0$, $g = 9.8$ m/s$^2 > 0$ and d$z > 0$, dP is necessarily negative; this means that the atmospheric pressure decreases as the height increases. Note that negative sign in Eq. (9.11) indicates decrease in pressure, not the direction of flows in momentum.

In such a case that the difference in height is a certain small range in which the density of air can be taken to be constant, then Eq. (9.11) may be rewritten as the result of integrating the pressure from P_0 to P_1 and the height from z_0 to z_1,

$$P_1 = P_0 - \rho g(z_1 - z_0). \tag{9.12}$$

Assuming $\rho = 1.2$ kg/m^3, $z_1 - z_0 = 20$ m, and $P_0 = 1000$ hPa, we find $P_1 = 997.7$ hPa. The difference between P_0 and P_1 corresponds to either 18 mm in the height of water or 1.8 mm in the height of mercury.

In the other case that the difference in the height is large so that the density of air cannot be assumed to be constant, we may use the density of air, which can be well characterized as ideal gas described in Chapters 7 and 8, as a function of temperature and pressure, that is $\rho = \dfrac{\mathfrak{M}P}{RT}$, where $\mathfrak{M} = 29 \times 10^{-3}$ kg/mol, R = 8.314 J/(mol·K) and T is absolute temperature of air [K]. Substitution of this relationship into Eq. (9.11) yields

$$\frac{dP}{P} = -\frac{\mathfrak{M}g}{RT}dz . \tag{9.13}$$

We all know that ambient air temperature decreases as we go up to high mountain areas. According to the measurement of air temperature with a balloon (Taylor 2005), it decreases almost linearly as the height increases as shown in Fig. 9.10, up to almost 10 km above the ground. The closed-square plots indicate measured values and the dashed line ovelapping the plots is the fitted line, whose slope is 6.49°C/ km (= 0.00649 K/m); this is called lapse rate and it is known to be independent of the seasonal change and almost constant anywhere all over the atmosphere near the ground surface. The other line having milder slope represents the theoretical air temperature profile obtained from thermodynamic calculation with the assumption that atmosphere is completely dry so that there is no condensation of water vapour, that is, no cloud formation.

In such a case that moist air ascends along the high mountains, the air is dehumidified in the course of cloud formation. Suppose that the dehumidified air further ascends and thereafter descends along those mountains. What happens in this series of air movement is that hot air is brought about in the local areas near the leeward foot of those mountains. This is called "Föhn wind" phenomenon; the line with the slope of 9.8°C/km (= 0.0098 K/m) in Fig. 9.9 corresponds to such a case.

The measured air temperature can be expressed as $T = T_o - \alpha z$, where T_o is air temperature near the gound [K] and α is the lapse rate at 0.00649 K/m. We further substitute this relationship into Eq. (9.13) and find the following equation that indicates the relationship between the atmospheric pressure and the height

$$\frac{dP}{P} = -\left(\frac{\mathfrak{M}g}{R}\right)\frac{dz}{T_o - \alpha z} . \tag{9.14}$$

By integrating this equation from P_o at z_o to P at z, we reach the following formula, with which the atmospheric pressure at an arbitrary height from the ground surface can be estimated

$$P = P_0\left(1 - \frac{\alpha z}{T_o}\right)^{\frac{\mathfrak{M}g}{R\alpha}}. \qquad (9.15)$$

In order to integrate Eq. (9.14), a mathematical characteristic hidden in the relationship between $\frac{1}{P}$ and dP as shown in Fig. 9.11 was used: this is called natural logarithm, which is expressed as either "ln" or "$\log_e$", which already appeared in a couple of sections in Chapters 7 and 8.

Figure 9.12 demonstrates the atmospheric pressure calculated from Eq. (9.15) by the triangular plots; the measured values by a balloon (Taylor 2005) are shown with circular plots. In the calculation, the values of P_o and T_o were assumed to be 1000 hPa and 298.15 K (= 25°C), respectively, taken from the measured data set. We find that the atmospheric pressure calculated from Eq. (9.15) fits very well with the whole of measured pressure values in the range between the ground surface and 10 km above. This implies that the momentum balance equation, Eq. (9.10), set up for a mass of air shown in Fig. 9.9 was reasonable.

For the atmosphere above 10 km, the measured air temperature stays almost constant as can be seen in Fig. 9.10 so that Eq. (9.13) may be integrated under the condition of constant temperature. Again, using the relationship demonstrated in Fig. 9.11, the result obtained is expressed as

$$P = P_{10}\exp\left(-\frac{\mathfrak{M}g}{RT_{10}}(z - 10^4)\right) \qquad (9.16)$$

where P_{10} and T_{10} are pressure and absolute temperature of the air at the height of 10 km. Note that symbol "exp" denotes Napier number, which is 2.7182818284···, usually expressed with symbol "e", as also shown in Fig. 9.11. Note that it is as important as number "π" since it relates to various attenuating natural phenomena to be observed including what can be seen in Fig. 9.12. Napier number is related to natural logarithm as shown in Fig. 9.11.

The right-hand side of Eq. (9.16) is the product of the pressure at the height of 10 km and Napier number powered by the value inside the large bracket; this value attenuates as the height increases. In Fig. 9.12, the pressure values calculated from Eq. (9.16) are also displayed with cross plots; they are again in good agreement with the measured pressure.

We can see that up until 5000 m or so, the atmospheric pressure decreases almost linearly as the height increases and thereafter the pressure attenuates exponentially. At the height of 30 km or so, the atmospheric pressure becomes nearly one hundredth. This implies that we may regard that the thickness of terrestrial atmosphere is about 30 km at the maximum, within which the uppermost ozone layer protects us from dangerous radiation such as ultra-violet rays (Eddy 2009). For this reason, we may regard that beyond this height is vacuum, which was first recognized by Pascal and others almost 400 years ago by performing their own very important experiments

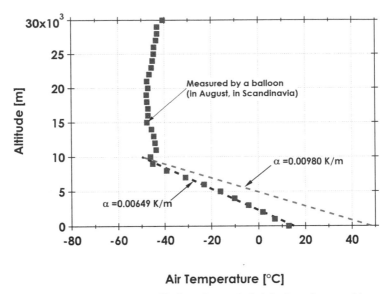

Air Temperature [°C]

Fig. 9.10: Atmospheric air temperature profile dependent on the height above the ground (measured data quoted from Taylor (2005)).

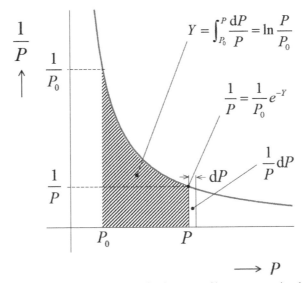

Fig. 9.11: Natural logarithm, which is the area under the curve of inverse proportionality, and its relation to Napier number "$e = 2.1718281\cdots$".

as described in Chapter 8. Although their experiments were rather primitive from the present-day viewpoint, within the height range of about 2 km near the ground surface, their finding was remarkable because we humans first came to recognize that there is vacuum space outside the Earth. More on the extra-terrestrial space will be described in Chapter 12.

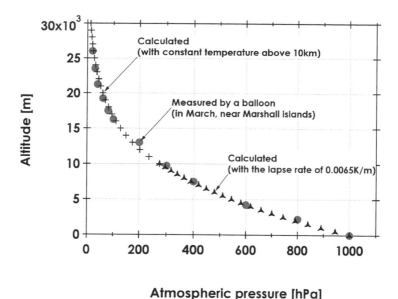

Fig. 9.12: Atmospheric pressure profile dependent on the height above the ground (measured data quoted from Tailor (2005)).

9.2.3 *Momentum balance of a raindrop and its fall*

What we have come to know through mathematical modelling described so far is the static pressure pattern in the vertical direction. We all know by experience that atmospheric pressure is not totally static, but dynamic as we experience the wind blowing here and there ceaselessly. This is basically due to so-called Bénard convection, first investigated by Bénard in 1900 (Bejan 2004). In the case of atmosphere, large cells of air are formed so that high-pressure and low-pressure zones emerge in various places all over the terrestrial surface. The variation of atmospheric pressure shown in Fig. 8.5 in Chapter 8 is due to this effect. Low-pressure zones emerge because of the ascending movement of air, with which clouds are formed and hence there is rainfall, while on the other hand high-pressure zones emerge because of the descending movement of air. Here let us focus on a raindrop and discuss how its fall can be modelled mathematically.

Figure 9.13 shows a model raindrop as a sphere in two cases: the left one represents the raindrop ascending and the right one descending. Of course, the latter is what we experience near the ground surface, but we need to consider both cases in order to confirm the rationality of momentum balance equation to be set up.

In a similar manner to what we did for a mass of atmospheric air, we first list up the inflow into and the outflow from the raindrop in terms of momentum. The inflow of momentum is caused by the inescapable gravitational force at the rate of $-mg$, where m is the raindrop mass [kg], g is gravitational accelerating rate at 9.8 m/s^2 and the negative sign is given to express the gravitation being downwards. For the period of time, dt, the inflow of momentum can be expressed as $-mg\mathrm{d}t$.

Since the raindrop is surrounded by the atmospheric air, a certain amount of momentum has to be passed onto the surrounding atmospheric air as the raindrop moves either upwards or downwards. This outflowing rate of momentum from the raindrop to the surrounding atmospheric air is known to be proportional to the raindrop velocity, v, with resistivity coefficient, γ (Yamamoto 2008). That is, this rate of outgoing momentum is given to be γv and the outgoing momentum in the period of dt is γvdt. Referring to the experimental facts in terms of various sizes of raindrops and their velocity (Gunn and Kinzer 1949, Mizuno 2000), we find that the resistivity coefficient, γ, becomes larger quadratically as the raindrop size increases as shown in Fig. 9.14.

Since $\gamma > 0$ and d$t > 0$, if the raindrop moves upwards, that is, $v > 0$, then γvd$t > 0$; if downwards, that is, $v < 0$, then γvd$t < 0$. In Fig. 9.13, the thin arrows inside the outflowing momenta indicate their respective directions depending on the movement of the raindrop. For both directions, the outflowing momentum is expressed in the same mathematical term as γvdt so that we can use it for either direction in setting up the momentum balance equation.

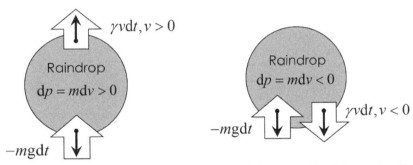

Fig. 9.13: An infinitesimal change in momentum of a raindrop caused by inflow and outflow of momentum.

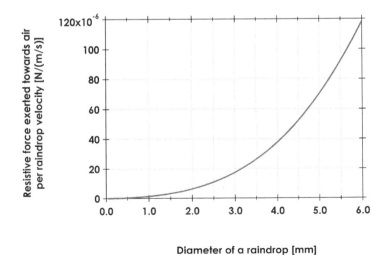

Diameter of a raindrop [mm]

Fig. 9.14: Resistive force exerted towards atmospheric air by a raindrop as per its velocity.

Assuming that the momentum of the raindrop changes from mv to $m(v + dv)$ in the infinitesimally short period of time, dt, we can express the momentum balance equation of a raindrop as

$$-mgdt = mdv + \gamma vdt. \tag{9.17}$$

Dividing both sides of this equation by m and denoting $\lambda = \gamma/m$,

$$-gdt = dv + \lambda vdt. \tag{9.18}$$

In order to derive the formula for calculating the raindrop velocity, v, from Eq. (9.18), in which there is an infinitesimally small increase of velocity, dv, in addition to velocity, v, itself, we need some mathematical operation. For this purpose, we first multiply an unknown function, f, to both sides of Eq. (9.18) and then make the mathematical operation demonstrated in the following equation

$$-f \cdot gdt = f \cdot dv + f \cdot \lambda vdt = f \cdot dv + (f\lambda dt)v = f \cdot dv + df \cdot v, \tag{9.19}$$

where $df = fdT$ and $dT = \lambda dt$. Referring to the relationship between two variables, f and v, in general as shown in Fig. 9.15, we can rewrite Eq. (9.19) as follows.

$$-fgdt = d(fv). \tag{9.20}$$

In the course of a raindrop falling down, its velocity must attenuate due to the friction between the raindrop and the atmospheric air. Therefore, the functional form of variable f should be the one that has such characteristic. Furthermore, it should also be in the form that requires the easiest mathematical operation for differentiation and integration. Such functional form is known to be $f = e^T$, since its either differential or integral form is exactly the same as the original form, that is, $df = fdT$ or $\int df = f$ as schematically demonstrated in Fig. 9.16. This function, $f = e^T$, has a kind of self-similarity similar to what we discussed earlier in the present chapter such as self-similar rectangles and decimal number system.

Substituting this relationship into Eq. (9.20) and using the relationship between T and t as $dT = \lambda dt$, we can come up with the following equation

$$-g \cdot e^T \cdot \frac{dT}{\lambda} = d(fv). \tag{9.21}$$

By integrating the left-hand side of Eq. (9.21) from 0 to T, and the right-hand side from $(fv)_0 = e^0 \cdot 0 = 0$ to fv, we can eventually reach the following final formula for calculating the raindrop velocity:

$$v = -\frac{g}{\lambda}\left(1 - e^{\lambda t}\right). \tag{9.22}$$

In Eq. (9.22), if $t = 0$, then $v = 0$; if $t = \infty$, then $v = -\frac{g}{\lambda}$. That is, the final velocity is expressed as $\frac{g}{\lambda}$ downwards. The mass of a raindrop is proportional to the third power of its diameter, D, since the raindrop can be assumed to be a sphere whose volume is expressed as $\frac{\pi}{6}D^3$. The resistivity coefficient, γ, shown in Fig. 9.14, may be regarded as proportional to the second power of D or the power of two and a half of D. Therefore, we can see the following relationship between the absolute value of raindrop velocity, $|v|$, and diameter, D.

$$|v| = \frac{g}{\lambda} = \frac{mg}{\gamma} \propto \frac{D^3}{D^2} = D, \text{ or } |v| = \frac{g}{\lambda} = \frac{mg}{\gamma} \propto \frac{D^3}{D^{5/2}} = \sqrt{D} \qquad (9.23)$$

where symbol "$\propto$" denotes proportionality. Equation (9.23) indicates that the raindrop velocity is proportional to its diameter or the squared root of diameter. This implies that large raindrops in heavy rain hit umbrellas, roofs, road surfaces and others on the ground very hard so that it is noisy. On the other hand, tiny raindrops in mist hit so softly that it is silent.

Figure 9.17 shows the variations of raindrop velocity in three cases, in which the raindrop diameter was assumed to be 5, 1 and 0.5 mm, respectively. One other case shown together with these three cases is that there is no atmosphere so that $\gamma = 0$; the velocity in this special case was calculated from $v = -gt$, which is obtained by solving Eq. (9.17) assuming $\gamma = 0$.

A smaller raindrop reaches the final velocity earlier than a larger one does. In three cases with atmospheric air, it takes about one, two, and four seconds or so for the raindrop of 0.5, 1, and 5 mm to reach their final velocity at about 3, 4 and 9 m/s, respectively. The height of clouds, where the raindrops originate, is about 2.5 km at the maximum. Therefore, we can say that the raindrops reach their final velocity values until they fall merely for a short distance of 40 m or so within several seconds.

If there is no atmosphere at all, as can be seen in the case of vacuum in Fig. 9.17, a raindrop reaches at 20 m/s in two seconds, regardless of its size. The velocity reaching after falling down for the distance of 2.5 km turns out to be about 220 m/s (= 792 km/h), which is almost the same order of the cruise speed of a jet airliner. It would be too dangerous for us to be outdoors since we would hardly survive if such high-velocity raindrops hit our body. Thanks to the presence of atmospheric air, we can walk outdoors in the rain. More on such safety in Chapter 12.

$$d(f \cdot v) = (f + df) \cdot (v + dv) - f \cdot v = f \cdot dv + df \cdot v$$

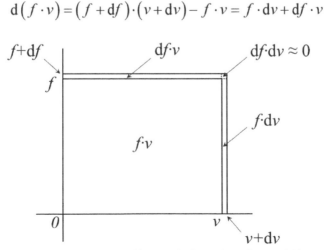

Fig. 9.15: An infinitesimal increase in the product of two variables.

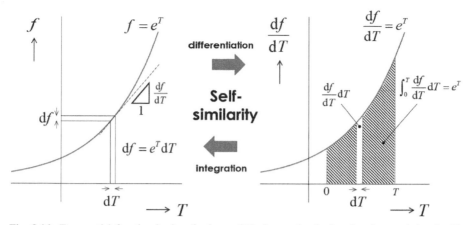

Fig. 9.16: Exponential function having the base of Napier number having the characteristic of self-similarity with respect to differentiation and integration.

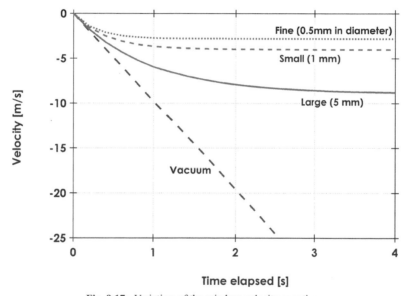

Fig. 9.17: Variation of the raindrop velocity over time.

9.3 Energy, entropy and exergy balances of a raindrop

Let us take a look at Eq. (9.17) again; the inflow and outflow of momentum which appear as $-mg\mathrm{d}t$ and $\gamma v\mathrm{d}t$, in the infinitesimal period of time, $\mathrm{d}t$, increased the momentum of a raindrop at the amount of $m\mathrm{d}v$. Parallel to this increase of momentum, the energy carried by the raindrop must also increase. For this increase of energy, there has to be the inflow of energy into the raindrop, which is the "work" expressed as the product of gravitational force, $-mg$, and the infinitesimal height, $\mathrm{d}h$, for which the raindrop falls, that is, $-mg\mathrm{d}h$.

The infinitesimal work, $-mgdh$, may be associated with the corresponding inflow of momentum, $-mgdt$, as follows.

$$-mgdh = -mgdh\left(\frac{dt}{dt}\right) = (-mgdt)v, \tag{9.24}$$

where $v = \dfrac{dh}{dt}$. Equation (9.24) suggests that multiplying the velocity, v, to respective sides of Eq. (9.17) can yield the energy balance equation standing in parallel to the momentum balance equation. That is,

$$-mgdh = mvdv + \gamma v^2 dt. \tag{9.25}$$

The whole of Eq. (9.25) can be integrated in terms of infinitesimal height, dh, from H_{cloud} to 0, in terms of infinitesimal velocity, dv, from 0 to v, and in terms of infinitesimal time, dt, from 0 to t. It is expressed as

$$-mg\int_{H_{cloud}}^{0} dh = m\int_{0}^{v} vdv + \gamma\int_{0}^{t} v^2 dt. \tag{9.26}$$

Some mathematical operation after substituting Eq. (9.22) for the formula of velocity, v, into the two terms of the right-hand side of Eq. (9.26) yields the following final equation of energy balance with respect to a raindrop

$$mgH_{cloud} = \frac{1}{2}mv^2 + Q, \tag{9.27}$$

where

$$H_{cloud} = \frac{g}{\lambda}t - \frac{g}{\lambda^2}\left(1 - e^{-\lambda t}\right), \tag{9.28}$$

$$Q = \frac{mg^2}{\lambda}\left(t + \frac{2}{\lambda}e^{-\lambda t} - \frac{1}{2\lambda}e^{-2\lambda t} - \frac{3}{2\lambda}\right). \tag{9.29}$$

We read Eq. (9.27) as follows: the amount of work is exerted as indicated in the left-hand side of this equation and its portion turns into kinetic energy carried by the raindrop as $\frac{1}{2}mv^2$ and the rest turns into heat as Q, which is dispersed into the atmosphere. This happened in the course of the raindrop falling down for the distance of H_{cloud} expressed by Eq. (9.28) as a function of time, t, and in due course, the friction between the raindrop and the atmospheric air causes the generation of heat at the amount of Q, which can be calculated from Eq. (9.29) as a function of time, t.

Figure 9.18 demonstrates one example of the calculation with respect to kinetic energy carried by a raindrop having the diameter of 5 mm and the amount of heat generated. The horizontal axis is the distance that the raindrop falls down for 30 m. It takes about three seconds to fall this distance because of its final velocity of about 9 m/s as shown in Fig. 9.17. As can be seen in Fig. 9.18, most of the work exerted into the raindrop turns into heat; almost 70% of the exerted work turns into heat until the raindrop falls down for 10 m, 87% for 30 m. As we discussed above, we are protected from the danger of raindrops by the atmospheric air; from the energetic

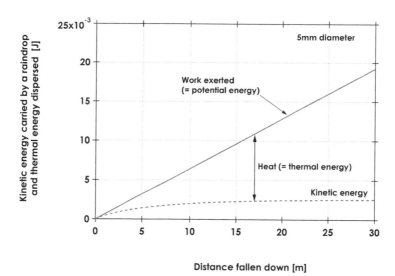

Fig. 9.18: Kinetic energy carried by a raindrop and thermal energy dispersion.

viewpoint, this is because much of energy exerted into the raindrop by gravitational work is dispersed as heat into the atmospheric air.

Discussion so far in this section has clarified the importance of thermal energy generation, but in order to quantify this more thoroughly, the entropy and exergy balance equations, which should be in parallel to energy balance equation, have to be set up. Let us do it here as the last step of the discussion on the fall of a raindrop.

As we discussed in Chapter 7, entropy is the concept not to be conserved, but to be generated. Therefore, we introduce entropy generation, S_g, in order to make it possible to set up the entropy balance equation. With this in mind, and also taking the concept of dispersality described in Chapter 7, we can set up the entropy balance equation to be in parallel to the energy balance equation, Eq. (9.27), as follows.

$$\left(\frac{1}{\infty}\right)mgH_{cloud} + S_g = \left(\frac{1}{\infty}\right)\frac{1}{2}mv^2 + \left(\frac{1}{T_o}\right)Q. \tag{9.30}$$

The corresponding temperature of work and kinetic energy is infinity, as described in Chapter 7, so that their dispersality is expressed as $1/\infty$, which is zero. We here assume that the temperature of heat sink is at atmospheric temperature near the ground surface. Equation (9.30) can be reduced to

$$S_g = \left(\frac{1}{T_o}\right)Q. \tag{9.31}$$

This equation implies that the entropy generation, S_g, is exactly proportional to heat generated, Q. We can derive the exergy balance equation by substituting this relationship into Eq. (9.27). That is

$$mgH_{cloud} - X_C = \frac{1}{2}mv^2, \tag{9.32}$$

where $X_C = S_g T_o$. X_C is the amount of exergy consumption in the fall of a raindrop. We read Eq. (9.32) as follows: the exergy input by gravitational work is partly consumed and thereby the rest is given to the raindrop as its kinetic exergy. Since $S_g T_o = Q$, we come to recognize that dispersed energy generated by friction is exactly the exergy consumption itself and also that such exergy consumption makes it possible for us to be outdoors safely under either heavy rain or mist near the ground surface.

9.4 Building envelope systems

Modelling physical systems mathematically can let us reveal their function, which is otherwise invisible or illegible. This is what has been demonstrated above. As the next step, we here review how the typical building envelope systems are modelled mathematically and confirm their function; what follows below is the basis of the discussion in 6.5, Chapter 6.

9.4.1 *Walls without considering heat capacity*

Suppose that there is an external wall composed of two materials as shown in Fig. 9.19. For simplifying our discussion, we first assume no heat capacity in these materials. Then, the thermal characteristics of the wall can be expressed with thermal resistivity and the thickness alone.

The external surface of this wall is, on the one hand, exposed to outdoor virtual surface, whose temperature is t_{ro}, and on the other hand, exposed to outdoor air, whose temperature is t_{ao}. Similarly, the internal surface is exposed to indoor virtual surface, whose temperature is t_{ri}, and indoor air, whose temperature is t_{ai}. Virtual surface temperature, either outdoors or indoors, may be called radiant temperature.

The values of two temperatures outside are to be given as local climatic conditions and those of two other temperatures indoors to be given as the results provided by a chosen combination of passive and active systems for built-environmental conditioning. These radiant and air temperature conditions together with thermal characteristics of the materials used for the wall are the determinant of the wall surface temperatures, t_{so} and t_{si}. In order to calculate these surface temperatures, what we have to do is to set up energy balance equation for two systems: one at the external surface and the other at the internal surface to be solved for the two surface temperatures, t_{so} and t_{si}.

Taking a look at Fig. 9.19, in which $t_{ao} < t_{so} < t_{si} < t_{ai}$ is assumed, and placing the energy input on the left-hand side and the energy output on the right, the energy balance equations for the external and for the internal surfaces as thermodynamic systems are expressed as follows, respectively.

[Absorption of external radiation] + [Conduction from the wall inside] =
[Emission of radiation toward outdoor space] + [Convection toward outside], (9.33)

[Absorption of internal radiation] + [Convection from indoor air] =
[Emission of radiation toward indoor space] + [Conduction into the wall]. (9.34)

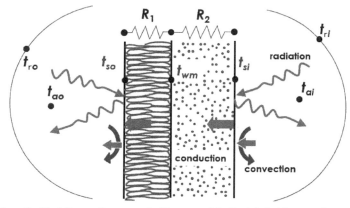

Fig. 9.19: A vertical building wall composed of two materials. Straight bold arrows for convection and conduction are from high temperature to low temperature.

Referring to what was described on long-wavelength radiation, convection, and conduction in Chapter 6, we come to realize that basic mathematical expression for radiant energy emission and absorption is intrinsically different from those for convection and conduction; radiation is proportional to fourth power of absolute temperature, while convection and conduction are proportional to space-wise temperature difference.

In order to make radiant energy emission and absorption be easier to handle in the calculation of two surface temperatures, we take the rates of radiant energy emission and absorption relative to those at outdoor temperature so that the fourth-powered temperature difference can be linearized. In so doing, we can introduce radiative heat transfer coefficient, which positions in parallel to convective heat transfer coefficient.

For the external surface, denoting the emission and absorption rates of radiant energy relative to the emission and absorption at outdoor air temperature as q_{ro_em} and q_{ro_ab}, respectively, they can be expressed as follows

$$q_{ro\_em} = h_{ro}(t_{so} - t_{ao}), \quad q_{ro\_ab} = h_{ro}(t_{ro} - t_{ao}). \tag{9.35}$$

where h_{ro} is radiative heat transfer coefficient at the external surface [W/(m²K)]. Taking the same procedure for the internal surface,

$$q_{ri\_em} = h_{ri}(t_{si} - t_{ao}), \quad q_{ri\_ab} = h_{ri}(t_{mi} - t_{ao}). \tag{9.36}$$

where q_{ri_em} and q_{ri_ab} are the relative rates of radiant energy emission and absorption at the internal surface, respectively, and h_{ri} is the radiative heat transfer coefficient at internal surface [W/(m²K)]. The formulation of Eqs. (9.35) and (9.36) is exactly consistent with what was done for deriving the simplified formula for radiant exergy, Eq. (7.44), in Chapter 7.

Equations (9.33) and (9.34) can then be expressed as

$$h_{ro}(t_{ro} - t_{ao}) + C(t_{si} - t_{so}) = h_{ro}(t_{so} - t_{ao}) + h_{co}(t_{so} - t_{ao}),$$
$$h_{ri}(t_{ri} - t_{ao}) + h_{ci}(t_{ai} - t_{si}) = h_{ri}(t_{si} - t_{ao}) + C(t_{si} - t_{so}) \tag{9.37}$$

$$C = \frac{1}{R_1 + R_2}. \tag{9.38}$$

where, h_{co} and h_{ci} are convective heat transfer coefficient along the external and internal surfaces, respectively [W/(m²K)], C is the conductance of the whole wall, and R_1 and R_2 are the resistance of the two materials in the wall shown in Fig. 9.19.

Let us regard the set of two surface temperatures, t_{so} and t_{si}, as a two dimensional quantity, that is, a two dimensional vector quantity. With this vector in mind, we can come up with a two-dimensional matrix comprising the thermal characteristics of the wall in question and also one other vector comprising two operative temperatures. With the matrix and the two vectors formulated, Eq. (9.37) can be rewritten as follows.

$$\begin{bmatrix} a_{11} & a_{12} \\ a_{21} & a_{22} \end{bmatrix} \cdot \begin{bmatrix} t_{so} \\ t_{si} \end{bmatrix} = \begin{bmatrix} t_{op\_o} \\ t_{op\_i} \end{bmatrix}, \tag{9.39}$$

The elements of the matrix on the left-hand side and the vector on the right-hand side of this equation are as follows:

$$a_{11} = \frac{h_{ro} + h_{co} + C}{h_{ro} + h_{co}}, a_{12} = \frac{-C}{h_{ro} + h_{co}}, a_{21} = \frac{-C}{h_{ri} + h_{ci}}, a_{22} = \frac{C + h_{ri} + h_{ci}}{h_{ri} + h_{ci}}.$$

$$t_{op\_o} = \frac{h_{ro}t_{ro} + h_{co}t_{ao}}{h_{ro} + h_{co}}, t_{op\_i} = \frac{h_{ri}t_{ri} + h_{ci}t_{ai}}{h_{ri} + h_{ci}} \tag{9.40}$$

Taking a look at the four elements of the matrix, we come to recognize that this matrix represents overall thermal characteristics of the wall in question, though the heat capacity is not included because of the assumption made in the beginning of this section. The reason why the elements of the vector in the right-hand side of Eq. (9.39) are called operative temperature is that, as can been seen in Eq. (9.40), each of them is the weighted value of radiant and air temperatures. They can be regarded as the stimuli for the wall in question and the responses are the two surface temperatures, t_{so} and t_{si}, to emerge in relation to the characteristic matrix, whose elements are expressed by Eq. (9.40).

Outdoor and indoor radiant temperatures are given as follows.

$$t_{ro} = f_{sky}t_{sky} + f_{gr}t_{gr} + \frac{a_w I_{TV}}{h_{ro}}, \text{ and } t_{ri} = \sum_{i=1}^{N} f_{si}t_{si}, \tag{9.41}$$

where t_{sky} is the apparent sky temperature, and a_w is solar absorptance of the external surface of the wall, whose typical values are as shown in Fig. 6.9, Chapter 6. The values of t_{sky} are dependent on outdoor air temperature and humidity; more will be described in Chapter 12. The factors such as f_{sky}, f_{gr} and f_{si} are weighting coefficients that can be determined in geometrical relation to the external or internal surfaces and their surrounding surfaces; they have to fulfil $f_{sky} + f_{gr} = 1$ and $\sum_{i=1}^{N} f_{si} = 1$, where N is the number of interior surfaces of ceiling, partitions, and floor.

If the temperature of the boundary of two materials, t_{wm}, has to be known, then it can be obtained from the following formula, which is derived from the energy balance equation of the whole wall under the condition of no effect of heat capacity

$$t_{wm} = \frac{R_2}{R_1 + R_2} t_{so} + \frac{R_1}{R_1 + R_2} t_{si}. \tag{9.42}$$

9.4.2 Overall heat transmission coefficient

Heat flow through the wall from internal surface to external surface can be expressed as

$$q_{total} = C(t_{si} - t_{so}). \tag{9.43}$$

Inverting the matrix in Eq. (9.39), that is, performing a mathematical operation equivalent to having a^{-1} for solving the value of x from a simple algebraic equation: $ax = b$, the resulting formulae to calculate the two surface temperatures, t_{so} and t_{si}, are expressed as follows.

$$t_{so} = \frac{a_{22}t_{op\_o} - a_{12}t_{op\_i}}{a_{11}a_{22} - a_{12}a_{21}}, t_{si} = \frac{-a_{21}t_{op\_o} + a_{11}t_{op\_i}}{a_{11}a_{22} - a_{12}a_{21}}. \tag{9.44}$$

Substituting Eq. (9.44) into Eq. (9.43) and recalling the relationships of $a_{11} + a_{12} = 1$ and $a_{21} + a_{22} = 1$ as can be known from the relationship expressed in Eq. (9.40), we reach the following equation

$$q_{total} = U(t_{op\_i} - t_{op\_o}). \tag{9.45}$$

where
$$U = \frac{C}{a_{11}a_{22} - a_{12}a_{21}} = \frac{1}{\dfrac{1}{h_{ro} + h_{co}} + R_1 + R_2 + \dfrac{1}{h_{ri} + h_{ci}}} \tag{9.46}$$

The value of U calculated from Eq. (9.46) is called overall heat-transmission coefficient, or simply U value. This is regarded to be an index that clarifies the level of thermal insulation of building walls in general. For example, two walls compared in the discussion made in 6.5.1, Chapter 6, have U values of 3.56 W/(m²K) and 0.34 W/(m²K), respectively. In engineering practice, this overall heat-transmission coefficient is calculated for walls, windows, roofs and floors, and then, taking the difference in temperature between outdoor air and indoor air instead of taking the operative temperature difference, thermal energy demand for a space heating and cooling system is estimated. The use of Eq. (9.46) for calculating U values and then Eq. (9.45) for estimating the thermal energy demand is convenient, but we should keep in mind that if such a way alone is taken to be the premise, then actual thermal behaviour within the indoor environment, the role of long-wavelength radiation in particular, could be slipped away from our minds.

In other words, we should carefully use so-called U values and also the estimation of heat loads based on them; upon necessity, we should come back to the calculation of surface temperature indoors as we discussed above in 9.4.1 and also take a careful look at the importance of radiant-exergy emission and absorption as was discussed in 6.5, Chapter 6 and also will be described in Chapter 10.

9.4.3 Radiant exergy absorption and consumption

As explained in the beginning of 9.4.1 using Fig. 9.19, it is important for us to have a look at a building wall being surrounded by outdoor and indoor virtual surfaces, whose temperatures are t_{ro} and t_{ri} being different from outdoor or indoor air temperature. This is because these virtual surfaces are the sources of radiation that affect more or less the surface temperatures of a wall in question. Furthermore, according to the thermodynamic discussion in Chapter 7, exergy is consumed not only in the course of convective or conductive thermal exergy flows, but also in the course of radiant exergy absorption by itself if the radiant source temperature is different from the temperature of a surface, where the absorption takes place.

In the cases of convection and conduction, exergy consumption takes place where there is thermal resistance, which is reciprocal of heat transfer coefficient or thermal conductance. In the case of radiation, on the other hand, what is equivalent to thermal resistance against convection and conduction is the absorptance. It is, in general, very much dependent on the wavelength of radiation, that is, the surface of materials acts differently, whether it is against short-wavelength or against long-wavelength radiation as explained in Chapter 6.

Here we discuss how the radiant exergy absorption and consumption can be modelled focussing on the case of a window having three components as shown in Fig. 9.20. The three components can be, for example, the combination of double panes and internal shading device, triple panes, or the combination of external shading device and double panes. Glass sheets and shading devices are usually very thin so that the effect of their heat capacities is negligible. The temperatures of these three components, t_1, t_2 and t_3 shown in Fig. 9.20 can be obtained by applying the method described in 9.4.1, provided that the absorption of solar energy is taken into consideration.

Assuming each of the three components to be a thermodynamic system, its exergy balance equation can be expressed as follows

[Solar (short-wavelength) radiant exergy absorption] +

[Long-wavelength radiant exergy absorption] – [Exergy consumption] =

[Long-wavelength radiant exergy emission] +

[Thermal exergy transfer by convection]   (9.47)

Exergy consumption emerges due to both the absorption of solar exergy and that of long-wavelength radiant exergy.

We all know through our everyday experience that the absorption of solar radiation by materials brings about "warmth" or "hotness". The energy-wise implication of this phenomenon is that solar radiant energy is converted into thermal energy to be transferred by long-wavelength radiation, convection or conduction. On

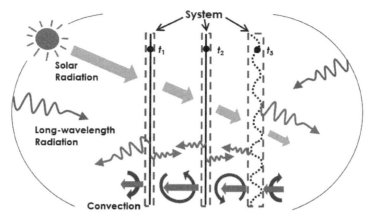

Fig. 9.20: A window comprising of three components, in each of which there are absorption of solar radiation and both emission and absorption of long-wavelength radiation.

the other hand, the entropy-wise implication is that some entropy is generated due to the absorption of solar radiation. These two statements are mathematically modelled as follows. In component 1,

$$a_1 I_{TV} = q_1, \quad a_1 s_{TV} + s_{g\_sol} = \left(\frac{1}{T_1}\right) q_1, \tag{9.48}$$

where a_1 is the solar absorptance of component 1, I_{TV} is the rate of solar energy incident on the whole of window [W/m²], q_1 is the rate of thermal energy generated due to the absorption of solar radiation [W/m²], s_{TV} is the rate of solar entropy incident on the whole of window [W/(m²K)], s_{g_sol} is the rate of entropy generation [W/(m²K)], and T_1 is the absolute temperature of component 1 [K]. Combining the two equations in Eq. (9.48) together with the environmental temperature, T_o [K], yields the following exergy balance equation:

$$a_1 x_{TV} - x_{C\_sol} = \left(1 - \frac{T_o}{T_1}\right) q_1. \tag{9.49}$$

where x_{TV} is solar radiant exergy [W/m²] and x_{C_sol} is exergy consumption rate [W/m²]; they are expressed as $x_{TV} = I_{TV} - s_{TV} T_o$ and $x_{C\_sol} = s_{g\_sol} T_o$, respectively. The values of I_{TV} and s_{TV} together with T_o can be given if the location, the season and the time of day are specified (Shukuya 2013).

Since the entropy generation rate, s_{g_sol}, cannot be zero, in other words, exergy has to be consumed for any macroscopic natural phenomena, the temperature of component 1 to be determined with the assumption of $s_{g\_sol} = 0$ is the upper limit temperature, T_{1_limit} [K].

$$T_1 < T_{1\_limit} = \frac{I_{TV}}{s_{TV}}. \tag{9.50}$$

The upper limit temperature, T_{1_limit}, is determined by the nature of solar radiation available alone. For example, in such a condition of $I_{TV} = 500$ W/m² and

s_{TV} = 0.15 W/(m²K), which resembles the solar availability on a clear sky day, T_{1_limit} ≈ 3,300 K (≈ 3000°C). As was shown in Fig. 6.11, Chapter 6, the apparent surface temperature of the Sun is estimated to be about 6000 K (≈ 5700°C). The solar radiant temperature right above the atmosphere, 30 km or so above the ground surface, is known to be 4500 K (≈ 4200°C) (Shukuya 2013). Therefore, we can say that solar radiation disperses in its travel from the Sun to the upper edge of the atmosphere, then at least down to the upper limit temperature, T_{1_limit} ≈ 3,300 K(≈ 3000°C) and usually much further down to a lower temperature, the order of 300 K (≈ 30°C). Such decrease of temperature emerges because of a large rate of solar exergy consumption, which can be expressed as follows by rewriting Eq. (9.49):

$$x_{C\_sol} = a_1 \left\{ \left(\frac{1}{T_1}\right) I_{TV} - s_{TV} \right\} T_o. \tag{9.51}$$

Equation (9.51) indicates that exergy consumption emerges as solar radiation is absorbed.

Similarly, as the surface of a material absorbs long-wavelength radiation, there also emerges exergy consumption. The formula for the calculation of exergy consumption rate due to the absorption of long-wavelength radiation, x_{C_lwr} [W/m²], which stands in parallel to x_{C_sol} to be calculated by Eq. (9.51), is derived from Eq. (9.47) as follows in the case of component 1, by substituting the exact formula for radiant exergy emission expressed in the form of Eq. (7.42) and also that for convective exergy transfer expressed as Eq. (7.51) altogether.

$$x_{C\_lwr} = \varepsilon \sigma T_1^3 \left\{ \frac{1}{3} + \left(\frac{T_{ro}}{T_1}\right)^3 \left(\frac{T_{ro}}{T_1} - \frac{4}{3}\right) \right\} T_o. \tag{9.52}$$

where ε is the emittance of the component 1 and σ is Stephan-Boltzmann constant (= 5.67 × 10⁻⁸ W/(m²K⁴)).

Table 9.2 summarizes all formulae for the exergy consumption rate in the window system shown in Fig. 9.20. The formulae other than Eqs. (9.51) and (9.52) are the ones to be derived from the discussion in Chapter 7. Numerical examples demonstrated in 6.5.3, Chapter 6, were in fact obtained from the calculation using the formulae given in Table 9.2.

9.4.4 Exergy balance within a wall with heat capacity

Building walls as shown in Fig. 9.19 have some heat capacity in reality and this causes time-wise temperature change in addition to space-wise temperature difference. As we all know through our everyday experience, the temperature of hot water decreases gradually and that of cold water increases, also gradually, as the time elapses. These are the delaying effect due to heat capacity of water. The same happens more or less in any building material having heat capacity. The temperature variations obtained from the sets of model experiment described in Chapter 6 are, more or less, due to the effects of heat capacity.

In order to take the effect of heat capacity into consideration, let us first model the wall as a chain of nodes having heat capacity as shown in Fig. 9.21 (Shukuya

Table 9.2: Formulae of exergy consumption rate due to the absorption of solar and long-wavelength radiation and also convection in a three-component window system.

Thermodynamic subsystem	Radiation	Convection
External surface and boundary layer	Eq. (9.52)	$\dfrac{h_{co}\left(T_1-T_o\right)^2}{T_1 T_o}T_o$
Component 1	Eq. (9.51)	---
Surfaces of components 1 and 2 and the space in between[*1]	$\dfrac{h_{r\_12}\left(T_2-T_1\right)^2}{T_2 T_1}T_o$	$\dfrac{h_{c\_12}\left(T_2-T_1\right)^2}{T_2 T_1}T_o$
Component 2	Eq. (9.51) [*2]	---
Surfaces of components 2 and 3 and the space in between[*3]	$\dfrac{h_{r\_23}\left(T_3-T_2\right)^2}{T_3 T_2}T_o$	$\dfrac{h_{c\_23}\left(T_3-T_2\right)^2}{T_3 T_2}T_o$
Component 3	Eq. (9.51)[*4]	---
Internal surface and boundary layer	Eq. (9.52)[*4]	$\dfrac{h_{ci}\left(T_{ia}-T_3\right)^2}{T_{ia} T_3}T_o$

[*1] h_{r_12} and h_{c_12} are radiative and convective heat transfer coefficients between the space.
[*2] Solar absorptance, emittance and temperature for component 2 instead of 1 are used.
[*3] h_{r_23} and h_{c_23} are radiative and convective heat transfer coefficients between the space.
[*4] Solar absorptance, emittance and temperature for component 3 instead of 1 are used.

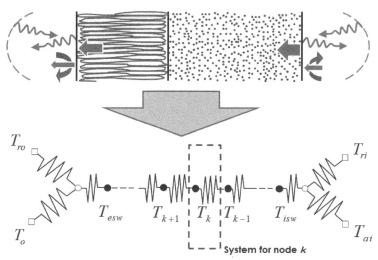

Fig. 9.21: Space-wise discretization of a building wall and its representation by a series of nodes connected with each other by thermal resistance indicated by corrugated lines in between.

2013). Such modelling enables us to calculate the variation of temperature within a building wall under unsteady-state conditions. Node k is surrounded by two nodes, $k-1$ and $k+1$. Their temperatures are denoted by T_k, T_{k-1}, and T_{k+1}, respectively. The diagram drawn with the symbols of closed circles and corrugated lines represents the whole of thermal conduction within the wall. The closed circles are the nodes having heat capacity and corrugated lines the thermal resistance of conduction, which equals the reciprocal of thermal conductance between two nodes. The nodes representing the exterior and interior surfaces can be assumed either to have heat capacity or not. Here we take the latter assumption. As we already discussed in 9.4.1, the node representing the exterior surface is connected with the whole of opposite surfaces that emit thermal radiation towards it and also with outdoor air for convective heat transfer. The same applies to the node at the interior surface.

Applying what was described in the previous sections, energy balance equations for node k can be expressed as follows, taking the terms representing the change in energy stored into consideration.

$$_{-}q_k dt = c_{pk} \rho_k l_k dT_k + {}_{+}q_k dt, \qquad (9.53)$$

where

$$_{-}q_k = {}_{-}C_k(T_{k-1} - T_k), \qquad (9.54)$$

$$_{+}q_k = {}_{+}C_k(T_k - T_{k+1}). \qquad (9.55)$$

The symbols used are as follows. Symbol $_{-}q_k$ is thermal energy flow rate by conduction from node $k-1$ to k, and $_{+}q_k$ is that from node k to $k+1$. Their unit is W/m². Symbol t denotes time in the unit of second and dt denotes an infinitesimally short period of time. $_{-}C_k$ and $_{+}C_k$ are thermal conductance between node $k-1$ and k and that between k and $k+1$ [W/(m²K)]. c_{pk} is specific heat capacity of node k [J/(kgK)], ρ_k is the density of material represented by node k [kg/m³], and l_k is the thickness of material represented by node k [m]. dT_k denotes an infinitesimally small increase in temperature at node k [K] within the infinitesimally short period of time, dt.

The entropy balance equation standing in parallel to the energy balance equation expressed by Eq. (9.53) is

$$\left(\frac{1}{T_{k-1}}\right) {}_{-}q_k dt + \sigma_k dt = \left(\frac{1}{T_k}\right) c_{pk} \rho_k l_k dT_k + \left(\frac{1}{T_k}\right) {}_{+}q_k dt \qquad (9.56)$$

where σ_k is the rate of entropy generation by the conduction of heat from node $k-1$ to node k. Equation (9.56) indicates that the entropy generated within the thermodynamic system defined by a closed dashed line in Fig. 9.21 flows into node k together with the entropy inflow coming from node $k-1$ and their sum is partly stored and partly outflows toward node $k+1$.

The diagram shown in Fig. 9.21 may be more easily understood by a schematic representation in Fig. 9.22. In this drawing, the nodes having heat capacity is analogically expressed as a series of vessels containing some amounts of water, whose respective heights are equivalent to the temperature of the respective nodes.

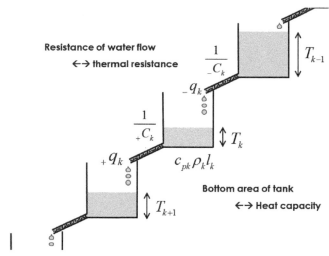

Fig. 9.22: Schematic representation of unsteady-state heat conduction with an analogy of unsteady-state flow of water.

The bottom-surface areas are equivalent to the heat capacity of the nodes. If the bottom-surface area is large, the height of water in the vessel tends to change slowly, which corresponds to a large heat capacity. The vessels are connected by tubes, in which quite a few pieces of gravel are filled so as to retard the flow of water from one vessel to another. The more the pieces of gravel are, the slower the flow of water. Thermally insulating materials are modelled with the tubes having many fine pieces of gravel that resist the flow of water from one vessel to another. Unsteady-state heat conduction is analogous to the flow of water through the vessels with the changes in the height of water in respective vessels, while on the other hand, the steady-state heat conduction is analogous to the flow of water through the vessels without a change in the height of water. The understanding of mathematical expression in Eqs. (9.53) and (9.56) with the help of the schematic diagrams in Fig. 9.22 may be recognized as an extension of the thought on structure (*Katachi*) to see and function to read (*Kata*).

As we performed algebraic operation from Eq. (9.48) to Eq. (9.49), we do the same with Eqs. (9.53) and (9.56) and the outdoor environmental temperature, T_o. We then reach the following exergy balance equation at node k

$$_-x_k - x_{ck} = x_{sk} + {_+}x_k, \tag{9.57}$$

where

$$_-x_k = \left(1 - \frac{T_o}{T_{k-1}}\right){_-}C_k(T_{k-1} - T_k), \tag{9.58}$$

$$x_{ck} = \sigma_k T_o, \tag{9.59}$$

$$x_{sk} = \frac{dX_{sk}}{dt} = \left(1 - \frac{T_o}{T_k}\right)c_{pk}\rho_k l_k \frac{dT_k}{dt}, \tag{9.60}$$

$$_+x_k = \left(1 - \frac{T_o}{T_k}\right)_+ C_k(T_k - T_{k+1}).$$ (9.61)

Equation (9.57) is read as follows referring to Fig. 9.21: a portion of exergy flowing across the right-hand side of the system into node k at the rate of $_-x_k$ is consumed at the rate of x_{ck}, which takes place through thermal resistance represented by the corrugated line inside the system, and thereby their difference is partly stored at node k with the rate of x_{sk} and partly flows out across the left-hand side of the system towards node $k+1$ at the rate of $_+x_k$.

We further approximate the infinitesimally short period of time, dt, with a finite difference of time, Δt, from $(n-1)\Delta t$ to $n\Delta t$ as shown in Fig. 9.23 and the finite increase in temperature of node i, that is a time-wise difference in temperature, ΔT_i, from $(n-1)\Delta t$ to $n\Delta t$. Note that n is integer, namely, 0, 1, 2, 3, $\cdots$.

The next problem to be considered is which time is taken, $(n-1)\Delta t$ or $n\Delta t$, for the space-wise temperature differences, $T_{k-1} - T_k$ and $T_k - T_{k+1}$, that appeared in Eqs. (9.54) and (9.55). If $(n-1)\Delta t$ is taken, the set of finite differential equations is called "explicit" type and if $n\Delta t$ is taken, then the resulting set is "implicit" type. We may take the average of "implicit" and "explicit", which is called Crank-Nicolson type.

Which of either "explicit", "implicit" or Crank-Nicolson type of equations is used for numerical calculation depends on which type suits better to the problem we have to solve. They require different manners of numerical calculation for solving the set of temperature at all nodes from $(n-1)\Delta t$ to $n\Delta t$.

The equations of "explicit" type set up for respective nodes are independent of each other so that the numerical calculation can simply be made from one equation to another for each time step, but there is one strict constraint, in which the sizes of space-wise and time-wise differences have to be small enough so as to make the results of calculation stable, that is, not to diverge.

On the other hand, the equations of "implicit" type or Crank-Nicolson type set up for respective nodes are dependent on each other so that these equations have to be solved simultaneously with algebraic operation of inverse matrix. This may require a bit of tedious calculation, but we do not need to worry about the possibility of diversion. Therefore, the space-wise and time-wise finite differentiation may be made wider than the explicit type.

If the "implicit" type of finite difference calculation is taken, the thermal conduction of $_-q_k$ and $_+q_k$ expressed in Eqs. (9.54) and (9.55) is approximated by those at $n\Delta t$. Denoting T_o and T_k at $n\Delta t$, as $T_o(n)$ to $T_k(n)$, then the rate of thermal exergy

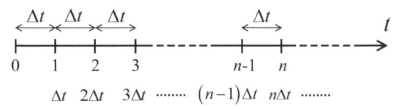

Fig. 9.23: Discretization of time t as a continuous variable into a discontinuous time series with the regular increment of time, Δt.

stored at node k, that was originally expressed as Eq. (9.60), can be approximated by the following finite difference equation.

$$x_{sk} \simeq \frac{\Delta X_{sk}}{\Delta t} = \left(1 - \frac{T_o(n)}{T_k(n)}\right) c_{pk} \rho_k l_k \frac{\{T_k(n) - T_k(n-1)\}}{\Delta t} \quad (9.62)$$

Taking a look at Eq. (9.62), the rate of thermal exergy stored, x_{sk}, can be either positive or negative depending on the relationships between three temperature values, $T_o(n)$, $T_k(n)$, and $T_k(n-1)$. This logic is exactly the same as that of convection described in 7.6.2, Chapter 7. Because of three variables, there are six combinations that are different from each other in terms of whether it implies "warm" or "cool" exergy and also whether it implies "increase" or "decrease". "Increase" or "decrease" in exergy stored as a quantity of state corresponds to "inflow" or "outflow" in exergy transferred by convection or conduction as a quantity of flow. We may regard a change in exergy stored to be time-wise flow, while exergy transfer by convection or conduction to be space-wise flow.

Table 9.3 shows the six combinations of the rate of exergy stored, x_{sk}, given by Eq. (9.62). In this table, the sign of Carnot factor denoted by d determines "warm" or "cool" exergy: positive is "warm" and negative "cool". The other factor denoted by e is the change in temperature of node k: positive is "increase" and negative "decrease". The diagram attached to Table 9.3 shows the six ranges determined by the combination of three temperatures, $T_o(n)$, $T_k(n)$, and $T_k(n-1)$.

Substitution of the rate of exergy stored, x_{sk}, calculated from Eq. (9.62) together with the rates of exergy inflow and outflow, x_k and $_{+}x_k$, for $n\Delta t$ into Eq. (9.57) leaves the exergy consumption rate due to conduction between node k and node $k+1$, x_{ck}, to be uniquely unknown so that its value can be calculated.

The variations of radiant exergy emission from the wall surface and exergy consumption rate within the wall together with the variation of interior wall surface temperature discussed in 6.5.2, Chapter 6, were all obtained from the above-mentioned method of calculation.

9.4.5 Differential exergy balance equation for heat conduction

The way of modelling a building wall having heat capacity described above was to first make space-wise discretization and then set up time-wise differential equation for energy and entropy balance. The time-wise differential equation was also finally discretized for numerical calculation to be made. This has to be consistent with not only time-wise but also space-wise differential equations to be discretized. For confirmation, let us discuss how the differential exergy balance equation for heat conduction is expressed by extending the differential energy balance equation for heat conduction that was established by J. B. Fourier (1768–1830).

Suppose that there is an axis of x, along which alone the heat flows; this is equivalent to considering a large flat wall, through which heat flows only in the direction of thickness. Let us denote the absolute temperature inside this wall as a function of position x and time t as

$$T = T(x, t). \quad (9.63)$$

Table 9.3: Increase or decrease in "warm" or "cool" exergy stored by node k.

Temperature	$d = 1 - \dfrac{T_o(n)}{T_k(n)}$	$e = \dfrac{T_k(n)}{-T_k(n-1)}$	$d \cdot e$	Warm/Cool	Increase/Decrease
I $\quad T_o(n) \leq T_k(n-1) < T_k(n)$	+	+	+	Warm	In
II $\quad T_o(n) < T_k(n) \leq T_k(n-1)$	+	−	−	Warm	De
III* $\quad T_k(n-1) < T_o(n) \leq T_k(n)$	+	+	+	Warm	In
IV* $\quad T_k(n) \leq T_o(n) < T_k(n-1)$	−	−	+	Cool	In
V $\quad T_k(n-1) \leq T_k(n) < T_o(n)$	−	+	−	Cool	De
VI $\quad T_k(n) < T_k(n-1) \leq T_o(n)$	−	−	+	Cool	In

* In the case of **III**, there is a change over from "cool" to "warm" exergy during the period of Δt. In the case of **IV**, a change over from "warm" to "cool" exergy.

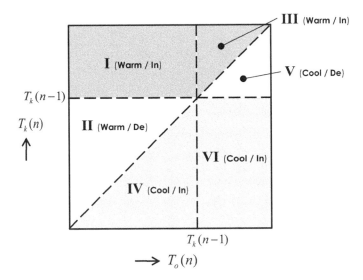

Referring to Fig. 9.24a, the infinitesimal increase of temperature, dT, can be expressed as

$$dT = \left(\frac{\partial T}{\partial x}\right)_t dx + \left(\frac{\partial T}{\partial t}\right)_x dt = \left(\partial T\right)_{space} + \left(\partial T\right)_{time}, \tag{9.64}$$

where $\left(\dfrac{\partial T}{\partial x}\right)_t$ is space-wise gradient; subscript "t" implies that time is held constant. The same rule applies to $\left(\dfrac{\partial T}{\partial t}\right)_x$ implying time-wise gradient with subscript "x" being held constant. This way of mathematical expression is consistent with what was used

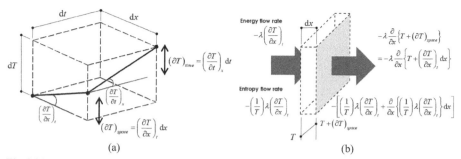

Fig. 9.24: (a) space-wise and time-wise infinitesimal increase of temperature and (b) thermal energy and entropy flow rates at position x and $x + \mathrm{d}x$.

in Eq. (8.14) and Fig. 8.10 in Chapter 8. The infinitesimal increase of temperature, $\mathrm{d}T$, is the sum of space-wise and time-wise increases, that is, $(\partial T)_{space}$ and $(\partial T)_{time}$. Note that "∂" indicates partial difference, here in this example, whether it is time-wise or space-wise.

Provided that the amount of thermal energy flowing across the area of 1 m² at location x during an infinitesimal period of time, $\mathrm{d}t$, is denoted as δQ_x, then it is expressed as

$$\delta Q_x = -\left\{ \frac{\lambda}{\mathrm{d}x}(\partial T)_{space} \right\} \mathrm{d}t = -\left\{ \frac{\lambda}{\mathrm{d}x}\left(\frac{\partial T}{\partial x} \right)_t \mathrm{d}x \right\} \mathrm{d}t = -\lambda \left(\frac{\partial T}{\partial x} \right)_t \mathrm{d}t, \quad (9.65)$$

where λ is thermal conductivity of solid material [W/(m·K)]. In Eq. (9.65), $\dfrac{\lambda}{\mathrm{d}x}$ indicates the conductance in the infinitesimal length of $\mathrm{d}x$, $(\partial T)_{space}$ is the infinitesimal space-wise temperature difference at time t as shown in Eq. (9.64). The reason for negative sign in Eq. (9.65) is that the heat flow direction is from $x + \mathrm{d}x$, at which the temperature is $T + (\partial T)_{space}$, to x, at which the temperature is T. If $(\partial T)_{space} > 0$, then $\delta Q_x < 0$, and if $(\partial T)_{space} < 0$, then $\delta Q_x > 0$. The last term of Eq. (9.65) was obtained by referring to Eq. (9.64). Note that the symbol "δ" indicates infinitesimal "flow", while "d" indicates infinitesimal "change".

At location $x + \mathrm{d}x$, the thermal energy flow $\delta Q_{x+\mathrm{d}x}$ can be expressed as

$$\delta Q_{x+\mathrm{d}x} = -\lambda \left[\frac{\partial \{ T + \partial(T)_{space} \}}{\partial x} \right] \mathrm{d}t = -\lambda \frac{\partial}{\partial x} \left\{ T + \left(\frac{\partial T}{\partial x} \right)_t \mathrm{d}x \right\} \mathrm{d}t$$
$$= -\left[\lambda \left(\frac{\partial T}{\partial x} \right)_t + \frac{\partial}{\partial x} \left\{ \lambda \left(\frac{\partial T}{\partial x} \right)_t \right\} \mathrm{d}x \right] \mathrm{d}t \quad (9.66)$$

The meaning of Eqs. (9.65) and (9.66) should become clearer by referring to Fig. 9.24b.

Defining a tiny thermodynamic system within the length of $\mathrm{d}x$, thermal energy to be stored in the volume of 1 m² multiplied by $\mathrm{d}x$ is expressed as $(c\rho \mathrm{d}x)\, \partial(T)_{time}$, where c is specific heat capacity [J/(kgK)], ρ is the density [kg/m³], and $(\partial T)_{time}$ is the infinitesimal time-wise temperature increase. This emerges if there is a difference between the inflow of thermal energy expressed by Eq. (9.65) and the outflow expressed by Eq. (9.66). This is again according to the law of energy conservation. Following the procedure so far described here in this chapter, thermal energy balance equation is expressed as

$$\delta Q_x = (c\rho dx)(\partial T)_{time} + \delta Q_{x+dx}. \tag{9.67}$$

Substituting Eqs. (9.65) and (9.66) into Eq. (9.67),

$$-\lambda \left(\frac{\partial T}{\partial x}\right)_t dt = (c\rho dx)(\partial T)_{time} - \left[\lambda\left(\frac{\partial T}{\partial x}\right)_t + \frac{\partial}{\partial x}\left\{\lambda\left(\frac{\partial T}{\partial x}\right)_t\right\}dx\right]dt. \tag{9.68}$$

This equation can further be reduced to the following equation by referring to the time-wise relationship shown in Eq. (9.64) and Fig. 9.24a

$$\lambda\frac{\partial}{\partial x}\left(\frac{\partial T}{\partial x}\right)_t = c\rho\left(\frac{\partial T}{\partial t}\right)_x, \text{ or abbreviating subscripts, } \lambda\frac{\partial^2 T}{\partial x^2} = c\rho\frac{\partial T}{\partial t}. \tag{9.69}$$

Equation (9.69) is exactly Fourier's differential equation for one-dimensional heat conduction.

As described in Chapter 7, the entropy flow is expressed as the product of the dispersality and thermal energy flow, with which a certain amount of entropy is necessarily generated. Denoting the rate of entropy generation within the volume of 1 m³ as σ [W/(m³·K)], the entropy balance equation to be standing in parallel to energy balance equation as Eq. (9.68) can be expressed as

$$-\left\{\left(\frac{1}{T}\right)\lambda\left(\frac{\partial T}{\partial x}\right)_t\right\}dt + \sigma dx dt = \left(\frac{1}{T}\right)(c\rho dx)(\partial T)_{time}$$
$$-\left[\left(\frac{1}{T}\right)\lambda\left(\frac{\partial T}{\partial x}\right)_t + \frac{\partial}{\partial x}\left\{\left(\frac{1}{T}\right)\lambda\left(\frac{\partial T}{\partial x}\right)_t\right\}dx\right]dt \tag{9.70}$$

This equation can be reduced to

$$\frac{\partial}{\partial x}\left\{\left(\frac{1}{T}\right)\lambda\left(\frac{\partial T}{\partial x}\right)\right\} + \sigma = \left(\frac{1}{T}\right)c\rho\frac{\partial T}{\partial t}. \tag{9.71}$$

Referring to the product rule shown in Fig. 9.15 and applying this rule to the product of two functions, $\frac{1}{T}$ and $\lambda\left(\frac{\partial T}{\partial x}\right)$, we can rewrite Eq. (9.71) as follows

$$\left\{-\frac{\lambda}{T^2}\left(\frac{\partial T}{\partial x}\right)^2 + \left(\frac{1}{T}\right)\lambda\frac{\partial^2 T}{\partial x^2}\right\} + \sigma = \left(\frac{1}{T}\right)c\rho\frac{\partial T}{\partial t}. \tag{9.72}$$

Taking the energy-balance relationship expressed by Eq. (9.69), we come to know how the entropy generation rate is expressed.

$$\sigma = \frac{\lambda}{T^2}\left(\frac{\partial T}{\partial x}\right)^2. \tag{9.73}$$

Since $\lambda > 0$, $T > 0$, and the square of space-wise temperature gradient is necessarily positive, the rate of entropy generation rate never becomes negative; here we have confirmed that there is no heat conduction in which no entropy is generated. Combining Eqs. (9.69) and (9.71) together with environmental temperature, T_o,

$$\lambda \frac{\partial}{\partial x}\left\{\frac{\partial T}{\partial x} - T_o\left(\frac{1}{T}\frac{\partial T}{\partial x}\right)\right\} - \sigma T_o = \left(1 - \frac{T_o}{T}\right)c\rho\frac{\partial T}{\partial t}. \qquad (9.74)$$

Since $\dfrac{1}{T}\dfrac{\partial T}{\partial x} = \dfrac{\partial}{\partial T}(\ln T)\dfrac{\partial T}{\partial x} = \dfrac{\partial}{\partial x}(\ln T)$, by referring to the characteristic of natural logarithm demonstrated in Fig. 9.11, we come to the final one-dimensional differential equation of exergy balance as

$$\lambda \frac{\partial^2}{\partial x^2}(T - T_o \ln T) - \sigma T_o = \left(1 - \frac{T_o}{T}\right)c\rho\frac{\partial T}{\partial t}. \qquad (9.75)$$

For three dimensional heat conduction, using the vector symbol ∇ (pronounced "nabla" or "del"), $\nabla = \left(\dfrac{\partial}{\partial x}, \dfrac{\partial}{\partial y}, \dfrac{\partial}{\partial z}\right)$, Eq. (9.75) can be extended as

$$\lambda\nabla^2(T - T_o \ln T) - \sigma T_o = \left(1 - \frac{T_o}{T}\right)c\rho\frac{\partial T}{\partial t}, \qquad (9.76)$$

$$\sigma = \frac{\lambda}{T^2}(\nabla T)^2. \qquad (9.77)$$

The first term of Eq. (9.76) is the net exergy inflow, its portion as σT_o is consumed, and the rest expressed by the right-hand side of this equation is stored. The set of Eqs. (9.76) and (9.77) is the most abstract expression of exergy consumption theorem. Space-wise and time-wise discretization of Eqs. (9.76) and (9.77) and numerical calculation of the resulting finite differential equations with respect to a building wall was discussed by Choi et al. (2018).

Abstract expressions based on mathematical formality as demonstrated by Eqs. (9.76) and (9.77) must look intimidating at first glance rather than beautiful, if it is introduced without any qualitative image of how nature functions. But provided that a certain qualitative image is always accompanied as demonstrated through the discussion in the previous sections, then it surely enriches the qualitative understanding of nature. This is what was pointed out in the very beginning of the present chapter.

References

Bejan A. 2004. Convective Heat Transfer. Third. Ed. John Wiley & Sons, Inc.

Choi W., Ooka R. and Shukuya M. 2018. Exergy analysis for unsteady-state heat conduction. International Journal of Heat and Mass Transfer 116: 1124–1142.

Devreese J. T. and Berghe G. V. 2007. 'Magic is No Magic': The wonderful world of Simon Stevin. WIT Press.

Eddy J. 2009. The Sun, the Earth, and near-Earth space—A guide to the sun-earth system. NASA NP-2009-1-066GFSC. ISBN0160838088.

Gunn R. and Kinzer G.D. 1949. The terminal velocity of fall for water droplets in stagnant air. Journal of Meteorology 6: 243–348.

Imai I. 2003. Introduction to a New Way of Physical Thinking. Iwanami Publishers (in Japanese).

Lloyd C. 2012. What on Earth Happened. Bloomsbury Publishers. London.

Maxwell J. C. 1891. Theory of Heat (with new introduction by Pesic P.) Dover Publishing Inc.

Mizuno H. 2000. Meteorology on clouds and rainfall. Asakura-Shoten Publishers (in Japanese).

Newton's cradle. https://en.wikipedia.org/wiki/Newton%27s_cradle#cite_ref-6 (retrieved 11th August 2017).

Shukuya M. 2013. Exergy—Theory and applications in the built environment. Springer-Verlag London.

Taylor F. W. 2005. Elementary Climate Physics. Oxford University Press.

Tooyama H. 1959. Introduction to Mathematics. Iwanami Publishers (in Japanese).

Tooyama H. 1970. Differentiation and Integration—their thought and method. Nippon-Hyoron Publishers (in Japanese).

Yamamoto Y. 2008. Mechanics and Differential Equations. Sugaku-Shobo Publishers (in Japanese).

Chapter 10
Human-Body Exergetic Behaviour

10.1 Sustenance of dynamic equilibrium

Everyone might have an experience where early in the morning a physiological need awakes you to go to the toilet. Recall the moment that after you come back from the toilet and slip into the bed because of your being still sleepy. You must have found yourself that your body was comfortably wrapped up by the warmth remaining within the bed. Why is it possible? Where does this warmth come from? It is, of course, from your own body, which never stops dissipating heat as long as we live. Too much self-evident it may sound, but still worth rethinking what really goes on in the whole process.

We all know that the intense solar radiation is the fundamental source for all living creatures existing on the Earth. This originates from thermonuclear fusion reaction, which has been occurring for several billion years in the Sun. The reason that it has lasted so long instead of blasting like a hydrogen bomb within a very short period of time is due to the tremendous gravitational force of the huge mass of the Sun. The round shape of the Sun can be regarded to be the result of two phenomena occurring simultaneously: one is thermonuclear explosion and the other enormous gravitational pressurization. We may regard that the Sun is forming its shape as the dynamic equilibrium with these two simultaneously occurring phenomena.

In due course, enormous amounts of thermal energy and entropy are being discharged ceaselessly into the Universal space as electro-magnetic wave and its tiny portion arrives on the terrestrial surface. This tiny portion of solar radiation reaching the Earth together with the ultimate coldness of the Universal space are the driving agents realizing all of the atmospheric, oceanic, and biological phenomena occurring here and there on the Earth.

In order to know the magnitude of energy emission being made by the Sun relative to that being made by each of us human, let us compare specific emission rates of the Sun and one single human body, that is, compare the rates per one unit of mass. The question is which of the two rates is larger? To help you make a guess, Fig. 10.1 gives five choices to be selected. Choose one of them; there is the right answer closest to the actual values among the five choices. The first one is that the rate of thermal energy discharged from the human body is only 0.0001 times that of

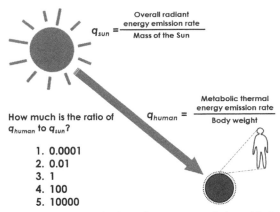

Fig. 10.1: Comparison of specific rate of solar radiant energy emission with that of thermal energy emission by human metabolism.

the Sun. The second is its 100 times, that is, 0.01 times larger. The third implies that the Sun and the human body is equivalent to each other in terms of the specific rate of thermal energy discharge. The fourth is that it is 100 times larger from the human body than from the Sun. The fifth is, even further large, 10000 times larger from the human body than from the Sun. Give your answer and write down it on your notebook or on a piece of paper nearby. Which one do you guess is right?

In fact, the right answer is "5" that implies the specific thermal energy emission rate of the human body being ten-thousand times larger than that of the Sun. Thanks to Kepler's law of planetary motion (1619), Galilei's gravitational acceleration on the Earth (1632), Newton's law of gravitation (1687) together with the careful determination of its proportional constant made by Cavendish (1798), and the distance between the Sun and the Earth to be determined by the method given in Chapter 4, we can estimate the masses of the Earth and the Sun as 5.96×10^{24} kg and 2.01×10^{30} kg, respectively. Extra-terrestrial solar radiation is at the rate of about 1370 W/m² so that with the distance of the Sun and the Earth, we can estimate the whole rate of energy emission by the Sun to be approximately 3.87×10^{17} GW. Therefore, the energy emission rate per one kilogram of the Sun turns out to be 0.2 mW/kg. The corresponding rate of human body is 1.75 W/kg assuming the average thermal energy emission rate for one person to be 125 W and his or her body weight to be 72 kg. The value of 1.75 W/kg is about 8700 times larger than 0.2 mW/kg; this is why the answer to the question in Fig. 10.1 is "5".

Although the human body keeps dissipating thermal energy at such a large specific rate, as confirmed with the question and answer raised in Fig. 10.1, it also keeps its core temperature at 37°C, which is required for the smooth bio-chemical reaction to keep proceeding as long as the human body is alive. For this purpose alone, there has to be the same rate of energy input to the human body in accordance with the law of energy conservation, otherwise it is not possible for the human body to keep its temperature at 37°C. This is why we eat food regularly, usually three times a day.

10.2 Exergy balance

If keeping the body-core temperature is the ultimate necessity, it might be sufficient enough to provide a heat source, whose temperature is a bit higher than the body-core temperature. But, the whole of our body surface is surrounded by the heat sink, into which thermal energy and entropy generated within the body has to be dumped. Therefore, there is no space for such heat source at all in our vicinity.

Moreover, the human body is subject to maintaining its structure made of about six trillion cells by constantly reconstructing quite a number of new cells while demolishing the corresponding old cells (Guggenheim 1991). In due course, the human body maintains its state of materially dynamic equilibrium in addition to thermally dynamic equilibrium. This implies that the food we eat consists of two types: one as fuel and the other as building blocks.

Here in this chapter, we focus on food as fuel represented by glucose, $C_6H_{12}O_6$ in chemical language, and how its exergy is consumed in order to maintain the state of thermally and materially dynamic equilibrium of the human body.

10.2.1 Chemical to thermal

Glucose molecules contained by a variety of food are absorbed through the stomach and intestines in the course of digestion and they are delivered to each of the human-body cells by the blood circulating all the time. The glucose molecules comprising a lot of carbon atoms are in fact very dangerous, though useful, because they are chemically very reactive. Therefore, the glucose molecules are converted into ATP (adenosine-triphosphate, $C_{10}H_{16}N_5O_{13}P_3$) as soon as they are absorbed by the human-body cells. This is performed by mitochondria, which exist in most of the cells as shown in Fig. 10.2. Muscular cells have a lot of mitochondria, especially those along the femurs and the upper arms.

A mitochondrion produces ATP molecules by combining the molecules of ADP (adenosine-diphosphate, $C_{10}H_{15}N_5O_{10}P_2$) with those of phosphoric acid (HPO_3). The chemical energy carried by ATPs is known to be about 35% of that originally carried by glucose molecules (Reece et al. 2011). This implies that 65% of the chemical energy turns into thermal energy. This conversion of energy from chemical to thermal, as the primary process, can be expressed as the exergy balance equation as follows.

$$[\text{Chemical exergy (glucose)}] - [\text{exergy consumption\_1}]$$

$$= [\text{chemical exergy (ATP)}] + [\text{``warm'' exergy\_1}] \qquad (10.1)$$

Equation (10.1) represents that a mitochondrion takes in chemical exergy contained by glucose molecules in order to produce ATP molecules that turn out to contain chemical exergy, while at the same time discharges "warm" exergy into the cellular spaces of the human body.

Let us suppose an average adult person who takes in about 1600 kcal of nutrition for one day. It is equivalent to the averaged energy intake at 77.5 W to be provided by 27.5 μmol/s of glucose molecules. Since the chemical exergy contained by glucose molecules is known to be 2978 kJ/mol (Shukuya 2013), the chemical exergy taken

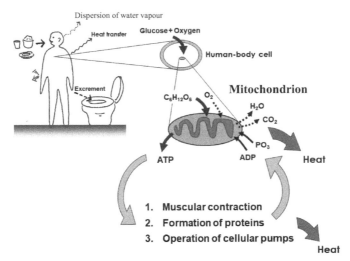

Fig. 10.2: Carbohydrate intake, its use for keeping dynamic equilibrium and subsequent heat generation.

in by the average person turns out to be 82.2 W. With this caloric intake, 27.1 W of chemical exergy is stored within ATP molecules; with this process, the exergy efficiency of ATP production is 33%. In due course, "warm" exergy is generated as its by-product and it amounts from 1 to 5 W depending on the environmental temperature. Substitution of these values into Eq. (10.1) lets us know that exergy consumption rate at mitochondria ranges between 50 and 54 W; this is 61 to 66% of the chemical exergy input.

The chemical exergy carried by ATP molecules is delivered to a variety of organelles and consumed for the respective functions to be performed by those organelles. Their functions may be categorized into three types as summarized in Fig. 10.2. The first is the contraction of fibre proteins, myosin and actin, both existing within muscular cells, with which various collective movements of the human body are realized. The second is the polymerization of protein molecules from amino acid molecules, which are in part newly absorbed and in part recycled. This is exactly the fundamental process required for maintaining the materially dynamic equilibrium of the whole human body. The third is the operation of sodium-potassium pump embedded within the cellular membranes. These pumps, which keep letting the sodium ions out and the potassium ions in, play a crucial role in maintaining the electric-charge difference between the internal and external spaces of the cellular membranes, relatively negative inside the cells. The dynamic equilibrium provided by the sodium-potassium pumps together with the channels, which are also embedded within the cellular membranes, is the basis for transferring the neural information, from sensory portals to the brain, and also from the brain to motor neurons.

As the result of exergy consumption for these three purposes, "warm" exergy is necessarily generated as a by-product and it amounts to a rate ranging between 0.6 and 2.8 W depending on the environmental temperature. The whole of these three processes, which follows the primary process for the production of ATP, is expressed as exergy balance equation as follows.

[Chemical exergy (ATP)] – [exergy consumption_2] = ["warm" exergy_2] (10.2)

Substituting 27.1 W for chemical exergy contained by ATP molecules into Eq. (10.2) and "warm" exergy of 0.6 or 2.8 W lets us know that exergy consumption_2 ranges from 24.3 to 26.5 W. This is about 90 to 98% of chemical exergy held by ATP molecules.

The exergy consumption processes discussed so far are summarized in Fig. 10.3 for a case of environmental temperature at 15°C. The total of the two-staged exergy consumption is 76.7 (= 51.5 + 25.2) W and the production of "warm" exergy is 5.5 (3.6 + 1.9) W. This "warm" exergy generated becomes the input for the process of thermally dynamic equilibrium at 37°C of the body-core temperature. From the exergetic viewpoint, mitochondria can be regarded as truly micro co-generation systems, since they produce both work and heat and let the surrounding cells and other organelles to function properly while providing thermally appropriate condition for themselves and for other organelles and cells.

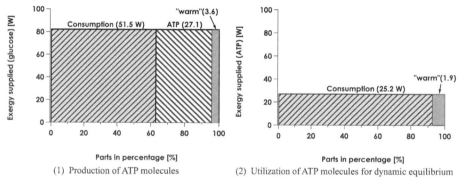

(1) Production of ATP molecules   (2) Utilization of ATP molecules for dynamic equilibrium

Fig. 10.3: Chemical exergy input with glucose, which results in the production of ATP molecules and "warm" exergy within the human body.

10.2.2 Three subsystems

The rate of "warm" exergy produced within the human body corresponds exactly to so-called metabolic-energy generation rate. The human body consumes this "warm" exergy in order to keep the thermally dynamic equilibrium. As shown in Fig. 10.4, let us assume that the human body consists of three sub-systems: the body core, the skin layer, and the clothing ensemble. The whole of human body discharges thermal energy and entropy through three paths of heat transfer, conduction, radiation and convection, and one other path as mass transfer, evaporation. The three subsystems at respective temperatures of T_{cr}, T_{sk} and T_{cl} [K] are in the state of thermally dynamic equilibrium, since these temperatures are, in general, different from the outdoor environmental temperature, T_o [K]. The pressures of the three subsystems are assumed to be the same as the surrounding atmospheric pressure, that is, the whole of the human body is in the state of mechanical equilibrium.

Within the body-core subsystem, there are inflow and outflow of moist air through the nose, trachea, and bronchi. Since the exhaled air is more humid than inhaled air, we assume that there exists metabolic water generation, whose amount

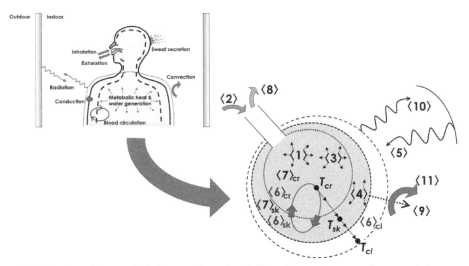

Fig. 10.4: Overall human-body heat transfer and modelling of three sub-systems for exergy balance.

corresponds to the discharged amount of water vapour in the course of respiration. This is in parallel to metabolic heat generation. The generated liquid water and the exhaled water vapour carry "warm" and "wet" exergies since their temperature and humidity are usually higher than the outdoor temperature and humidity.

In addition to these moist-air input and output as mass transfer, there are also the inflow and outflow of blood between the body-core and the skin-layer subsystems as indicated in Fig. 10.4. This is active transport of heat by blood circulation. There is also passive transport of heat, that is, heat conduction from the body-core subsystem at the temperature of T_{cr} to the skin-layer subsystem at T_{sk}, since the skin-layer subsystem encloses the whole of body-core system and both subsystems are continuous with each other. With both active and passive transports, there is exergy delivery and transfer.

The skin-layer subsystem discharges liquid water more or less in the course of sweat secretion and lets it evaporate into the surrounding space, while admitting the heat from the body core and releasing its portion as sensible heat into the clothing ensemble. This sensible heat released by the skin-layer subsystem is transferred partly by conduction and partly by radiation and convection to the clothing-ensemble subsystem. The clothing-ensemble subsystem absorbs long-wavelength radiation emitted by the surrounding surfaces of walls, windows, ceiling, and floor and also short-wavelength radiation depending on the time and spatial position of the human body, while at the same time it discharges heat by radiation and convection into the surrounding space.

In general, the exergy balance equation for a system can be expressed as follows, as described in Chapters 7 and 9: [exergy input] – [exergy consumption] = [exergy stored] + [exergy output]. We apply this general rule to each of the three subsystem and express their respective exergy balance equations. For this purpose, we here use the symbols shown in Fig. 10.4, from $\langle 1 \rangle$ to $\langle 5 \rangle$, then $\langle 6 \rangle_{cr}$, $\langle 6 \rangle_{sk}$, $\langle 6 \rangle_{cl}$, $\langle 7 \rangle_{cr}$, $\langle 7 \rangle_{sk}$, and further from $\langle 8 \rangle$ to $\langle 11 \rangle$, altogether fourteen symbols to express either exergy input, consumption, stored, or output.

$\langle 1 \rangle$, $\langle 3 \rangle$ and $\langle 4 \rangle$ are metabolic exergy generation (in Fig. 10.4, they are expressed with radiating arrows); $\langle 2 \rangle$, $\langle 8 \rangle$ and $\langle 9 \rangle$ are the transport of exergy by moist air; $\langle 5 \rangle$ and $\langle 10 \rangle$ are radiant exergy emissions; $\langle 11 \rangle$ is convective exergy transfer; $\langle 6 \rangle_{cr}$, $\langle 6 \rangle_{sk}$, and $\langle 6 \rangle_{cl}$ represent exergy consumption within respective subsystems; and $\langle 7 \rangle_{cr}$ and $\langle 7 \rangle_{sk}$ are the changes in exergy stored by body-core and skin-layer subsystems, respectively. There is no term as $\langle 7 \rangle_{cl}$, since clothing-ensemble subsystem is assumed to have no heat capacity.

In addition to these fourteen exergy terms, we need to consider the following four thermal exergy flows. The first two of them, X_{bld_sk} and X_{bld_cr}, are "warm" exergy carried by blood circulated from skin-layer to body-core and that from body-core to skin-layer subsystems, respectively. The other two are by passive transport: "warm" exergy flow by conduction from the body-core towards the skin-layer subsystem, X_{cond_cr}; and that from the skin-layer towards the clothing-ensemble subsystem, X_{cl_sk}.

The details of these eighteen, fourteen plus four, terms are given in Table 10.1. The variables used in respective formulae in Table 10.1 are given in Table 10.2. All of these formulae are derived from what was discussed in Chapters 7 and 8. Formulae $\langle 1 \rangle$, X_{cond_cr}, $\langle 7 \rangle_{cr}$, $\langle 7 \rangle_{sk}$, X_{cl_sk}, and $\langle 11 \rangle$ have the same forms as those appeared in 7.6.2. Formulae $\langle 2 \rangle$, $\langle 3 \rangle$, $\langle 4 \rangle$, $\langle 8 \rangle$ and $\langle 9 \rangle$ consist of two parts: thermal and moisture-related parts. The former parts are given by the same form as Eq. (7.33) and the latter parts are derived from the procedure discussed in 8.4.1, 8.4.2, and 8.5. Formulae $\langle 5 \rangle$ and $\langle 10 \rangle$ are derived from the procedure described in 7.6.1 and $\langle 6 \rangle_{cr}$, $\langle 6 \rangle_{sk}$, and $\langle 6 \rangle_{cl}$ are derived from what was described in 7.5. X_{bld_sk} and X_{bld_cr} are also given in the same form as Eq. (7.33).

Note that energy and entropy to be emitted at the environmental temperature are considered for deriving the radiant-exergy formulae of $\langle 5 \rangle$ and $\langle 10 \rangle$ as discussed in Chapter 7. Note also that an amount of dry air to be dispersed with the evaporated sweat is considered for deriving the moist-air exergy formulae of $\langle 4 \rangle$ and $\langle 9 \rangle$ as discussed in Chapter 8. Figure 10.5 depicts briefly these thoughts on radiant and moist-air exergy calculation.

We assume that the boundary-surface temperature of human-body system is represented by the average clothing-ensemble temperature. Therefore, as can be seen in formulae $\langle 10 \rangle$ and $\langle 11 \rangle$ in Table 10.1, thermal exergy outflow by radiation and convection from the human body includes the clothing-ensemble temperature. On the other hand, the water-vapour pressure to be taken for the calculation of outgoing "wet" or "dry" exergy of moist air from the skin-layer subsystem, formula $\langle 9 \rangle$ in Table 10.1, is assumed to be the water-vapour pressure of room air, to which the evaporation of liquid water as sweat takes place.

Exergy balance equations of the respective three subsystems can be expressed with these eighteen terms so far, explained as follows. At the body-core subsystem,

$$\left\{ \langle 1 \rangle + \langle 2 \rangle + \langle 3 \rangle + X_{bld\_sk} \right\}_{in} - \langle 6 \rangle_{cr}$$
$$= \langle 7 \rangle_{cr} + \left\{ \langle 8 \rangle + X_{cond\_cr} + X_{bld\_cr} \right\}_{out} \tag{10.3}$$

Table 10.1: The mathematical formulae of the respective terms in Eqs. (10.3)–(10.5).

$\langle 1 \rangle$:	The rate of "warm" exergy generation by metabolism	$(1-\dfrac{T_o}{T_{cr}})q_{met}$
$\langle 2 \rangle$:	The rate of "warm/cool" and "wet/dry" exergies carried by the inhaled humid air	$V_{in}\left[\begin{array}{l}\left\{c_{pa}(\dfrac{\mathfrak{M}_a}{R\,T_{ra}})(P-p_{vr})+c_{pv}(\dfrac{\mathfrak{M}_{vo}}{R\,T_{ra}})p_{vr}\right\}\left\{(T_{ra}-T_o)-T_o\ln\dfrac{T_{ra}}{T_o}\right\}\\ +\dfrac{T_o}{T_{ra}}\left\{(P-p_{vr})\ln\dfrac{P-p_{vr}}{P-p_{vo}}+p_{vr}\ln\dfrac{p_{vr}}{p_{vo}}\right\}\end{array}\right]$
$\langle 3 \rangle$:	The rate of "warm" and "wet" exergies of liquid water generated in the core by metabolism	$V_{w-core}\rho_w\left[\begin{array}{l}c_{pw}\left\{(T_{cr}-T_o)-T_o\ln\dfrac{T_{cr}}{T_o}\right\}\\ +\dfrac{R}{\mathfrak{M}_{vo}}T_o\ln\dfrac{p_{vs}(T_o)}{p_{vo}}\end{array}\right]$
X_{bld_sk}:	The rate of "warm" exergy delivered from skin layer to body core	$C_{bld}\left\{(T_{sk}-T_o)-T_o\ln\dfrac{T_{sk}}{T_o}\right\}$
X_{cond_cr}:	The rate of "warm" exergy conduction from body core to skin layer	$(1-\dfrac{T_o}{T_{cr}})K(T_{cr}-T_{sk})$
X_{bld_cr}:	The rate of "warm" exergy delivered from body core to skin layer	$C_{bld}\left\{(T_{cr}-T_o)-T_o\ln\dfrac{T_{cr}}{T_o}\right\}$
$\langle 4 \rangle$:	The rate of "warm/cool" and "wet/dry" exergies of the sum of liquid water generated in the body shell by metabolism and dry air to let the liquid water disperse	$V_{w-sk}\rho_w\left[\begin{array}{l}c_{pw}\left\{(T_{sk}-T_o)-T_o\ln\dfrac{T_{sk}}{T_o}\right\}\\ +\dfrac{R}{\mathfrak{M}_{vo}}T_o\left\{\ln\dfrac{p_{vs}(T_o)}{p_{vo}}+\dfrac{P-p_{vr}}{p_{vr}}\ln\dfrac{P-p_{vr}}{P-p_{vo}}\right\}\end{array}\right]$
$\langle 5 \rangle$:	The rate of "warm/cool" radiant exergy absorbed by the whole of skin and clothing surfaces	$f_{eff}f_{cl}\displaystyle\sum_{j=1}^{N}a_{pj}\varepsilon_{cl}h_{rb}\dfrac{(T_j-T_o)^2}{(T_j+T_o)}$
$\langle 6 \rangle_{cr}$:	Exergy-consumption rate within body core	$\sigma_{cr}T_o$
$\langle 6 \rangle_{sk}$:	Exergy-consumption rate within skin layer	$\sigma_{sk}T_o$
$\langle 6 \rangle_{cl}$:	Exergy-consumption rate within clothing ensemble	$\sigma_{cl}T_o$
$\langle 7 \rangle_{cr}$:	The rate of "warm" exergy stored in the body core	$Q_{cr}(1-\dfrac{T_o}{T_{cr}})\dfrac{dT_{cr}}{dt}$
$\langle 7 \rangle_{sk}$:	The rate of "warm" exergy stored in the skin layer	$Q_{sk}(1-\dfrac{T_o}{T_{sk}})\dfrac{dT_{sk}}{dt}$
X_{cl_sk}:	The rate of "warm" exergy conduction from skin layer to clothing ensemble	$(1-\dfrac{T_o}{T_{sk}})\left(\dfrac{1}{R_{cl}}\right)(T_{sk}-T_{cl})$
$\langle 8 \rangle$:	The rate of "warm" and "wet" exergies carried by the exhaled moist air	$V_{out}\left[\begin{array}{l}\left\{c_{pa}(\dfrac{\mathfrak{M}_a}{R\,T_{cr}})(P-p_{vs}(T_{cr}))+c_{pv}(\dfrac{\mathfrak{M}_{vo}}{R\,T_{cr}})p_{vs}(T_{cr})\right\}\left\{(T_{cr}-T_o)-T_o\ln\dfrac{T_{cr}}{T_o}\right\}\\ +\dfrac{T_o}{T_{cr}}\left\{(P-p_{vs}(T_{cr}))\ln\dfrac{P-p_{vs}(T_{cr})}{P-p_{vo}}+p_{vs}(T_{cr})\ln\dfrac{p_{vs}(T_{cr})}{p_{vo}}\right\}\end{array}\right]$

Table 10.1 contd. ...

...Table 10.1 contd.

$\langle 9 \rangle$:	The rate of "warm/cool" exergy of the water vapour originating from the sweat and "wet/dry" exergy of the humid air containing the evaporated sweat	$V_{w-sk}\rho_w \left[c_{pv}\left\{ (T_{cl}-T_o)-T_o\ln\frac{T_{cl}}{T_o}\right\} + \dfrac{R}{\mathfrak{M}_{w}}T_o\left\{ \ln\frac{p_{vr}}{p_{vo}}+\frac{P-p_{vr}}{p_{vr}}\ln\frac{P-p_{vr}}{P-p_{vo}}\right\} \right]$
$\langle 10 \rangle$:	The rate of "warm/cool" radiant exergy discharged from the clothing ensemble	$f_{eff}f_{cl}\varepsilon_{cl}h_{rb}\dfrac{(T_{cl}-T_o)^2}{(T_{cl}+T_o)}$
$\langle 11 \rangle$:	The rate of "warm/cool" exergy transferred by convection from the clothing ensemble	$f_{cl}h_{ccl}(T_{cl}-T_{ra})(1-\dfrac{T_o}{T_{cl}})$

At the skin-layer subsystem,

$$\left\{ \langle 4 \rangle + X_{cond\_cr} + X_{bld\_cr}\right\}_{in} - \langle 6 \rangle_{sk}$$

$$= \langle 7 \rangle_{sk} + \left\{ X_{cl\_sk}+\langle 9 \rangle + X_{bld\_sk}\right\}_{out} \tag{10.4}$$

At the clothing-ensemble subsystem,

$$\left\{ \langle 5 \rangle + X_{cl\_sk}\right\}_{in} - \langle 6 \rangle_{cl} = \left\{ \langle 10 \rangle + \langle 11 \rangle\right\}_{out} \tag{10.5}$$

Combining all of these three equations brings about the whole human-body exergy balance equation, which was originally given through the series of previous studies (Shukuya 2013).

$$\left\{ \sum_{i=1}^{5}\langle i \rangle\right\}_{in} - \left\{ \langle 6 \rangle_{cr}+\langle 6 \rangle_{sk}+\langle 6 \rangle_{cl}\right\} = \left\{ \langle 7 \rangle_{cr}+\langle 7 \rangle_{sk}\right\} + \left\{ \sum_{i=8}^{11}\langle i \rangle\right\}_{out} \tag{10.6}$$

In order to make it possible to calculate all of the formulae given in Table 10.1, we need to assume the following eight variables as the human-body and surrounding boundary conditions: (1) metabolic energy generation rate, q_{met}; (2) average surrounding surface temperature (mean radiant temperature), $T_{MRAD}\left(=\sum_{j=1}^{N}f_jT_j\right)$; (3) room air temperature, T_{ra}; (4) room air relative humidity, φ_{ra}; (5) resistance of clothing ensemble, R_{cl}; (6) surrounding air velocity, v_a; (7) outdoor air temperature, T_o; and (8) outdoor air relative humidity, φ_o.

Among these eight variables, four variables, T_{MRAD}, T_{ra}, φ_{ra}, and v_a specify the indoor condition, and two variables, q_{met} and R_{cl}, specify the state of human body. All of them affect, more or less, the values of internal thermal characteristics: C_{bld}, V_{w-sk}, Q_{cr} and Q_{sk}, and three subsystem temperatures: T_{cr}, T_{sk}, and T_{cl}. Therefore, these internal thermal characteristics, together with three subsystem temperatures, have to be simultaneously determined by solving the set of nonlinear energy-balance equations for the three subsystems. The algorithm for this calculation is described in next section.

Once all these values, C_{bld}, V_{w-sk}, Q_{cr}, Q_{sk}, T_{cr}, T_{sk} and T_{cl}, are given, then we can determine each value of the formulae in Table 10.1, except the three values of exergy

Table 10.2: The mathematical symbols used in Table 10.1.

Each of the terms in Table 10.1 is expressed for one squared-meter of human-body surface.

The symbols used in the formulae from the top to the bottom in Table 10.1 denote as follows.

T_o :   Outdoor air temperature as environmental temperature for exergy calculation [K].

T_{cr} :   Body-core temperature [K]. Its infinitesimal change is dT_{cr}.

q_{met} :   Metabolic energy generation rate [W/m^2].

V_{in} :   Volumetric rate of inhaled air [(m^3/s)/m^2]. It is given by $V_{in} \approx 1.2 \times 10^{-6} q_{met}$.

C_{pa} :   Specific heat capacity of dry air [J/(kg K)] (= 1005).

$\mathfrak{M}_a$:   Molar mass of dry air [kg/mol] (= 28.97 × 10^{-3}).

R :   Gas constant [J/(mol K)] (= 8.314).

T_{ra} :   Room air temperature [K].

P :   Atmospheric air pressure [Pa] (= 101325).

p_{vr} :   Water-vapour pressure in the room space [Pa].

C_{pv} :   Specific heat capacity of water vapour [J/kg K] (= 1846).

$\mathfrak{M}_m$:   Molar mass of water molecules [kg/mol] (= 18.015 × 10^{-3}).

p_{vo} :   Water-vapour pressure of the outdoor air [Pa].

V_{w-core} :   Volumetric rate of liquid water generated in the body core, which turns into water vapour and is exhaled through the nose and the mouth [(m^3/s)/ m^2].
It is given by $V_{w-core} \approx V_{in}(0.029 - 0.049 \times 10^{-4} p_{vr})$.

ρ_w :   Density of liquid water [kg/m^3] (= 1000).

C_{pw} :   Specific heat capacity of liquid water [J/(kg K)] (= 4186).

$p_{vs}(T_o)$:   Saturated water-vapour pressure at outdoor air temperature [Pa].

C_{bld} :   Blood circulation rate between body core and skin layer [W/(m^{2}K)].

T_{sk} :   Skin temperature [K]. Its infinitesimal change is dT_{sk}.

K :   Conductance between body core and skin layer [W/(m^{2}K)] (= 5.28).

Table 10.2 contd. ...

...Table 10.2 contd.

$V_{w\text{-}sk}$:   Volumetric rate of liquid water generated at the skin layer as sweat [(m³/s)/ m²]. It can determined from the procedure described in the next section 10.3.

f_{eff} :   The ratio of the effective area of human body for radiant-heat exchange to the surface area of human body with clothing (= 0.696 ~ 0.725).

f_d :   The ratio of human body area with clothing to the naked human body area (= 1.05 ~ 1.5); the thicker the cloth is, the larger the value of f_d is.

q_{pj} :   Absorption coefficient between the human body surface and the surrounding surface denoted by j [dimensionless]. It can be assumed in most cases to be equal to configuration factor, the ratio of diffuse radiation incident on the human body to the diffuse radiation emitted from surface j.

ε_{cl} :   Emittance of clothing surface [dimensionless]. Its value is usually higher than 0.9.

h_{rb} :   Radiative heat-transfer coefficient of the perfect-black surface [W/(m²K)] (= 5.7 ~ 6.3).

T_j :   Temperature of wall surface j [K].

σ_{cr} :   Entropy-generation rate within the body core [W/(m²K)].

σ_{cr} :   Entropy-generation rate within the skin layer [W/(m²K)].

σ_{cr} :   Entropy-generation rate within the clothing ensemble [W/(m²K)].

Q_{cr} :   Heat capacity of body core [J/(m²K)].

Q_{sk} :   Heat capacity of skin layer [J/(m²K)].

t :   Time [s]. Its infinitesimal change is dt.

R_{cl} :   Overall resistance of clothing ensemble [m²K/W].

T_{cl} :   Clothing-ensemble temperature [K].

V_{out} :   Volumetric rate of exhaled air [(m³/s)/m²]. It is assumed to be equal to V_{in}.

$p_{vs}(T_{cr})$:   Saturated water-vapour pressure at body-core temperature [K].

h_{ccl} :   Convective heat-transfer coefficient over clothed body-surface [W/(m²K)].

---

consumption rates, $\langle 6 \rangle_{cr}$, $\langle 6 \rangle_{sk}$, and $\langle 6 \rangle_{cl}$. In order to determine $\langle 6 \rangle_{cr}$ and $\langle 6 \rangle_{sk}$, we solve Eq. (10.3) for $\langle 6 \rangle_{cr}$ and Eq. (10.4) for $\langle 6 \rangle_{sk}$, both by substituting all other terms other than $\langle 6 \rangle_{cr}$ and $\langle 6 \rangle_{sk}$ into these equations.

In the case of $\langle 6 \rangle_{cl}$, we can determine its value by the following manner. Exergy consumption within the clothing ensemble system emerges in two courses

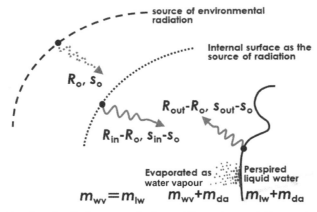

Fig. 10.5: Consideration of radiant energy and entropy at the environmental temperature, R_o and s_o, and an amount of dry air, m_{da}, to be mutually dispersed with evaporated sweat.

of phenomena: one is the heat conduction from skin-layer subsystem to clothing-ensemble system and the other is the absorption of radiation at the clothing-ensemble subsystem. Denoting these two types of exergy consumption as $X_{C\_cl\_cond}$ and X_{C_rad}, respectively,

$$\langle 6 \rangle_{cl} = X_{C\_cl\_cond} + X_{C\_rad}. \tag{10.7}$$

Referring to the discussion we had in 7.6.2, $X_{C\_cl\_cond}$ can be given by the following equation

$$X_{C\_cl\_cond} = \left(\frac{1}{R_{cl}} \right) \frac{\left(T_{sk} - T_{cl} \right)^2}{T_{sk} T_{cl}} T_o. \tag{10.8}$$

On the other hand, X_{C_rad} can be given by the following equation.

$$X_{C\_rad} = \varepsilon \sigma T_{cl}^3 \left\{ \frac{1}{3} + \left(\frac{T_{MRAD}}{T_{cl}} \right)^3 \left(\frac{T_{MRAD}}{T_{cl}} - \frac{4}{3} \right) \right\} T_o, \tag{10.9}$$

referring to Eq. (9.52) in 9.4.3.

Appendix 10.A shows the VBA codes for the calculation of human-body exergy balance based on the internal thermal characteristics and three-subsystem temperatures to be obtained from the procedure described in the next section.

10.3 Internal thermal characteristics and three-subsystem temperatures

10.3.1 Energy balance equations to be solved

Determination of respective temperatures of the three subsystems, T_{cr}, T_{sk}, and T_{cl}, together with sweat secretion rate, V_{w-sk}, heat capacities of body core, Q_{cr}, and skin layer, Q_{sk}, blood-flow rate between body core and skin layer, C_{bld}, is made by solving the following so-called "two-node model" of energy balance equations developed originally by Gagge et al. (1971, 1973, 1986) for the calculation of body-core, skin-layer and clothing ensemble temperature of the aforementioned three subsystems.

In general, energy balance equations are expressed in the form of [Energy input] = [Energy stored] + [Energy output]. This also applies to the energy balance equations in the two-node model. At the "core" subsystem,

[Metabolic thermal energy generation as input] =

[Thermal energy stored in the body core] +

[Thermal energy transferred into the surrounding space by breathing] +

[Thermal energy transported by blood circulation towards skin layer]. (10.10)

At the "skin-layer" subsystem,

[Thermal energy transported from the body core to skin layer] =

[Thermal energy stored in the skin layer] +

[Heat transfer to the clothing ensemble] +

[Thermal energy dispersed by the evaporation of water]. (10.11)

Since the second term of the right-hand side of Eq. (10.11), [Heat transfer to the clothing ensemble], has to be equal to the sum of heat transfer to the surrounding space by radiation and convection,

[Heat transfer to the clothing ensemble] =

[Net radiant heat transfer from the clothing surface] +

[Convective heat transfer from the clothing surface]. (10.12)

According to Gagge et al. (1971), Eqs. (10.10), (10.11) and (10.12) are expressed with mathematical symbols as follows

$$q_{met} = Q_{cr}\frac{dT_{cr}}{dt} + (c_{res} + e_{res}) + K^*(T_{cr} - T_{sk}), \tag{10.13}$$

$$K^*(T_{cr} - T_{sk}) = Q_{sk}\frac{dT_{sk}}{dt} + \frac{1}{R_{cl}}(T_{sk} - T_{cl}) \\ + w(F_{pcl}f_{cl}l_r h_c)\{p_{sks}(T_{sk}) - p_a(T_{ra})\} \tag{10.14}$$

$$\frac{1}{R_{cl}}(T_{sk} - T_{cl}) = f_{cl}h_r(T_{cl} - T_{MRAD}) + f_{cl}h_{ccl}(T_{cl} - T_{ra}). \tag{10.15}$$

These equations are expressed for one-squared metre surface area of the human body. All variables appearing in these three equations are given as follows.

q_{met} is the metabolic energy emission rate [W/m²] and given by the following equation.

$$q_{met} = 58.2M_{et}(1 - \eta) + q_{shiv}, \tag{10.16}$$

where M_{et} is dimensionless rate of metabolic energy emission: $M_{et} = 1.0$ for sedentary posture, in which the average human body emits heat at the rate of 58.2 W/m²;

M_{et} = 0.7 for lying on bed; M_{et} = 2.0 for walking at 3 km/h of speed; and so on. η is the ratio of external work to the total metabolic thermal energy emission rate: $\eta = 0$ for $M_{et} < 1.4$, $\eta = 0.1$ for $1.4 \leq M_{et} < 3$, $\eta = 0.2$ for $3 \leq M_{et}$. q_{shiv} is the rate of shivering thermogenesis, which occurs under severe cold environmental condition and is given by an empirical formula as

$$q_{shiv} = 19.4(T_{sk\_set} - T_{sk})(T_{cr\_set} - T_{cr}), \tag{10.17}$$

where T_{sk_set} = 33.7 + 273.15 [K], T_{cr_set} = 36.8 + 273.15 [K]. These values are the set-point temperature kept in the hypothalamus that orders the human body to take physiological reaction for thermal homeostasis. If $T_{sk\_set} - T_{sk} \leq 0$ or $T_{cr\_set} - T_{cr} \leq 0$, then the value of q_{shiv} = 0. In other words, only when $T_{sk} < T_{sk\_set}$ and $T_{cr} < T_{cr\_set}$, the shivering thermogenesis occurs. This implies that the average skin layer temperature lower than 33.7°C together with the body-core temperature lower than 36.8°C is below the critical limit for the human body.

Since Eqs. (10.13) and (10.14) are expressed for the body surface area of 1 m², the variables, Q_{cr} and Q_{sk}, are in the unit of J/(m²K). dT_{cr} and dT_{sk} are infinitesimal increments in the body-core and skin layer temperatures. The blood flow rate between the body core and the skin layer varies with environmental conditions and it affects the values of Q_{cr} and Q_{sk}, both of which are determined by the ratio of skin-layer mass to the total body mass, α_{sk}, as follows

$$Q_{cr} = (1 - \alpha_{sk})\frac{M_{body}}{A_{body}}c_{p\_body}, \tag{10.18}$$

$$Q_{sk} = \alpha_{sk}\frac{M_{body}}{A_{body}}c_{p\_body}, \tag{10.19}$$

where M_{body} is body mass [kg], A_{body} is human body surface area [m²], and c_{p_body} is specific heat capacity of human body, which is 3490 J/(kgK).

The value of α_{sk} is determined by the following empirical formula.

$$\alpha_{sk} = 0.0418 + \frac{0.745}{v_{bl}^* + 0.585}, \tag{10.20}$$

where v_{bl}^* is blood flow rate between the body core and the skin layer [L/(m²h)]. Equation (10.20) indicates that the value of α_{sk} becomes larger as the blood flow rate decreases. This results in a decrease of skin-layer temperature. The variation of blood flow rate between the body core and the skin layer is in order for thermoregulation to keep the body-core temperature remain unchanged. In other words, the human body has a characteristic that it can control the sizes of body core and skin layer in response to the change of thermal environmental conditions in his or her vicinity. This feature is reflected as the variable body-core and skin-layer heat capacities.

Human-body surface area, A_{body}, can be calculated either from Dubois' formula, A_{body} = $0.007184H_{body}^{0.725}M_{body}^{0.425}$ or from Fujimoto's formula, A_{body} = $0.008883H_{body}^{0.663}M_{body}^{0.444}$. Two variables in these formulae, H_{body} and M_{body}, are body height in the unit of centimetre and body mass in kilogram.

The blood flow rate between body core and skin layer, v_{bl}^*, varies depending on the condition characterized by the body-core and the skin-layer temperatures, T_{cr} and T_{sk}, themselves and two set-point temperatures, T_{cr_set} and T_{sk_set}, described for Eq. (10.17).

$$v_{bl}^* = \frac{6.3 + c_{dil}\left(T_{cr} - T_{cr\_set}\right)}{1 + \sigma_{tr}\left(T_{sk\_set} - T_{sk}\right)}, \tag{10.21}$$

For the conditions of $(T_{cr} - T_{cr\_set}) \leq 0$, the value of numerator is taken to be 6.3, and for the conditions of $(T_{sk\_set} - T_{sk}) \leq 0$, the value of denominator is taken to be 1.

Two factors, c_{dil} and σ_{tr}, in Eq. (10.21) are c_{dil} = 75 ~ 225, σ_{tr} = 0.25 ~ 0.75, respectively. According to Gagge et al. (1973), which values of c_{dil} and σ_{tr} are taken do not have much influence on the final result of body-core and skin-layer temperatures under a constant environmental condition, but do influence the manner in which the body-core and skin-layer temperatures reach their final values; this implies that, for unsteady-state human-body exergy analysis, we need to be careful in choosing the values of c_{dil} and σ_{tr}. According to a sensitivity analysis made by Schweiker and Shukuya (2015), the values of c_{dil} = 100 and σ_{tr} = 0.25 bring about reasonable results of calculated skin temperature which fit very well with a set of measured skin temperature. In the previous human-body exergy analyses for steady-state conditions, the values of c_{dil} and σ_{tr} were chosen to be c_{dil} = 200, σ_{tr} = 0.5.

The sum of c_{res} and e_{res} represents the rate of thermal-energy discharge due to breathing and each of them is given by the following empirical formula

$$c_{res} = 0.0014(58.2M_{et})(34 - t_{ra}), \tag{10.22}$$

$$e_{res} = 0.0173(58.2M_{et})\{5.87 - 0.01\varphi_a p_{sa}(t_{ra})\}, \tag{10.23}$$

where t_{ra} is the air temperature (= inhaled air temperature) [°C], φ_{ra} is the surrounding air relative humidity [%], and $p_{sa}(t_{ra})$ is the saturated water vapour pressure at temperature t_{ra} in the unit of kPa and is given as

$$p_{sa}(t_{ra}) = 0.001e^{25.89 - \frac{5319}{t_{ra} + 273.15}}. \tag{10.24}$$

Equation (10.24) is the formula derived in Chapter 8, Eq. (8.19).

K^* is the sum of passive and active heat-transfer coefficients between body core and skin layer and is expressed as follows

$$K^* = K + C_{bld} = K + 1.163v_{bl}^*. \tag{10.25}$$

K is passive heat-transfer coefficient (thermal conductance), which is 5.28 W/(m²K), and C_{bld} is active heat-transfer coefficient. K is necessary for the calculation of X_{cond_cr} and C_{bld} for X_{bld_cr} and X_{bld_sk} as shown in Table 10.1. The blood circulation rate, v_{bl}^* is, as denoted in relation to Eq. (10.20), in the unit of L/(m²h). The factor of 1.163, which appeared in front of v_{bl}^*, is to convert the unit of L/(m²h) to that of W/(m²K).

R_{cl} is the clothing thermal resistance [m²K/W]. Thermal conductance of the clothing U_{cl} [W/(m²K)] is

$$U_{cl} = \frac{1}{R_{cl}} = \frac{1}{0.155 \cdot I_{clo}}. \tag{10.26}$$

I_{clo} represents the level of clothing insulation in the unit of "clo". One "clo" equals 0.155 (m²K)/W. I_{clo} is zero for a naked human body, 0.3 for the human body wearing a T-shirt and short pants, 0.5 for a long-sleeved shirt and light trousers, 1.0 for a thick jacket or sweater and wintry trousers, and 2.0 for those clothes with a thick coat for wintry weather.

How much of skin surface gets wetted is determined by the set of following equations

$$w = 0.06 + 0.94 \frac{E_{rsw}}{E_{max}}, \tag{10.27}$$

where w is called skin wettedness (wetness factor), E_{rsw} is thermal-energy discharge rate of secreted sweat, and E_{max} is the maximum limit of thermal-energy discharge rate by evaporation from the skin surface including the effect of clothing; both are in the unit of W/m².

The thermal-energy discharge rate by evaporation of secreted sweat, E_{rsw}, is expressed by the following empirical formula

$$E_{rsw} = 170\left(t_b - t_{b\_set}\right)e^{\frac{t_{sk} - t_{sk\_set}}{10.7}}, \tag{10.28}$$

where $t_b = \alpha_{sk}t_{sk} + (1 - \alpha_{sk})t_{cr}$, and $t_{b\_set} = \alpha_{sk}t_{sk\_set} + (1 - \alpha_{sk})t_{cr\_set}$: t_{sk_set} and t_{cr_set} are the set-point temperature values memorized within the hypothalamus and $t_{sk\_set} = 33.7°C$ and $t_{cr\_set} = 36.8°C$, both of which also appeared in Eq. (10.17) in relation to shivering thermogenesis, and t_{sk} and t_{cr} are skin-layer and body-core temperatures in the unit of Celsius. If $t_b - t_{b\_set} \leq 0$, then $E_{rsw} = 0$. In other words, the sweat secretion occurs when $t_{b\_set} < t_b$, in which case $t_{sk\_set} < t_{sk}$.

Even if there is no sweat perceived, that is, $E_{rsw} = 0$, the skin wetness factor, w, does not become zero. The minimum value of w being 0.06 implies that some amount of water is always evaporated from the whole of skin surface and its equivalent skin surface area as being totally wet is about 6% of the whole skin surface area, while 94% is totally dry. Namely, there is passive discharge of heat due to evaporation from the skin surface even when the skin surface is perceived not wet. On the other hand, sweat secretion can be regarded to be performed for active discharge of heat due to evaporation.

The maximum limit of thermal-energy discharge rate by evaporation, E_{max}, is expressed by

$$E_{max} = \left(F_{pcl}f_{cl}l_r h_{ccl}\right)\left\{p_{sks}\left(T_{sk}\right) - p_a\left(T_{ra}\right)\right\}, \tag{10.29}$$

where F_{pcl} is permeation efficiency factor of the clothing, which is dimensionless, and given by

$$F_{pcl} = \frac{1}{R_{e,cl}f_{cl}l_r h_{ccl} + 1}, \tag{10.30}$$

where $R_{e,cl}$ is clothing moisture resistance [m²kPa/W] and f_{cl} is the ratio of overall clothing surface area to naked body surface area, which is given as $f_{cl} = 1 + 0.15I_{clo}$ or $f_{cl} = 1 + 0.3I_{clo}$. I_r is the ratio of moisture transfer coefficient to convective heat transfer coefficient, which is called Lewis Relationship and can be taken to be 16.5°C/kPa as a constant under the atmospheric condition of 101.3 kPa as also described in Chapter 8.

h_{ccl} is convective heat transfer coefficient [W/(m²K)] and can be calculated from the following empirical formulae depending on the given conditions

$$\begin{cases} h_{ccl} = 5.66\left(\dfrac{q_{met}}{58.2} - 0.85\right)^{0.39}, \\ h_{ccl} = 8.6v_a^{0.53} \end{cases} \tag{10.31}$$

where v_a is air velocity in the vicinity of the human body [m/s]. The larger value of h_{ccl} calculated from the two formulae is chosen.

The clothing moisture resistance, $R_{e,cl}$, is related to the clothing insulation with the proportional constant of $(l_r i_c)^{-1}$

$$R_{e,cl} = \frac{R_{cl}}{l_r i_c}, \text{ that is, } i_c = \frac{R_{cl}}{l_r R_{e,cl}}, \tag{10.32}$$

where i_c is the ratio of thermal resistance to moisture resistance with respect to clothing. The smaller value of i_c implies that the clothing is less permeable. According to the previous studies done by Nishi and Gagge (1970), F_{pcl} can be determined by the following formula

$$F_{pcl} = \frac{1}{1 + 0.143I_{clo}f_{cl}h_{ccl}}. \tag{10.33}$$

This equation implies that $i_c = 1.084$ is assumed. There is also another formula that may be used for less permeable clothing.

$$F_{pcl} = \frac{1}{1 + 0.388I_{clo}f_{cl}h_{ccl}}. \tag{10.34}$$

This equation implies that $i_c = 0.4$ is assumed.

$p_{sks}(T_{sk})$, a variable appeared inside the brace of Eq. (10.29), is saturated water-vapour pressure at the skin temperature, T_{sk}, and $p_a(T_{ra})$, the other variable that appeared in Eq. (10.29), is partial water-vapour pressure of the surrounding air, which is equal to $0.01\varphi_a P_{sa}(t_{ra})$, where φ_{ra} is surrounding-air relative humidity in the unit of per cent and $p_{sa}(t_{ra})$ is the saturated water vapour pressure at temperature t_{ra} [°C].

Referring to Eq. (10.27), if the value of E_{rsw} is larger than that of E_{max}, then w turns out be $1.0 \le w$; this is a case that the whole body gets soaked and the amount of secreted sweat corresponding to $\left(\dfrac{E_{rsw}}{E_{max}} - 1.0\right)$ does not contribute to decreasing the skin surface temperature. Therefore, the skin wetness factor, w, is $0.06 \le w \le 1.0$.

The last term of Eq. (10.14) may be rewritten as follows, using the relationships given in Eqs. (10.27) and (10.29)

$$w\left(F_{pcl}f_{cl}l_rh_{ccl}\right)\left\{p_{sks}\left(T_{sk}\right)-p_a\left(T_{ra}\right)\right\}=wE_{max}=\left(0.06E_{max}+0.94E_{rsw}\right). \quad (10.35)$$

The last formula in Eq. (10.35) again confirms that the skin surface is considered to be composed of two portions: one portion, 6% of the whole body surface area, is always fully wet and the other portion, 94%, becomes either fully dry or wet depending on the internal and external conditions of the human body.

The volumetric rate of liquid water secreted from the skin surface, V_{w_sk} [(m^{3}g/s)/m^2], to be used for the calculation of formulae $\langle 4 \rangle$ and $\langle 8 \rangle$ in Table 10.1, is given by

$$V_{w\_sk}=\frac{wE_{max}}{\rho_w L_H}, \quad (10.36)$$

where ρ_w is the density of liquid water (= 1000 kg/m^3) and L_H is the latent heat of evaporation for liquid water; we may assume that $L_H = 2450 \times 10^3$ J/kg, which is for 30°C.

Radiative heat transfer coefficient, h_r, can be determined by

$$h_r=f_{eff}\varepsilon_{cl}h_{rb}\approx4.41\times\varepsilon_{cl}, \quad (10.37)$$

where f_{eff} is the ratio of apparent body-surface area to the clothed body-surface area, ε_{cl} is overall emittance of the clothed body-surface, and h_{rb} is radiative-heat transfer coefficient of black-body surface [W/(m^{2}K)], which was introduced in Chapter 7, Eq. (7.44).

Equation (10.15) may be rewritten as follows in the form for the calculation of clothing-ensemble temperature

$$T_{cl}=\frac{\dfrac{1}{R_{cl}}T_{sk}+f_{cl}h_rT_{MRAD}+f_{cl}h_{ccl}T_a}{\dfrac{1}{R_{cl}}+f_{cl}h_r+f_{cl}h_{ccl}}. \quad (10.38)$$

Substitution of Eq. (10.38) into Eq. (10.15) yields

$$\frac{1}{R_{cl}}\left(T_{sk}-T_{cl}\right)=F_{cl}f_{cl}\left(h_r+h_{ccl}\right)\left(T_{sk}-T_{op}\right), \quad (10.39)$$

where
$$F_{cl}=\frac{1}{1+R_{cl}f_{cl}\left(h_r+h_{ccl}\right)}=\frac{1}{1+0.155\times I_{clo}f_{cl}\left(h_r+h_{ccl}\right)}, \quad (10.40)$$

$$T_{op}=\frac{h_rT_{MRAD}+h_{ccl}T_a}{h_r+h_{ccl}}. \quad (10.41)$$

F_{cl} is sensible heat-transfer effectiveness, which is in parallel to F_{pcl} representing permeation efficiency factor that is associated with latent heat-transfer effectiveness. T_{op} is called "operative temperature", which implies the overall average temperature of an imaginary uniform environmental space.

Equation (10.14) may be rewritten as follows by replacing its second term on the right-hand side with Eq. (10.39) as

$$K^* \left(T_{cr} - T_{sk} \right) = Q_{sk} \frac{\mathrm{d} T_{sk}}{\mathrm{d} t} + F_{cl} f_{cl} \left(h_r + h_{ccl} \right) \left(T_{sk} - T_{op} \right)$$
$$+ w \left(F_{pcl} f_{cl} l_r h_{ccl} \right) \left\{ p_{sks} \left(T_{sk} \right) - p_a \left(T_{ra} \right) \right\} \tag{10.42}$$

10.3.2 Calculation of body-core and skin-layer temperatures

The set of Eq. (10.13) and Eq. (10.42) to be solved for T_{cr} and T_{sk} is so-called "two-node model". This set of differential equations is nonlinear, since the variables q_{shiv}, Q_{cr}, K^*, Q_{sk}, and w are dependent on the values of T_{cr} and T_{sk}. Because of this non-linearity, the easiest way to solve these equations is to approximate them with the respective finite-differential equations. Let us take time, t, as a continuous variable to be in a series of discrete time interval Δt, that is, 0, Δt, $2\Delta t$, $3\Delta t$, $\cdots$, $(n-1)\Delta t$, $n\Delta t$, $(n+1)\Delta t$, $\cdots$. During the period of time from $(n-1)\Delta t$ to $n\Delta t$, the finite increase of T_{cr} and T_{sk}, ΔT_{cr} and ΔT_{sk}, are expressed as $\Delta T_{cr} = T_{cr}(n) - T_{cr}(n-1)$ and $\Delta T_{sk} = T_{sk}(n) - T_{sk}(n-1)$, where $T_{cr}(n)$ and $T_{cr}(n-1)$ implies T_{cr} at the time of $n\Delta t$ and $(n-1)\Delta t$, respectively, and $T_{sk}(n)$ and $T_{sk}(n-1)$ at the time of $n\Delta t$ and $(n-1)\Delta t$, respectively. ΔT_{cr} and ΔT_{sk} are the approximation of $\mathrm{d} T_{cr}$ and $\mathrm{d} T_{sk}$.

Then, the next step that we need to do is to make all terms other than $\mathrm{d} T_{cr}$ and $\mathrm{d} T_{sk}$ be approximated at the time, either $n\Delta t$ or $(n-1)\Delta t$. Theoretically speaking, either is possible: if the former is taken, then the equations are called "implicit" type of finite difference equations, and if the latter is taken, then "explicit" type.

Here we take the explicit type of finite differentiation, since the calculation is easier to proceed than "implicit" type. Then, the finite differential equation to be calculated for $T_{cr}(n)$ is expressed as

$$T_{cr}(n) = \left\{ 1 - \frac{\Delta t \cdot K^*(n-1)}{Q_{cr}(n-1)} \right\} T_{cr}(n-1) + \frac{\Delta t}{Q_{cr}(n-1)} \mathbf{Z}_{cr}, \tag{10.43}$$

where
$$\mathbf{Z}_{cr} = q_{met}(n-1) + q_{shiv}(n-1)$$
$$- \left\{ c_{res}(n-1) + e_{res}(n-1) \right\} + K^*(n-1) T_{sk}(n-1) \tag{10.44}$$

$K^*(n-1)$, $Q_{cr}(n-1)$, $q_{met}(n-1)$, $q_{shiv}(n-1)$, $c_{res}(n-1)$, and $e_{res}(n-1)$ represent the values of K^*, Q_{cr}, q_{met}, q_{shiv}, c_{res}, and e_{res} at the time $(n-1)\Delta t$. The factor in front of $T_{cr}(n-1)$ in the 1st term of right-hand side of Eq. (10.43) has to satisfy the condition, $0 \leq 1 - \dfrac{\Delta t \cdot K^*(n-1)}{Q_{cr}(n-1)}$, that is, the value of Δt has to be determined so as to make the following equation be valid:

$$\Delta t \leq \frac{Q_{cr}(n-1)}{K^*(n-1)}. \tag{10.45}$$

The finite differential equation to be calculated for $T_{sk}(n)$ is

$$T_{sk}(n) = \left\{ 1 - \frac{\Delta t \cdot K^*(n-1)}{Q_{sk}(n-1)} - \frac{\Delta t \cdot f_{cl}F_{cl}(h_r + h_{ccl})}{Q_{sk}(n-1)} \right\} T_{sk}(n-1)$$

$$- \frac{\Delta t \cdot w(n-1)(f_{cl}l_r h_{ccl})F_{pcl}}{Q_{sk}(n-1)} p_{sks}(n-1) + \frac{\Delta t}{Q_{sk}(n-1)} Z_{sk}$$, (10.46)

where

$$Z_{sk} = K^*(n-1)T_{cr}(n-1) + f_{cl}(h_r + h_c)clF_{cl}T_{op}(n-1)$$
$$+ w(n-1)(f_{cl}l_r h_{ccl})F_{pcl}p_a(n-1)$$. (10.47)

$Q_{sk}(n-1)$, $w(n-1)$, $p_{sks}(n-1)$, $T_{op}(n-1)$, and $p_a(n-1)$ represent the values of Q_{sk}, w, q_{shiv}, $P_{sks}(T_{sk})$, T_{op}, and $p_a(T_a)$ at the time $(n-1)\Delta t$. In Eq. (10.46), the constraint parallel to the one for Eq. (10.43) is

$$0 \le 1 - \frac{\Delta t \cdot K^*(n-1)}{Q_{sk}(n-1)} - \frac{\Delta t \cdot f_{cl}F_{cl}(h_r + h_{ccl})}{Q_{sk}(n-1)}$$
$$- \frac{\Delta t \cdot w(n-1)(f_{cl}l_r h_{ccl})F_{pcl}}{Q_{sk}(n-1)} \cdot \frac{p_{sks}(n-1)}{T_{sk}(n-1)}$$. (10.48)

Table 10.3 summarizes the whole steps of calculation, with which all of the internal thermal characteristics and three subsystem temperatures necessary for the calculation of the exergy balance of the human body can be determined.

10.3.3 Extended operative temperature—ET* and SET*

The whole of thermal energy dispersion from the human body into the immediate environmental space, q_{out}, which originally appeared in the second and third terms of the right-hand side of Eq. (10.14), can be once again expressed as follows, since we have come to know that Eq. (10.14) may be rewritten as Eq. (10.42)

$$q_{out} = F_{cl}f_{cl}(h_r + h_{ccl})(T_{sk} - T_{op}) + w(F_{pcl}f_{cl}l_r h_{ccl})\{p_{sks}(T_{sk}) - p_a(T_a)\}.(10.49)$$

Equation (10.49) may be rewritten as

$$q_{out} = f_{cl}(h_r + h_{ccl})F_{cl} \left[\begin{array}{c} \left\{ T_{sk} + w\dfrac{h_{ccl}F_{pcl}}{(h_r + h_{ccl})F_{cl}}l_r p_{sk}(T_{sk}) \right\} \\ - \left\{ T_{op} + w\dfrac{h_{ccl}F_{pcl}}{(h_r + h_{ccl})F_{cl}}l_r p_a(T_{ra}) \right\} \end{array} \right] .$$ (10.50)

The 2nd term inside the squared bracket of the above equation can be regarded as an extended operative temperature. This is the uniform temperature of an imaginary space, in which the human body would discharge thermal energy by radiation, convection and evaporation at the same rate as it does in actual environmental space of T_{MRAD}, T_{ra}, φ_{ra} and v_a. This 2nd term inside

Table 10.3: The 26 steps for the calculation of three subsystem temperatures.

1)	Take the given values of T_{cr} and T_{sk} at the time $(n-1)\Delta t$, that is, $T_{cr}(n-1)$ and $T_{sk}(n-1)$.
2)	Take the values of M_{et}, η, I_{clo}, and v_a at the time $(n-1)\Delta t$.
3)	q_{shiv} at the time $(n-1)\Delta t$ is determined from Eq. (10.17).
4)	q_{met} at the time $(n-1)\Delta t$ is determined from Eq. (10.16)
5)	f_{cl} at the time $(n-1)\Delta t$ is determined by $f_{cl} = 1 + 0.15I_{clo}$.
6)	h_r at the time $(n-1)\Delta t$ is determined from Eq. (10.37).
7)	h_{ccl} at the time $(n-1)\Delta t$ is determined from Eq. (10.31).
8)	F_{cl} is determined from Eq. (10.40) using the results of 5), 6) and 7).
9)	F_{pcl} is determined from Eq. (10.33) (or from Eq. (10.34)) using the results of 5) and 7).
10)	Take the given values of T_{MRAD}, T_a and φ_a at the time $(n-1)\Delta t$.
11)	$p_a(n-1)$ is determined from the equation, $0.01\varphi_{ra}p_{sa}(t_{ra})$, by substituting t_{ra}, which is equal to $T_{ra} - 273.15$.
12)	$c_{res}(n-1)$ and $e_{res}(n-1)$ are determined from Eq. (10.22) and Eq. (10.23), respectively.
13)	v^*_{bl} at the time $(n-1)\Delta t$ is determined from Eq. (10.21).
14)	α_{sk} at the time $(n-1)\Delta t$ is determined from Eq. (10.20) using the result of 13).
15)	K^* at the time $(n-1)\Delta t$ is determined from Eq. (10.25) using the result of 13).
16)	$Q_{cr}(n-1)$ and $Q_{sk}(n-1)$. are determined from Eq. (10.18) and Eq. (10.19), respectively.
17)	$T_{cr}(n)$ is calculated from Eq. (10.43) with Eq. (10.44).
18)	$p_{sks}(n-1)$ is determined from Eq. (10.24) substituting the value of $t_{sk}(n-1)$ in the place of t_{ra}, where $t_{sk}(n-1) = T_{sk}(n-1) - 273.15$.
19)	$E_{max}(n-1)$ is determined from Eq. (10.29) using the results of 11) and 18).
20)	$E_{rsw}(n-1)$ is determined from Eq. (10.28) using $T_{sk}(n-1)$ and $T_{cr}(n-1)$.
21)	$w(n-1)$ is determined from Eq. (10.27) using the results of 19) and 20).
22)	$T_{sk}(n)$ is calculated from Eq. (10.46) with Eq. (10.47).

Table 10.3 contd. ...

...Table 10.3 contd.

23)	Take the given values of T_{MRAD}, T_a and φ_a at the time $n\Delta t$.
24)	$T_{cl}(n)$ is calculated from Eq. (10.38).
25)	Prepare the next step of calculation by making $T_{cr}(n)$ obtained in 17) be $T_{cr}(n-1)$ and $T_{sk}(n)$ obtained in 22) be $T_{sk}(n-1)$.
26)	Return to 2).

the squared bracket may be rewritten with the extended operative temperature, T_{eop}, as follows

$$T_{op} + w \cdot i_m \cdot l_r p_a (T_{ra}) = T_{eop} + w \cdot i_m^* \cdot l_r p_a (T_{eop}), \tag{10.51}$$

where

$$i_m = \frac{h_{ccl} F_{pcl}}{(h_r + h_{ccl}) F_{cl}} \quad \text{and} \quad p_a (T_{ra}) = 0.01 \varphi_{ra} p_{sa} (T_{ra}), \tag{10.52}$$

$$i_m^* = \frac{h_{ccl}^* F_{pcl}^*}{(h_r + h_{ccl}^*) F_{cl}^*} \quad \text{and} \quad p_a (T_{eop}) = 0.01 \varphi_{ra}^* p_{sa} (T_{eop}). \tag{10.53}$$

If i_m^* is assumed to be the same as i_m and $p_a(T_{eop})$ is given for $\varphi_{ra}^* = 50\%$, then the value of T_{eop}, which makes both sides of Eq. (10.51) be equal to each other, is called ET^*. If i_m^* is determined so as to be for h_{ccl}^* with $v_a = 0.1$ and $M_{et} = 1.1$, F_{pcl}^* and F_{cl}^* with $I_{clo} = 0.6$, and $p_a(T_{eop})$ for $\varphi_{ra}^* = 50\%$, then the value of T_{eop} is called SET^*.

ET^* is the energy-balance-wise equivalent temperature of an imaginary environmental space, where $T_{MRAD} = T_a$, $\varphi_a = 50\%$, and SET^* is that of another imaginary environmental space, where $T_{MRAD} = T_a$, $\varphi_a = 50\%$ with the condition of seated posture, for which $v_a = 0.1$ m/s, $I_{clo} = 0.6$, and $M_{et} = 1.1$.

In order to determine the value of T_{eop}, it is necessary to solve Eq. (10.51) for T_{eop} as the unknown variable. Since Eq. (10.51) is not linear due to the fact that the saturated vapour pressure is related to the temperature in an exponential form, it requires an iterative calculation to have the value of T_{eop}. This can be done following the steps described in Table 10.4.

10.3.4 Combination of T_{MRAD} and T_{ra} for energy-balance-wise neutral conditions

We can combine the energy balance equations for three subsystems, originally expressed by Eqs. (10.13), (10.14) and (10.15), and thereby come up with the following equation

$$q_{met} = q_{Load} + (c_{res} + e_{res}) + \varsigma h_r (T_{sk} - T_{MRAD}) + \varsigma h_{ccl} (T_{sk} - T_{ra}) + \Gamma_{per}, \tag{10.54}$$

Table 10.4: The steps for the calculation of ET* and SET* as extended operative temperature.

1) Using T_{MRAD}, T_{ra}, $\varphi_{ra} = 50\%$ and also skin wettedness (wetness factor), w, obtained from the calculation described in Table 10.3, calculate the value of the left-hand side of Eq. (10.51). Take this value to be C.

2) Suppose that there is a function, $Y(T_{eop})$, in the form of

$Y = T_{eop} + w \cdot i^{*}_{m} \cdot l_{r}p_{a}(T_{eop}) - C$ and its differential quotient with respect

to T_{eop} as $Z(T_{eop}) = \dfrac{dY}{dT_{eop}} = 1 + w \cdot i^{*}_{m} \cdot l_{r} \times 0.5 \times e^{25.89 - \frac{5319}{T_{eop}}}\left(\dfrac{5319}{T_{eop}^{2}}\right).$

3) Assume an initial value of T_{eop} to be T_{eop_0}.

4) Calculate the values of $Y(T_{eop\_0})$ and $Z(T_{eop\_0})$.

5) Then, using the relationship: $Z\left(T_{eop\_0}\right) = \dfrac{Y\left(T_{eop\_0}\right)}{T_{eop\_0} - T_{eop\_1}}$, which represents

the tangent of function $Y(T_{eop\_0})$ at T_{eop_0}, calculate the value of T_{eop_1} from

the following equation: $T_{eop\_1} = T_{eop\_0} - \dfrac{Y\left(T_{eop\_0}\right)}{Z\left(T_{eop\_0}\right)}.$

6) Calculate $\left| T_{eop\_1} - T_{eop\_0}\right|.$

7) If $\left| T_{eop\_1} - T_{eop\_0}\right| \le \delta$, where $\delta = 0.05 \sim 0.1$, then the value of T_{eop_1} is taken to be the final value of T_{eop}. Otherwise, take T_{eop_1} to be a renewed value of T_{eop_0} and return to 4).

where $\qquad q_{Load} = Q_{cr}\dfrac{dT_{cr}}{dt} + Q_{sk}\dfrac{dT_{sk}}{dt}$, $\varsigma = \dfrac{f_{cl}}{1 + R_{cl}f_{cl}\left(h_{r} + h_{ccl}\right)}$,

$$\Gamma_{per} = wF_{pcl}\left(f_{cl}l_{r}h_{ccl}\right)\left\{p_{sks}\left(T_{sk}\right) - p_{a}\left(T_{a}\right)\right\}.$$

Under the energy-balance-wise neutral conditions, according to Rohles and Nevins (1971) and also Fanger (1970) (Parsons 2002), the skin-layer temperature, T^{*}_{sk}, and the rate of thermal energy dispersion due to the evaporation of sweat secreted from the sweat glands, q^{*}_{per}, are given by

$$T^{*}_{sk} = 35.7 - 0.028q_{met}, \tag{10.55}$$

$$q^{*}_{per} = 0.42(q_{met} - 58.15) \ge 0. \tag{10.56}$$

The evaporation of water from the skin layer other than the sweat glands may be expressed by the following empirical formula

$$q_{skin\_dif} = 3.05 \times 10^3 \{5733 - 6.99q_{met} - p_a(T_a)\}. \tag{10.57}$$

Provided that the sum of q_{per}^* and q_{skin_dif} is expressed as Γ_{per}^*, then Γ_{per}^* corresponds to the term Γ_{per} in Eq. (10.54). Substituting $\Gamma_{per}^* = q_{per}^* + q_{skin\_dif}$ for Γ_{per} and also making $q_{Load} = 0$, Eq. (10.54) turns out to be the following equation.

$$q_{met} = (c_{res} + e_{res}) + \varsigma h_r \left(T_{sk}^* - T_{MRAD}\right) + \varsigma h_{ccl} \left(T_{sk}^* - T_{ra}\right) + \Gamma_{per}^*. \tag{10.58}$$

We may rewrite Eq. (10.58) into an explicit form for the calculation of T_{MRAD}. That is,

$$T_{MRAD} = \frac{1}{\varsigma h_r} \left\{ \varsigma \left(h_r + h_{ccl}\right) T_{sk}^* - \varsigma h_{ccl} T_a - q_{met} + \left(c_{res} + e_{res}\right) + \Gamma_{per}^* \right\}. \tag{10.59}$$

Let us review all the variables appearing on the right-hand side of Eq. (10.59). In so doing, let us express, for example, ς as a function of R_{cl} and h_{ccl} in the form of $\varsigma = \varsigma(R_{cl}, h_{ccl})$. Since we can express $R_{cl} = R_{cl}(I_{cl})$ and $h_{ccl} = h_{ccl}(q_{met}, v_a)$, $\varsigma = \varsigma(R_{cl}, h_{ccl})$ turns out to be expressed as $\varsigma = \varsigma(I_{cl}, q_{met}, v_a)$.

Following this rule, we may write $T_{sk}^* = T_{sk}^*(q_{met})$, $c_{res} = c_{res}(q_{met}, T_{ra})$, $e_{res} = e_{res}(q_{met}, T_{ra}, \varphi_{ra})$, and $\Gamma_{per}^* = \Gamma_{per}^*(q_{met}, p_a) = \Gamma_{per}^*(q_{met}, \varphi_{ra}, T_{ra})$. Then we find that T_{MRAD} can be expressed as

$$T_{MRAD} = T_{MRAD}(q_{met}, T_{ra}, \varphi_{ra}, v_a, I_{clo}). \tag{10.60}$$

Equation (10.60) indicates that we can calculate a value of mean radiant temperature, T_{MRAD}, which provides the energy-balance-wise neutral state of human body with the given combination of T_{ra} together with q_{met}, φ_{ra}, v_a, and I_{clo}.

10.4 Exergy-consumption rate patterns

Based on the theory so far described, we can decipher how the human body behaves exergetically depending on the indoor and outdoor thermal conditions. Here in this section, let us discuss some numerical examples obtained from the calculation under quasi steady-state thermal conditions.

10.4.1 Winter cases

Figure 10.6 shows the relationship between the whole human-body exergy consumption (HBXC) rate, which is exactly expressed by the 2nd term of the left-hand side of Eq. (10.6), and mean radiant temperature (MRT) with a parameter of air temperature being at 14, 18, 22 and 24°C. The values of exergy consumption rate here are those at the moment after a one-hour period of calculation performed at sixty-second intervals. The interval of time can be chosen arbitrarily as far as the two conditions of constraint defined by Eqs. (10.45) and (10.48) are satisfied. The calculation with the interval of time shorter than three minutes were confirmed to give almost the same results as those shown in Fig. 10.6, but if the interval of time is too long, let's say five minutes, then the calculation becomes unstable.

Outdoor air temperature and relative humidity were assumed to be 5°C and 50%, respectively. Indoor thermal conditions, other than mean radiant temperature and air temperature, are as follows: air velocity at 0.1 m/s and relative humidity at 50%. The state of human body is at 1.1 Met, which is mostly for sedentary posture and

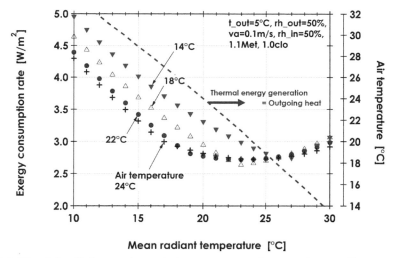

Fig. 10.6: The relationship between human-body exergy consumption rate and mean radiant temperature in winter case. The dashed line drawn represents the combination of mean radiant temperature and air temperature in the case of thermal energy generation being equal to outgoing heat.

occasionally for standing posture, and at 1.0 clo, which represents typical winter clothes.

The dashed line drawn from upper left to lower right represents the condition of energy-balance-wise neutral state of human body as the combination of air temperature shown on the right-hand side of vertical axis and mean radiant temperature given on the horizontal axis. This line is drawn based on the relationship expressed by Eq. (10.60).

The whole HBXC rate decreases as the MRT increases, but it starts increasing gradually at certain values of mean radiant temperature, which differ by the values of air temperature. With air temperature either at 18, 22 or 24°C, the HBXC rate reaches the minimum in the range from 21 to 24°C of MRT. For MRT lower than 20°C, the air temperature has to be increased in order to lower the HBXC rate, but the reduction seems to saturate as the air temperature is raised over 22°C. With this in mind, we can say that increasing MRT makes more sense than increasing air temperature.

According to the energy-balance-wise thought, for example, 16°C of MRT combined with 28°C of air temperature is the same as 22°C with 22°C or 26°C with 18°C. But, it does not mean that the human thermal sensation emerging from these combinations of MRT and air temperature is the same as each other. Let us ask ourselves whether people perceive the three indoor thermal environmental conditions provided by these combinations of MRT and air temperature equally or not? In a well-designed so-called passive solar house, the occupants can perceive pleasant warmth with a sufficient amount of solar heat gain even if the room air temperature is rather low, while on the other hand, in a poorly insulated house, even if the room air temperature is raised up to the value to be more than sufficient, the occupants are still likely to perceive discomfort, for example, due to draught. Once having recalled such indoor thermal conditions in our minds, we may recognize that the difference in sensation and perception may well be related to the difference in the values of HBXC rate rather than the energy balance of the human body.

As was shown in Fig. 7.20, the rate of "warm" radiant exergy emission increases as the wall surface temperature increases. This implies that as the availability of "warm" radiant exergy increases, the HBXC rate becomes smaller. We may say that comfortable warmth must originate from the exposure of human body to sufficient "warm" radiant exergy. Nonetheless, more than enough exposure to radiant exergy is too much, since the HBXC rate increases as MRT increases over 24°C.

In Fig. 10.6, as mentioned in the beginning, the assumption of clothing insulation is constant at 1.0 clo regardless of the increase in MRT. But, in reality, we humans have a habit of reducing the clothing insulation, for example, by taking off a jacket or a sweater, as we are surrounded for one hour or so by the surfaces whose temperature is much higher than the air temperature. If such an action is taken into account, for example, changing the "clo" values from 1.0 to 0.6, then the HBXC rate of 2.84 W/m^2 at MRT of 28°C with the air temperature of 22°C goes down to 2.33 W/m^2; the reduction is 18%. This suggests that the reason why we take off a jacket is to reduce the exergy consumption rate within the human body to get less thermal stress and thereby restore thermal comfort.

Let us take another look at the relationship between the HBXC rate and indoor thermal condition from a different angle as shown in Fig. 10.7. Here, the horizontal axis represents air temperature. Two kinds of circular plots are for MRT at 18°C and 22°C, respectively. We can see that there is the value of air temperature resulting in the smallest HBXC rate with a given value of MRT. Again, we confirm that such minimum value of HBXC rate becomes smaller as MRT increases. If the air temperature becomes higher than the value with which the minimum exergy consumption rate is provided, the HBXC rate tends to increase, but its degree of increase looks saturating as the air temperature reaches 25°C or 30°C.

The reason for such tendency is that the skin wetness indicated with triangular plots starts increasing beyond its minimum at 0.06 as indicated in the right-hand side vertical axis. Again we can confirm that taking off a jacket or a sweater is due to avoiding a further sweat secretion and thereby to keep the dissipation of heat at a reasonable rate. From the exergetic viewpoint, this is exactly in order to reduce the HBXC rate. The increase of skin wetness as a physiological reaction is to make the difference in temperature between the body-core and the skin-layer subsystems be larger for evaporative cooling at the skin surface; this results in an increase of exergy consumption rate. The perspiration of sweat can be regarded as physiological adaptive behaviour; this may be, in other words, regarded as the evaporative cooling system embodied within the human body.

Let us next move on to take a detailed look at the four components of exergy consumption rate. Figure 10.8 shows how these four components vary with the change of MRT. The exergy consumption at the body-core subsystem is almost constant regardless of MRT and it is the smallest among the four components in the range of MRT lower than 20°C. In the range of MRT lower than 23°C, the exergy consumption rate at the skin-layer subsystem stays constant, but in the range of MRT higher than 23°C, it becomes gradually larger as the values of MRT increases. This

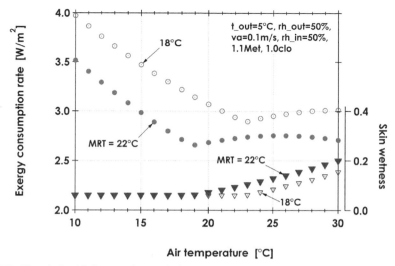

Fig. 10.7: The relationship between human-body exergy consumption rate and air temperature in winter case. The circular plots indicate HBXC rates and the triangular plots skin wetness.

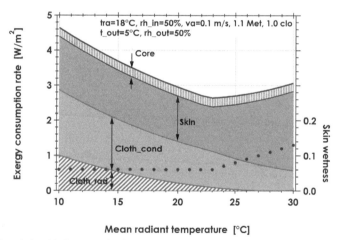

Fig. 10.8: The relationship between the four components of HBXC rate and MRT in a winter case. The round plots indicate the skin wetness.

trend is consistent with the increase of skin wetness in the range of MRT higher than 23°C.

The exergy consumption due to the conduction between the skin layer and clothing ensemble decreases monotonically as MRT increases. The same applies to the exergy consumption due to the absorption of radiant exergy available from the surrounding. The exergy consumption caused by the absorption of radiant exergy is reduced by a factor of ten from the MRT at 10°C to that at 23°C; this is due to an increase of "warm" radiant exergy available from the surrounding surface in the order from 300 mW/m² to 3000 mW/m², which can be known from Fig. 7.20 or from the values to be calculated from Eq. (7.42) or Eq. (7.44).

An increase in surrounding surface temperature results in the corresponding increase in the clothing-ensemble temperature and thereby the exergy consumption taking place between the skin layer and the clothing ensemble is reduced. The enhancement of thermal insulation in building envelope systems has been considered, in general, to be aimed at the reduction of energy use for heating, but it is worth recognizing that appropriate thermal insulation of building envelope systems makes the HBXC rate minimized and thereby to restore thermal comfort rationally. This is taken to be the first priority rather than simply reducing the so-called energy use for heating.

10.4.2 Summer cases

Let us discuss the summer cases in a similar manner to winter cases. Figure 10.9 shows the relationship between HBXC rate and MRT under a summer condition of outdoor air temperature and relative humidity at 32°C and 60%. Four groups of plots: circular, cross, triangular, and plus, are for air velocity of 0.1, 0.3, 0.5 and 0.8 m/s, respectively. The reason for taking different air-velocity values as parameter instead of air temperature is that the air movement plays a crucial role in summer conditions for the creation of pleasant coolness. The air velocity at 0.1 m/s or below corresponds to the condition of still air in a room with the windows closed, while the air velocity fluctuating between 0.3 m/s and 1.0 m/s corresponds to that of a room with its windows open for natural ventilation or with a fan being on. The room air temperature and relative humidity are assumed to be at 26°C and 60% and the state of human body to be at 1.1 Met and at 0.4 clo representing a typical summer clothes.

All of the four sets of plots for HBXC rate have quadratic shape. Their respective minimum values shift from 2.3 W/m² at 26°C of MRT for 0.1 m/s, via 2.05 W/m² at 28°C for 0.5 m/s, to 2.05 W/m² at 30°C of MRT for 0.8 m/s. The dashed line indicates the combination of MRT and air temperature with air velocity at 0.1 m/s; this line represents such cases that outgoing thermal energy flow from the human body becomes equal to thermal energy generation within the human body. If setting up such a condition alone is the mission of built-environmental conditioning, then the MRT at 20°C with the air temperature at 31°C is the same as the MRT at 30°C with the air temperature at 23°C. But, in reality, as we discussed in relation to winter cases, the thermal perception in those conditions must be different from each other.

Suppose that there is a room in which a lot of people are seated, like a large lecture hall. Let us assume that MRT increases for a large number of people, or for the air-conditioning system working less properly than to be expected. Sooner or later, we can easily observe quite a few people taking a behaviour of fanning. This is of course the behaviour to increase the air velocity and its purpose is considered to decrease their own HBXC rates, though he or she is unconscious.

Figure 10.10 shows the same relationship between the HBXC rate and MRT as Fig. 10.9, but this is for the cases of air temperature being at 30°C instead of 26°C. The overall quadratic shaped patterns for the four cases with different air velocities are basically the same as what can be seen in Fig. 10.9, but the difference in HBXC rate due to the change in air velocity is less significant. The minimum values of HBXC rate emerge at some values between 28°C and 30°C of MRT. The minimum

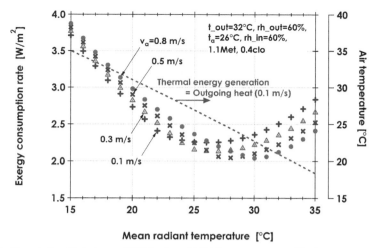

Fig. 10.9: The relationship between HBXC rate and MRT in summer case with air temperature at 26°C. The dashed line represents the combination of MRT and air temperature in the condition of thermal energy generation being equal to outgoing heat.

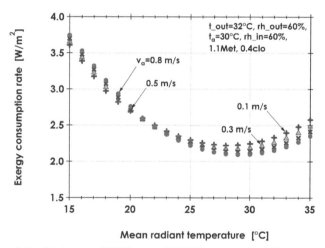

Fig. 10.10: The relationship between HBXC rate and MRT in summer case with air temperature at 30°C switched from 26°C in Fig. 10.9.

HBXC rate decreases with the enhancement of air movement: 2.22 W/m² for 0.1 m/s and 2.1 W/m² for 0.8 m/s.

Comparing Fig. 10.10 with Fig. 10.9, we find that the rise of indoor air temperature makes the HBXC rate slightly smaller. If the MRT is made equal to or a bit lower than 30°C, the air velocity of about 0.8 m/s makes the HBXC rate be about 2.1 W/m². A further increase in air temperature and relative humidity higher than 30°C and 60% assumed for Fig. 10.10 tends to result in smaller values of HBXC rate, let's say, lower values than 2.0 W/m². It does not mean that such low values of HBXC rate brings about the human body with less thermal stress, since the human body generates "warm" exergy as the result of ceaseless "chemical" exergy consumption as was discussed in 10.2.1 and thereby the entropy produced in this course of exergy consumption has to be kept being discarded into his or her immediate environmental space, otherwise he or she cannot keep his or her life. For this reason, there is a certain threshold of HBXC rate. This issue will be further discussed in the next section.

Figure 10.11 shows the four components of HBXC rate in the case of room air temperature at 30°C and air velocity at 0.4 m/s. The exergy consumption rates at skin-layer subsystem account for the largest portion since the condition of high air temperature requires the human body to take the evaporative cooling based on the sweat secretion as can be seen with the skin wetness in the right-hand side of vertical axis. The skin wetness is about 0.13 at the MRT of 29°C, with which the whole of HBXC rate becomes the smallest at 2.07 W/m².

The minimization of the whole HBXC rate down to the values of 2.07 W/m² is realized by the increase in exergy consumption rate due to the absorption of "cool" radiant exergy. In this example, since the outdoor air temperature is assumed to be 32°C, any surface whose temperature is lower than 32°C emits "cool" radiant exergy. Therefore, at the MRT of 29°C, the human body is exposed to the "cool" radiant exergy, which is in the order of 20 to 60 mW/m², as expected from Fig. 7.20 or from Eq. (7.44) described in Chapter 7.

In the range of MRT from 29°C to 32°C, as the MRT increases, the exergy consumption rate due to the absorption of "cool" radiant exergy becomes smaller. But this results in the whole HBXC rate becoming rather larger. This suggests the supremacy of radiant cooling to be in harmony with natural ventilation. For example, natural ventilation to be performed during night time together with the storage of coolness to be emerged as "cool" radiant exergy from the interior surfaces of building structure having appropriate heat capacity during the following daytime to come will

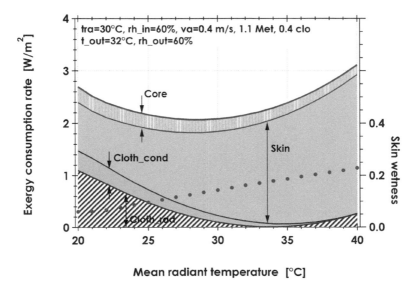

Fig. 10.11: The relationship between the four components of HBXC rate and MRT in summer case with room air temperature at 30°C and air velocity at 0.4 m/s. The round plots indicate the skin wetness.

become an attractive sound technology to be further developed in various places all over the world.

The intrinsic nature of "cool" radiant exergy is different from that of "warm" radiant exergy, but their roles are not different in the sense that either of them decreases the whole HBXC rate. In hot summer, increasing the absorption of "cool" radiant exergy by lowering the radiant temperature makes the whole of HBXC rate decreased, while on the other hand in winter, increasing the availability of "warm" radiant exergy by raising the radiant temperature makes the whole of HBXC rate decreased.

10.5 Thermal history and perception

The characteristics of human-body exergy consumption rate demonstrated in the previous section was based on constant environmental conditions. But, in reality, the indoor and outdoor environmental conditions never stay constant but always vary from time to time. Furthermore, people are likely not to keep staying in a small confined space, but move around from one place to another, the thermal conditions of which are, in general, different from each other. In all these courses, as described in Chapter 3, the cyclic process from sensation via perception to behaviour takes place. A variety of behaviours to control thermal conditions for himself or herself and built-environmental space, such as drinking cold water, taking off or on a jacket, opening or closing a window, switching on or off a fan or mechanical air conditioning unit, emerge in response to the changes of environmental conditions indoors and outdoors. Here in this section, let us take a look at the nature of such ever-changing environmental conditions and its relation to human-body exergetic behaviour together with thermal perception for hot and humid summer cases in particular.

10.5.1 Unsteady-state exergy consumption rate and thermal preference

Figure 10.12 outlines a two-stage subjective experiment performed in order to clarify the above-mentioned feature (Nagai et al. 2016, Shukuya et al. 2018). As the 1st stage, thirty-eight persons of the age 20 to 21 years old, were asked to carry a set of compact sensors for measuring grey-coloured globe temperature, air temperature, and relative humidity as shown in the bottom left of Fig. 10.12. The purpose of this measurement is to know their overall thermal environmental conditions for respective one-week periods.

Figure 10.13 shows two examples of the variation of measured globe temperature values in the vicinities of two persons, S and W, whose average values of globe temperature are the highest at 29.6°C and the lowest at 25.1°C among 38 persons, respectively. We can see that the globe temperature varies from time to time. The range of globe temperature is from 24 to 38°C for Subject_S, whose average is the highest, and from 20 to 36°C for Subject_W, whose average is the lowest. The outdoor air temperature measured at the weather station in Yokohama varies diurnally between 26 and 32°C and the variation looks rather milder than the globe temperature measured in the vicinities of the two persons. These two persons live in geographically same regions, but as their measured globe temperatures for a one-week period demonstrate, they are as though living in different climatic regions.

Two-stage experiment

The 1ˢᵗ stage
■ Thermal history

Continuous measurement of
● Grey-coloured globe temperature;
● Air temperature;
● Relative humidity
for one week was made prior to the 2ⁿᵈ stage.

38 persons (male 21, female 17) participated.

The 2ⁿᵈ stage
■ Thermal perception

21 persons visited three thermally different rooms and stayed in one after another for 15 minutes each (6ᵗʰ–8ᵗʰ August, 2014).

They were asked to answer
● Preference of thermal condition;
● Cognitive temperature.

All together 622 votes were obtained.

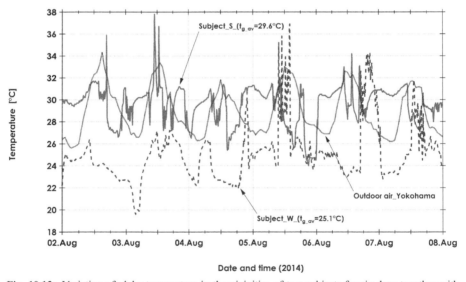

Fig. 10.12: A two-stage experiment to clarify the dynamic nature of thermal perception.

Fig. 10.13: Variation of globe temperature in the vicinities of two subjects for six days together with outdoor air temperature measured at the nearby weather station.

Figure 10.14 shows all of the respective average globe-temperature and water-vapour-concentration values in the vicinities of 38 subjects; among them, 22 subjects are with their average globe temperature higher than the whole average, 27.6°C, and 16 subjects are with lower. The whole average of water vapour concentration is 15.4 g/m^3 and 25 subjects are with their respective average of water vapour concentration higher than the whole average, 15.4 g/m^3, and 13 subjects are with lower. We denote those 22 subjects, who experienced the thermal environment having higher temperature, as Sb_H and those 16 subjects, who experienced the thermal environment having relatively lower temperature, as Sb_L.

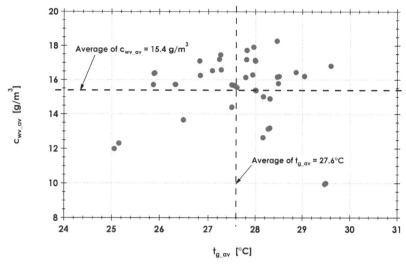

Fig. 10.14: The averages of grey-coloured globe temperature and water-vapour concentration measured in the vicinities of 38 subjects for a one-week period.

Figure 10.15 shows the frequency distribution of grey-colored globe temperature and water vapour concentration of these two groups of subjects, Sb_H and Sb_L, respectively. There is an obvious difference, 2°C, in the median values of measured globe temperature; it is at the rank of 27.5°C ($27 \leq t_g < 28$) in the group Sb_H, while on the other hand, at the rank of 25.5°C ($25 \leq t_g < 26$) in the group Sb_L. There is not such a clear difference in the case of water vapour concentration. Therefore, 38 persons were not separated with respect to water-vapour concentration.

All of 38 subjects were invited to the 2nd stage of experiment and 21 out of them could participate in it; 11 are Sb_H and 10 are Sb_L. As shown in Fig. 10.12, they visited three thermally different rooms from one to another and stayed for 15 minutes each. The three rooms are as follows: one with window closed and no mechanical cooling unit operated (denoted as WC), another with natural ventilation by keeping windows opened (NV), and the last with window closed and mechanical cooling unit operated (MC). All together they were exposed to these varying thermal environments for the period of two and a half hours, and during the participation they were asked to answer their preference and also cognitive temperature (Saito and Tsujihara 2015, Saito 2017).

Answering their preference and cognitive temperature was made four times in each room: firstly, right after entering a room; secondly, five minutes later; thirdly, further five minutes later; and lastly, right before leaving the room. The preference asked was whether they prefer lowering the room temperature, prefer keeping it as it is, or prefer raising the room temperature. The cognitive temperature is a single value of temperature that they imagine in accordance with their experiencing conditions. It is considered to reflect not only the present state of thermal conditions that the subjects are experiencing, but also their respective thermal history that they had unconsciously built up until then, as was demonstrated in Fig. 10.13, by the variation of thermal environmental conditions.

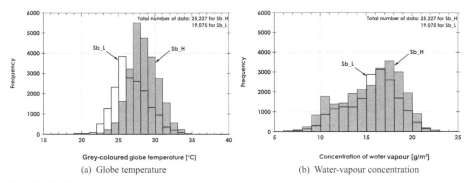

(a) Globe temperature   (b) Water-vapour concentration

Fig. 10.15: Frequency distribution of globe temperature and water vapour concentration measured in the vicinities of 38 subjects for a one-week period. Sb_H and Sb_L denote two groups of subjects, whose respective averages of globe temperature are higher or lower than the whole average.

This 2nd stage experiment was performed for three days, from 6th to 8th of August, 2014 and altogether 622 votes were obtained. Figure 10.16 demonstrates the variation of thermal environmental condition for the participants indoors together with the outdoor environmental conditions during the series of visit from WC to NV, WC to MC, and so forth on 6th August, 2014. It was a typical hot and humid summer day in Yokohama area. Other two days, 7th and 8th August, were more or less the same as the first day. For one session of the experiment, six, seven or eight subjects participated in and they made visit in the course of WC → NV → WC → MC → and so on as indicated in the graphs.

In WC, the MRT and room air temperature were almost the same as each other and they are 2 to 3°C lower than outdoor air temperature. In NV, the MRT was slightly higher than room air temperature when the first visit was made, but they were almost the same as each other when the second visit was made. Room air temperature and the MRT in NV was the highest in three rooms, since the outdoor air, whose temperature is higher than the room air temperature in WC, was taken in by natural ventilation.

In WC, no window was opened and no mechanical cooling was on so that the air velocity was the lowest in three rooms, while in NV the air velocity fluctuated in the range from 0.4 to 0.9 m/s because of the windows being kept open. In MC, the air velocity was observed to be slightly higher than that in WC and the room air temperature was the lowest in three rooms. It was about 27°C, which is approximately 5 to 6°C lower than outdoor air temperature and 2 to 3°C lower than the room air temperature in WC. Throughout all sessions, relative humidity, indoors and outdoors, is more or less within the range between 50 to 57%. In room MC, the relative humidity is the lowest at about 50%. As shown in the middle graph, ET* calculated using the set of measured data is the highest in NV, the lowest in MC and in between in WC.

The bottom graph shows the variation of human-body exergy consumption rate calculated using the measured data mentioned above. Since the measurement was made at one-minute intervals, the human-body exergy balance calculation was also

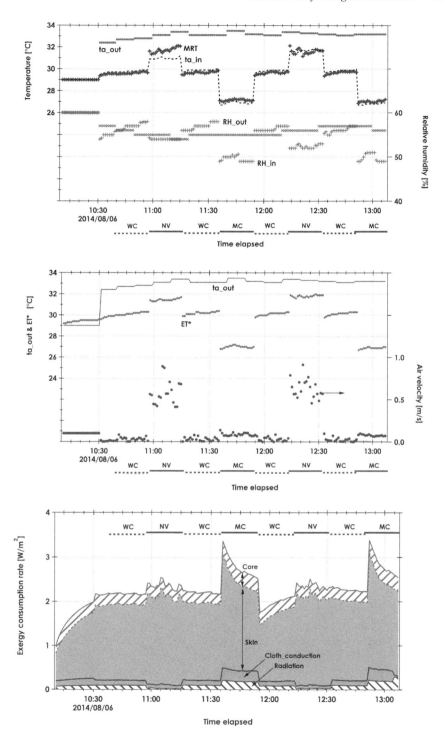

Fig. 10.16: Variation of thermal environmental condition in one session of the 2nd stage experiment and variation of the human-body exergy consumption rate calculated.

made at one-minute intervals. For this calculation, the metabolic energy generation rate and the clothing insulation were assumed to be 1.1 Met and 0.3 clo. The first twenty-minute period in Fig. 10.16 is the period for preliminary calculation so that the actual results to be seen are from 10:30 on.

The human-body exergy consumption rate varies from time to time depending on the thermal conditions of three rooms. A sharp rise of HBXC rate emerges right after they enter room MC from room WC. On the other hand, a sharp drop of HBXC rate emerges right after they enter room WC from room MC. In room NV, the whole HBXC rate varies in response to the fluctuation of air velocity. The attenuating feature appeared in the variation of HBXC rate is exactly the reflection of unsteady-state calculation.

Since this is a summer case, the exergy consumption emerged at the skin layer occupies the largest portion within the whole HBXC rate. This is particularly so in room NV since the exergy consumption due to the conduction between the skin layer and the clothing ensemble and also the exergy consumption due to the absorption of radiation becomes the smallest for the MRT and air temperature being the highest. In room MC, the exergy consumption due to conduction and absorption of radiation becomes large for the lowest values of the MRT and room air temperature. These features, together with the low relative humidity, about 50%, makes the whole HBXC rate in room MC be the largest.

All of the 622 preference votes obtained from the participants in the 2nd stage experiment held for three days were allotted in the bins of HBXC rate such as $1.0 \leq X_C < 1.1$, $1.1 \leq X_C < 1.2$, $\cdots$, and so on, where X_C denotes the whole HBXC rate [W/m^2], and then, the percentage of subjects who prefer lowering the room temperature in each bin was calculated. This calculation was made for all participants together and also for two groups of the participants, Sb_H and Sb_L, separately. Figure 10.17 shows the results in the case of all participants since there was no significant difference for Sb_H and Sb_L. As can been seen, the smaller the HBXC rate, the more the preference for lowering the room temperature. The percentage of people prefering to lower the room temperature, p_L [%], can be expressed by the following logistic curve as a function of HBXC rate, X_C, with the coefficient of determination being 0.724 and the level of significance being less than 1%.

$$p_L = \left(\frac{1}{1 + e^{3.940 X_c - 7.399}} \right) \times 100 \tag{10.61}$$

For the condition at which the HBXC rate becomes 1.5 W/m^2, approximately 80% of subjects prefer lowering the room temperature. On the other hand, for the conditions at which the HBXC rate is 2.2 W/m^2, approximately 20% of the subjects prefer lowing the room temperature. It implies that 80% of the subjects would not necd a change of thermal condition. With this statistical evidence in mind, let us again take a look at the bottom graph of Fig. 10.16. We find that the HBXC rate ranging from 2.2 to 2.5 W/m^2 is realized in room NV. This range of HBXC rate coincides with the minimum values of HBXC rate shown in Fig. 10.10.

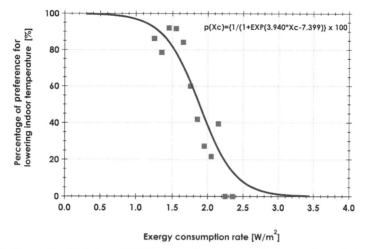

Fig. 10.17: The relationship between HBXC rate and the percentage of subjects who prefer lowering the room temperature.

10.5.2 *Radiant exergy input rate and avoidance of discomfort*

Since, as described earlier, the cognitive temperature as the other subjective appraisal of thermal condition was asked in addition to the preference, whether or how it relates to HBXC rate was examined as shown in Fig. 10.18. The examination was made for Sb_H and Sb_L separately. Since the cognitive temperature is a kind of subjective indicator, the plots scatter much against a certain value of HBXC rate, but we can see some patterns in the plots as a whole. As the two graphs show, the smaller is the HBXC rate, the higher is the cognitive temperature. This tendency looks similar to each other for both Sb_H and Sb_L and it is consistent with what can be seen in Fig. 10.17. The values of cognitive temperature become higher as the HBXC rate becomes lower than 2.5 W/m^2.

The same way of comparison was also made with respect to radiant exergy input rate, radiant exergy emission rate, and outgoing exergy rate by convection, and it was found that there is hardly any correlation at all between the cognitive temperatue and radiant exergy emission rate and also between the cognitive temperature and outgoing exergy rate by convection. But, there was some distinctive pattern suggesting a correlation between the cognitive temperature and the radiant exergy input rate for Sb_H, but not for Sb_L as shown in Fig. 10.19. For Sb_H, whether "cool" or "warm" radiant exergy is available looks like affecting the values of cognitive temperature, while on the other hand, for Sb_L, it does not look so.

Since the thermo-sensory portals of human body is embedded within the outermost portion of skin layer as was described in Chapter 3, radiant exergy input, whether it is "warm" or "cool", could be the primary stimulant to the brain, which leads to the perception and the follow-up consciousness to emerge. Assuming it to be so, Sb_H is sufficiently sensitive to the availability of radiant exergy, but Sb_L is not.

According to a previous investigation on the role of radiant exergy emission rate from the surrounding surfaces in naturally ventilated rooms (Tokunaga et al.

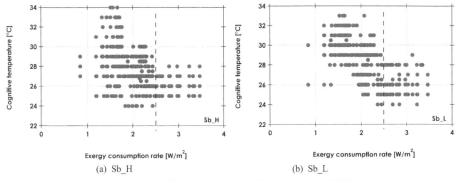

Fig. 10.18: Cognitive temperature and its relation to HBXC rate.

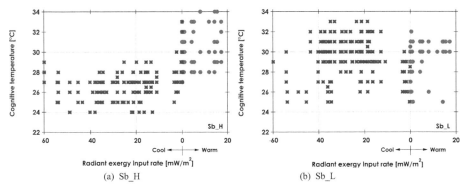

Fig. 10.19: Cognitive temperature and its relation to radiant exergy input rate.

2005, Shukuya et al. 2006, Shukuya 2013), the avoidability of discomfort is very much influenced by the availability of "warm" or "cool" radiant exergy. This result is depicted in Fig. 10.20 as the preference lowering the indoor temperature; the circular plots represent the experimental result given by Tokunaga et al. (2005) and the curve represents their logistic regression. Shown together is the perecentage of Sb_H and Sb_L, who prefer lowering the room temperature. Since the number of votes obtained were much smaller than in the previous studies done by Tokunaga et al. (2005) the size of bins for radiant exergy emission rate were made larger: e.g., from 0 to 20 mW/m², 20 to 40 W/m² and so on.

In the thermal condition in which "warm" radiant exergy is available, it is very likely that the preference for lowering the room temperature emerges regardless of either Sb_H or Sb_L. But under the thermal condition in which "cool" radiant exergy is available, the emergence of preference looks rather different, depending on whether one belongs to Sb_H or Sb_L. The subjects of Sb_H look very responsive to the turnover of radiant exergy from "warm" to "cool", but those of Sb_L look much less responsive. These tendencies look consistent with what was observed in Fig. 10.19, that is, Sb_H tends to vote lower cognitive temperature with the availability of "cool" radiant exergy, but Sb_L does not. This is considered due to the subjects of Sb_L having been exposed very often to the thermal environmental

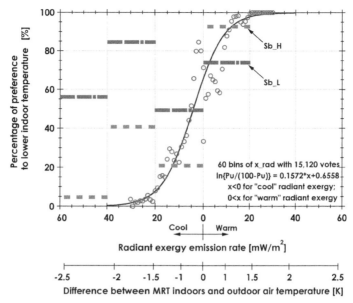

Fig. 10.20: The relationship between radiant exergy emission rate from the surrounding surfaces and the percentage of subjects who prefer lowering the room temperature.

condition with low temperature. They cannot be responsive to the condition of MRT being 1 or 2°C lower than outdoor air temperature. On the contrary, those belonging to Sb_H are rather highly sensitive to the change in MRT, as was indicated by the cognitive temperature given by them as shown in Fig. 10.19.

What can be seen in Figs. 10.19 and 10.20 again confirms the importance of controlling radiant temperature described in the previous section 10.4, and also discussed in Chapter 6, because such rational control of radiation within the built environment not only optimizes the exergy consumption for space cooling together with the human-body exergy consumption, but also revitalizes the human cyclic process from sensation via perception to behaviour to function towards the direction of well being. This is exactly what we need to develop for rational technology to be the well-balanced passive and active system solution.

References

ASHRAE Handbook of Fundamentals 2005. Chapter 8: Thermal Comfort. 8.1–8.8.

Fanger P. O. 1970. Thermal Comfort. Danish Technical Press.

Gagge A. P., Stolwijk J. A. J. and Nishi Y. 1971. An effective temperature scale based on a simple model of human physiological regulatory response. ASHRAE Transactions 77(1): 247–262.

Gagge A. P., Nishi Y. and Gonzalez R. R. 1973. Standard effective temperature—A single temperature index of temperature sensation and thermal discomfort. Proceedings of the CIB Commission W45 Symposium, London 1972. HMSO, 229–250.

Gagge A. P., Fobelets A. P. and Berglund L. G. 1986. A standard predictive index of human response to the thermal environment. ASHRAE Transactions 92(2B): 709–731.

Guggenheim K. Y. 1991. Rudolf Schönheimer and the concept of the dynamic state of body constituents. Journal of American Institute of Nutrition 121: 1701–1704.

Nagai R., Rijal H. B. and Shukuya M. 2016. Research on the effect of long-term thermal history on thermal perception and human-body exergy balance. Proceedings of Annual Meeting of Architectural Inst. of Japan, 309–310.

Nishi Y. and Gagge A. P. 1970. Moisture permeation of clothing: A factor governing thermal equilibrium and comfort. ASHRAE Transactions 76(Part 1): 137–145.

Parson K. 2002. Human Thermal Environments—The Effects of Hot, Moderate, and Cold Environments on Human Health, Comfort, and Performance. CRC Press.

Reece J. B., Urry L. A., Cain M. L., Wasserman S. A., Minorsky P. V. and Jackson R. B. 2011. Campbell—Biology. Global Edition. Pearson, 220–223.

Rohles F. H. and Nevins R. G. 1971. The nature of thermal comfort for sedentary man. ASHRAE Transactions 77(1): 239–246.

Saito M. and Tsujihara M. 2015. Thermal adaptation of elementary students in summer based on cognitive temperature scale in the case of Sapporo and Kumamoto. Proceedings of Annual Meeting of Architectural Inst. of Japan, 471–474.

Saito M. 2017. Radiative environmental design for active behaviour and high comfort of occupants. Energy and Natural Resources 38(4): 208–212 (in Japanese).

Schweiker M. and Shukuya M. 06th November 2015. Personal communication.

Schweiker M., Kolark J., Dovjak M. and Shukuya M. 2016. Unsteady-state human-body exergy consumption rate and its relation to subjective assessment of dynamic thermal environments. Energy and Buildings 116: 164–180.

Shukuya M. 2002. Exergetic way of thinking and symbiotic architecture. Proceedings of the 32nd Symposium on Heat Transfer Problems in Buildings, Architectural Institute of Japan, 51–57 (in Japanese).

Shukuya M., Tokunaga K., Nishiuchi M., Iwamatsu T. and Yamada H. 2006. Thermal radiant exergy in naturally-ventilated room space and its role on thermal comfort. Proceedings of Healthy Buildings 2006 Vol. IV: 257–262.

Shukuya M. 2009. Exergy concept and its application to the built environment. Building and Environment 44(7): 1545–1550.

Shukuya M. (ed.). 2010. Theory on exergy and environment—revised version. Inoue Shoin Publishers Ltd. (in Japanese).

Shukuya M. 2013. Exergy—Theory and applications in the built environment. Springer-Verlag London.

Shukuya M., Nagai R. and Rijal H. B. 2018. An exergetic investigation on the effect of long-term thermo-physical exposure on thermal perception. Proceedings of 10th Windsor Conference: Rethinking Comfort, pp. 1076–1084.

Tokunaga K., Iwamatsu T., Hiraga T., Nishiuchi M., Yamada H. and Shukuya M. 2005. An experimental study on thermal comfort caused by solar-heat control and ventilation in summer season—the relationship between radiant environment and comfort. Proceedings of Annual Meeting of Architectural Inst. of Japan, 607–608.

Appendix

10.A VBA codes for the calculation of human-body exergy balance

Table 10.A.1 shows an example of the VBA code for the calculation of C_{bld}, V_{w-sk}, Q_{cr}, Q_{sk}, T_{cr}, T_{sk}, and human-body exergy balance. The program shown here is to perform the calculation for a period of time of $N\Delta t$, with the time step of Δt, with the input of constant human-body and surrounding thermal conditions. In the case of unsteady conditions, the code can be modified so that the input variables are given inside the loop of time-steps. Table 10.A.2 lists up all of VBA function codes used in the VBA code shown in Table 10.A.1.

Table 10.A.1: VBA code for the calculation of human-body exergy balance.

```
Sub Tcr_Tsk_HbEx_balance()
'
'   This is a program for the calculation of human-body exergy balance under unsteady-state conditions.
'                   1st ver.  MS 13 February, 2013
'                   MS 11 May, 2014// 13th July, 2014// 18th February, 2015//06 November, 2015
'                       //01Feb2017
'
    TK0 = 273.15: tsk_set = 33.7: tcr_set = 36.8: lr = 16.5 * 10 ^ (-3)      'constants
    cdil = 200: sigma_tr = 0.5: Csw = 170                                    'constants
    M_air = 28.97 * 0.001: M_water = 18.015 * 0.001: R_gas = 8.314   'constants
    Row = 1000#: Roa = 1.2                                                   'constants
    Cp_body = 3490#: cpa = 1005#: cpv = 1846#: cpw = 4186#           'constants
    PO = 101325#                                    'Atmospheric pressure [N/m2=J/m3]
'
    Icl = Sheet10.Cells(6, 2): va = Sheet10.Cells(4, 2): Met = Sheet10.Cells(5, 2)     'Input
    qmet = Meta_therm(Met)                                                   '--------Function
    fcl = 1 + 0.15 * Icl: Rcl = 0.155 * Icl
'
    tmr = Sheet10.Cells(1, 2): ta = Sheet10.Cells(2, 2): phia = Sheet10.Cells(3, 2)     'Input
    TKa = ta + TK0: TKmr = tmr + TK0
'
    to_env = Sheet10.Cells(1, 11): TKo_env = to_env + TK0            'Outdoor air temperature
    phia_env = Sheet10.Cells(2, 11): pv_env = p_vapor(to_env, phia_env)     '--------Function
    Sheet10.Cells(3, 11) = pv_env
'
    frad = Sheet10.Cells(1, 6): eps = Sheet10.Cells(2, 6): i_c = Sheet10.Cells(3, 6)     'Input
    Hbody = Sheet10.Cells(7, 2): Mbody = Sheet10.Cells(8, 2)                 'Input
    Abody = 0.008883 * Hbody ^ 0.663 * Mbody ^ 0.444: Sheet10.Cells(9, 4) = Abody: r_MA = Mbody / Abody
'
    Cal_hour = Sheet10.Cells(15, 2) * 60 * 60: Dt_cal = Sheet10.Cells(16, 2) * 60          'Input
'
    hr = 6.13 * frad * eps: Sheet10.Cells(1, 4) = hr
    hc = hcv_g(va, Met): Sheet10.Cells(2, 4) = hc                            '--------Function
    top = (hr * tmr + hc * ta) / (hr + hc)
    FFCL = 1 / (1 + 0.155 * Icl * fcl * (hr + hc)): FPCL = 1 / (1 + 0.155 * Icl * fcl * hc / i_c)
    CL = 1 / (1 + FPCL / ((hr + hc) * FFCL))
    im = hc * FPCL / ((hr + hc) * FFCL)
    Sheet10.Cells(10, 2) = fcl: Sheet10.Cells(11, 2) = FFCL: Sheet10.Cells(12, 2) = FPCL: Sheet10.Cells(13, 2) = im
'
    Icl_star = 0.6: fcl_star = 1 + 0.15 * Icl_star
    hc_star = hcv_g(0.1, 1#)
    FFCL_star = 1 / (1 + 0.155 * Icl_star * fcl_star * (hr + hc_star))
    FPCL_star = 1 / (1 + 0.155 * Icl_star * fcl_star * hc_star / i_c)
    im_star = hc_star * FPCL_star / ((hr + hc_star) * FFCL_star): Sheet10.Cells(14, 2) = im_star
    C_res = 0.0014 * (58.2 * Met) * (34 - ta)
'
    pva = p_vapor(ta, phia)                         '--------Module_1
    E_res = 0.0173 * (58.2 * Met) * (5.87 - pva / 1000)
'
    tcr = Sheet10.Cells(5, 4): tsk = Sheet10.Cells(6, 4)        'Initial values of body-core and skin-layer temperature
'
    jcount = 0: time_elp = 0
```

Table 10.A.1 contd. ...

...Table 10.A.1 contd.

' A series of calculation step by step

11 If Cal_hour < time_elp Then

' GoTo 10   '→ ending the calculation

' Else

q_shiv = Mshiv(tcr_set, tsk_set, tcr, tsk)   '--------Function
Vbl_s = vbl_cdil_str(cdil, sigma_tr, tcr_set, tsk_set, tcr, tsk)   '--------Function
alfa_sk = 0.0418 + 0.745 / (Vbl_s + 0.585): K_s = 5.28 + 1.163 * Vbl_s

C_bld = 1.163 * Vbl_s: K_cr_sh = 5.28

' Qcr = (1 - alfa_sk) * r_MA * Cp_body: Qsk = alfa_sk * r_MA * Cp_body

Sheet10.Cells(jcount + 12, 12) = q_shiv: Sheet10.Cells(jcount + 12, 13) = Vbl_s
Sheet10.Cells(jcount + 12, 14) = alfa_sk
Sheet10.Cells(jcount + 12, 15) = K_s: Sheet10.Cells(jcount + 12, 16) = Qcr: Sheet10.Cells(jcount + 12, 17) = Qsk

tcr_n = (1 - Dt_cal * K_s / Qcr) * tcr + Dt_cal / Qcr * (qmet + q_shiv - (C_res + E_res) + K_s * tsk)
psks = p_vapor(tsk, 100#)   '--------Function
tb = alfa_sk * tsk + (1 - alfa_sk) * tcr: tb_set = alfa_sk * tsk_set + (1 - alfa_sk) * tcr_set
Ersw = Csw * (tb - tb_set) * Exp((tsk - tsk_set) / 10.7)

Sheet10.Cells(jcount + 12, 9) = Ersw: Sheet10.Cells(jcount + 12, 10) = Emax

w = 0.06 + 0.94 * Ersw / Emax
    If w < 0.06 Then
        w = 0.06
    End
If 1 < w Then
        w = 1#
    End If
    DTQ = Dt_cal / Qsk
    tsk_n = (1 - DTQ * K_s - DTQ * fcl * FFCL * (hr + hc)) * tsk - DTQ * w * fcl * lr * hc * FPCL * psks + _
        DTQ * (K_s * tcr + fcl * (hr + hc) * FFCL * top + w * fcl * lr * hc * FPCL * pva)
    tcl_n = ((1 / Rcl) * tsk_n + fcl * hr * tmr + fcl * hc * ta) / (1 / Rcl + fcl * hr + fcl * hc)

' Exergy balance

' Thermal exergy generation by metabolism

TKcr_n = tcr_n + TK0: TKsk_n = tsk_n + TK0: TKcl_n = tcl_n + TK0
Meta_energy = qmet + q_shiv
X_met = Meta_energy * (1# - TKo_env / TKcr_n): xwc = WC_X_check(TKcr_n, TKo_env)
Sheet10.Cells(jcount + 12, 19) = X_met: Sheet10.Cells(jcount + 12, 20) = xwc

' Inhaled humid air

Vin = 1.2 * 10 ^ (-6) * Meta_energy
cpav = cpa * (M_air / (R_gas * TKa)) * (PO - pva) + cpv * (M_water / (R_gas * TKa)) * pva
X_inhale_wc = Vin * WC_Ex(cpav, TKa, TKo_env): xwc = WC_X_check(TKa, TKo_env)   '--------Function
Sheet10.Cells(jcount + 12, 21) = X_inhale_wc: Sheet10.Cells(jcount + 12, 22) = xwc

X_inhale_wd = Vin * WD_Ex(TKa, TKo_env, pva, pv_env): xwd = WD_X_check(pva, pv_env)   '--------Function
Sheet10.Cells(jcount + 12, 23) = X_inhale_wd: Sheet10.Cells(jcount + 12, 24) = xwd

---

Table 10.A.1 contd. ...

...Table 10.A.1 contd.

'
'   Liquid water generated in the core by metabolism to be dispersed into the exhaled air
'

Vw_core = Vin * (0.029 - 0.049 * 10 ^ (-4) * pva)
X_lw_core_wc = Vw_core * Roa * WC_Ex(cpw, TKcr_n, TKo_env): xwc = WC_X_check(TKcr_n, TKo_env) '--------Function
Sheet10.Cells(jcount + 12, 25) = X_lw_core_wc: Sheet10.Cells(jcount + 12, 26) = xwc

pvs_env = p_vapor(to_env, 100#)
X_lw_core_wet = Vw_core * Roa * R_gas / M_water * TKo_env * Log(pvs_env / pv_env)
Sheet10.Cells(jcount + 12, 27) = X_lw_core_wet: Sheet10.Cells(jcount + 12, 28) = "WET"

'
'   Liquid water generated in the shell by metabolism to be secreted as sweat
'

Vw_shell_row = w * Emax / (2450# * 1000#)
X_lw_shell_wc = Vw_shell_row * WC_Ex(cpw, TKsk_n, TKo_env): xwc = WC_X_check(TKsk_n, TKo_env) '--------Function
Sheet10.Cells(jcount + 12, 29) = X_lw_shell_wc: Sheet10.Cells(jcount + 12, 30) = xwc

X_lw_shell_wd = Vw_shell_row * WD_Ex_lw(TKo_env, pvs_env, pva, pv_env): xwd = WD_X_check(pva, pv_env) '--------Function
Sheet10.Cells(jcount + 12, 31) = X_lw_shell_wd: Sheet10.Cells(jcount + 12, 32) = xwd

'
'   Radiant exergy input
'

X_in_rad = fcl * hr * (TKmr - TKo_env) ^ 2 / (TKmr + TKo_env): xwc = WC_X_check(TKmr, TKo_env)
Sheet10.Cells(jcount + 12, 33) = X_in_rad: Sheet10.Cells(jcount + 12, 34) = xwc

Xin_total = X_met + X_inhale_wc + X_inhale_wd + X_lw_core_wc + X_lw_core_wet + X_lw_shell_wc + X_lw_shell_wd + X_in_rad
Sheet10.Cells(jcount + 12, 35) = Xin_total

'
'   Exergy stored
'

X_st_core = Qcr * (1# - TKo_env / TKcr_n) * (tcr_n - tcr) / Dt_cal
X_st_shell = Qsk * (1# - TKo_env / TKsk_n) * (tsk_n - tsk) / Dt_cal
Sheet10.Cells(jcount + 12, 36) = X_st_core: Sheet10.Cells(jcount + 12, 37) = X_st_shell

'
'   Exhaled humid air
'

pvss_cr = p_vapor(tcr_n, 100#)
cpav = cpa * (M_air / (R_gas * TKcr_n)) * (PO - pvss_cr) + cpv * (M_water / (R_gas * TKcr_n)) * pvss_cr
X_exhale_wc = Vin * WC_Ex(cpav, TKcr_n, TKo_env): xwc = WC_X_check(TKcr_n, TKo_env)   '--------Function
Sheet10.Cells(jcount + 12, 38) = X_exhale_wc: Sheet10.Cells(jcount + 12, 39) = xwc

X_exhale_wd = Vin * WD_Ex(TKcr_n, TKo_env, pvss_cr, pv_env): xwd = WD_X_check(pvss_cr, pv_env) '--------Function
Sheet10.Cells(jcount + 12, 40) = X_exhale_wd: Sheet10.Cells(jcount + 12, 41) = xwd

'
'   Water vapor originating from the sweat and humid air containing the evaporated sweat
'

X_sweat_wc = Vw_shell_row * WC_Ex(cpv, TKcl_n, TKo_env): xwc = WC_X_check(TKcl_n, TKo_env) '--------Function
Sheet10.Cells(jcount + 12, 42) = X_sweat_wc: Sheet10.Cells(jcount + 12, 43) = xwc

X_sweat_wd = Vw_shell_row * WD_Ex_lw(TKo_env, pva, pva, pv_env): xwd = WD_X_check(pva, pv_env) '--------Function
Sheet10.Cells(jcount + 12, 44) = X_sweat_wd: Sheet10.Cells(jcount + 12, 45) = xwd

'
'   Radiant exergy output
'

X_out_rad = fcl * hr * (TKcl_n - TKo_env) ^ 2 / (TKcl_n + TKo_env): xwc = WC_X_check(TKcl_n, TKo_env)
Sheet10.Cells(jcount + 12, 46) = X_out_rad: Sheet10.Cells(jcount + 12, 47) = xwc

'

Table 10.A.1 contd. ...

...Table 10.A.1 contd.

```
'
'   Exergy transfer by convection
'
    X_out_conv = fcl * hc * (TKcl_n - TKa) * (1# - TKo_env / TKcl_n): xwc = WC_X_check(TKcl_n, TKo_env)
    Sheet10.Cells(jcount + 12, 48) = X_out_conv: Sheet10.Cells(jcount + 12, 49) = xwc
'
    Xout_total = X_st_core + X_st_shell + X_exhale_wc + X_exhale_wd + X_sweat_wc + X_sweat_wd + X_out_rad + X_out_conv
    Sheet10.Cells(jcount + 12, 50) = Xout_total
'
'   4 components of exergy consumption rate
'
    X_bld_sh = WC_Ex(C_bld, TKsk_n, TKo_env)                    '----------------------Module_4
    Xin_123bldsk = (X_met + X_inhale_wc + X_inhale_wd + X_lw_core_wc + X_lw_core_wet) + X_bld_sh   '-------------Input to the core

    Xcond_cr = (1# - TKo_env / TKcr_n) * K_cr_sh * (TKcr_n - TKsk_n)
    X_bld_cr = WC_Ex(C_bld, TKcr_n, TKo_env)                    '----------------------Module_4
    Xst7_out8_condcr_bldcr = X_st_core + X_exhale_wc + X_exhale_wd + Xcond_cr + X_bld_cr           '-------------Stored at core and
output
'
    Xin_4condcr_bldcr = X_lw_shell_wc + X_lw_shell_wd + Xcond_cr + X_bld_cr                         '-------------Input to the shell
    X_clsh = (1# - TKo_env / TKsk_n) * (1# / Rcl) * (TKsk_n - TKcl_n)
    Xst7sh_clsh_bldsh = X_st_shell + X_clsh + X_sweat_wc + X_sweat_wd + X_bld_sh                    '-------------Stored at shell and output

    Xconsumption_cr = Xin_123bldsk - Xst7_out8_condcr_bldcr: Sheet11.Cells(3 + ii, 3) = Xconsumption_cr
    Xconsumption_sh = Xin_4condcr_bldcr - Xst7sh_clsh_bldsh: Sheet11.Cells(3 + ii, 4) = Xconsumption_sh
'
    Xconsumption_cl_cond = (1# / Rcl) * (TKsk_n - TKcl_n) ^ 2 / (TKsk_n * TKcl_n) * TKo_env
    Sheet11.Cells(3 + ii, 5) = Xconsumption_cl_cond
    Xin_5clth = X_in_rad + X_clsh
    Xout_clth = X_out_rad + X_out_conv
    Xconsumption_clth = Xin_5clth - Xout_clth
'   Xconsumption_rad = Xconsumption_clth - Xconsumption_cl_cond
    Xconsumption_rad = 0.95 * 5.67 * 10 ^ (-8) * TKcl_n ^ 3 * (1# / 3# + (Tkmr / TKcl_n) ^ 3 * (Tkmr / TKcl_n - 4# / 3#)) * TKo_env
'   Sheet11.Cells(3 + ii, 10) = Xconsumption_rad_1
'
    Sheet11.Cells(3 + ii, 6) = Xconsumption_rad
    Sheet11.Cells(3 + ii, 7) = Xconsumption_cl_cond + Xconsumption_rad
    Sheet11.Cells(3 + ii, 8) = Sheet11.Cells(3 + ii, 7) + Xconsumption_sh

    X_consumption = Xconsumption_cr + Xconsumption_sh + Xconsumption_cl_cond + Xconsumption_rad
    Sheet10.Cells(jcount + 12, 51) = X_consumption: Sheet11.Cells(3 + ii, 2) = X_consumption
    Sheet11.Cells(3 + ii, 9) = w'
    ET_star = ET(top, ta, phia, w, im, 50#, im): Sheet10.Cells(jcount + 12, 11) = ET_star       '--------Function
'
    Sheet10.Cells(8, 16) = ET_star

    jcount = jcount + 1: time_elp = time_elp + Dt_cal
    tcr = tcr_n
    tsk = tsk_n

    GoTo 11        'Go back to the top of next time step
'
    End If
'
10  Sheet10.Cells(jcount + 12, 3) = "END": Sheet10.Cells(jcount + 12, 4) = "END"
    SET_star = ET(top, ta, phia, w, im, 50#, im_star): Sheet10.Cells(8, 17) = SET_star         '--------Function
'
End Sub
```

Table 10.A.2: VBA functions used for the calculation of human-body exergy balance.

```
Function hcv_g(v, Met)
'Calculation of convective heat transfer coefficient
'Equations used by Gagge et al. (1986)
  qmet = Met * 58.2
  QQ = qmet / 58.2 - 0.85
  If QQ < 0 Then
    hc1 = 0
  Else
    hc1 = 5.66 * QQ ^ 0.39
  End If
  hc2 = 8.6 * v ^ 0.53
  hcv_g = hc1
  If hc1 < hc2 Then hcv_g = hc2
End Function
```
```
Function p_vapor(t, phi)
'Calculation of water vapor pressure in Pa
phi_air = phi / 100#
TK = t + 273.15
p_vapor = phi_air * Exp(25.89 - 5319 / TK)
End Function
```
```
 Function vbl_cdil_str(cdil, str, tcr_set, tsk_set, tcr, tsk)
'Calculation of skin blood flow
'vbl is in the unit of "litre/(m2h)".
  Signal_cr = tcr - tcr_set
  If 0 <= Signal_cr Then GoTo 10
  Signal_cr = 0
10 Signal_sk = tsk_set - tsk
  If 0 <= Signal_sk Then GoTo 11
  Signal_sk = 0
11 vbl_cdil_str = (6.3 + cdil * Signal_cr) / (1 + str * Signal_sk)
  If vbl_cdil_str < 0.5 Then vbl_cdil_str = 0.5
  If 90 < vbl_cdil_str Then vbl_cdil_str = 90
End Function
```
```
Function Mshiv(tcr_set, tsk_set, tcr, tsk)
'Calculation of thermal-energy generation by shivering
'By Stolwijk and Hardy
Signal_cr = tcr_set - tcr: If Signal_cr < 0 Then Signal_cr = 0
Signal_sk = tsk_set - tsk: If Signal_sk < 0 Then Signal_sk = 0
Mshiv = 19.4 * Signal_cr * Signal_sk
End Function
```
```
Function Meta_therm(Met)
'
'Calculation of thermal energy emission rate taking
'      the rate of external work into consideration
'
eff = 0#
If Met >= 1.4 And Met < 3# Then
    eff = 0.1
ElseIf Met >= 3# Then
    eff = 0.2
End If
Meta_therm = 58.2 * Met * (1# - eff)
End Function
```

Table 10.A.2 contd. ...

...*Table 10.A.2 contd.*

Function WC_Ex(cp, T1, TOO)

' Calculation of thermal exergy contained by a body with heat capacity of "cp"

'

   WC_Ex = cp * ((T1 - TOO) - TOO * Log(T1 / TOO))

'

End Function
==

Function WC_X_check(T1, TOO)

'

'   Judgement of exergy regarding "warm" or "cool"

'

    If T1 < TOO Then
         WC_X_check = "COOL"
    Else
         WC_X_check = "WARM"
    End If
End Function
==

Function WD_Ex(T1, TOO, pv1, pvo)

'

' Calculation of material(wet/dry) exergy contained by one cubic-meter of moist air

'

   POT = 101325#
   WD_Ex = TOO / T1 * ((POT - pv1) * Log((POT - pv1) / (POT - pvo)) + pv1 * Log(pv1 / pvo))

'

End Function
==

Function WD_Ex_lw(TOO, pvso, pv1, pvo)

'

' Calculation of material(wet/dry) exergy contained by one kilogram of liquid water

'

   R_gas = 8.314: M_water = 18.015 * 0.001: POT = 101325#
   WD_Ex_lw = R_gas / (M_water) * TOO * (Log(pvso / pvo) + (POT - pv1) / pv1 * Log((POT - pv1) / (POT - pvo)))

'

End Function
==

Function WD_X_check(p1, POO)

'

'   Judgement of exergy regarding "wet" or "dry"

'

    If p1 < POO Then
         WD_X_check = "DRY"
    Else
         WD_X_check = "WET"
    End If
End Function

Table 10.A.2 contd. ...

...Table 10.A.2 contd.

Function ET(top, ta, pha, w, im, pha_ET, im_star)
'
' Calculation of Effective temperature
'
' top: operative temperature, ta: air temperature, pha: air humidity[%], w: skin wettedness, im: permeability factor
' pha_ET: air humidity of imaginary environment[%],
' im_star: permeability factor for imaginary clothing in assumed environment
'
  Delta = 0.1: lr = 16.5 * 10 ^ (-3): TK0 = 273.15
  C = top + w * im * lr * 0.01 * pha * Exp(25.89 - 5319 / (TK0 + ta))
  tes0 = ta
12 Y = tes0 + w * im_star * lr * 0.01 * pha_ET * Exp(25.89 - 5319 / (TK0 + tes0)) - C
  Z = 1 + w * im_star * lr * 0.01 * pha_ET * Exp(25.89 - 5319 / (TK0 + tes0)) * 5319 / ((TK0 + tes0) ^ 2)
  tes1 = tes0 - Y / Z
  If Abs(tes1 - tes0) > Delta Then GoTo 10
  GoTo 11
10 tes0 = tes1: GoTo 12
11 ET = tes1
End Function

Chapter 11

Flow and Circulation of Matter

11.1 Development of active transport

The underlying feature of heat and mass transfer phenomena taking place here and there within the built environment is the diffusion caused either by the space-wise difference in temperature or by chemical potential (water-vapour pressure). The diffusion itself emerges spontaneously wherever there is a difference either in temperature or in chemical potential (water-vapour pressure) and how much of diffusion occurs simply depends on the physical characteristics of the matter involved. In due course, exergy consumption (entropy generation) is necessarily accompanied as was discussed in previous chapters.

Provided that such spontaneous dispersion is called "passive transport", then there is one other way to be called "active transport" with a variety of sizes, from the ones as tiny as internal organs inside a biological cell, to the one as large as atmosphere on the Earth. Existing in between are various mechanical heating and cooling systems for built-environmental conditioning. They are composed of the units uniquely necessary for active transport, that is, pumps and fans, which transport the mass of water or air for exergy transfer. Heat-transfer units, through which thermal exergy transfer takes place, need to be installed together with appropriate units for transport.

Pumps and fans are of course man-made, but their origin of formation may well be found in various examples in the nature; a chicken embryo as shown in Fig. 11.1 is a typical example. The embryo, which first looks as if being a mere tiny round-shaped spot, proceeds its proper development in case that a thermally appropriate environmental condition is kept. In the very beginning, it does not have any pipe networks, but it gradually self-organizes them as shown in the lower drawing of Fig. 11.1. The embryo imports its necessary nutrients from the yolk sac through the network of pipes called vitelline artery and vein, while at the same time it takes in oxygen molecules from the environmental space and exhausts carbon dioxides into the environmental space through the tiny micro-scaled openings embedded within the eggshell, via the network of pipes called umbilical artery and vein.

The emergence of active-transport network is of crucial importance in the course of development since the amounts of nutrients required and wastes to be discarded become very large in comparison to the relative size of the embryotic body being grown. "Blood" as the carrier of nutrients and wastes is generated together with "arteries" and "veins" as pipes and "heart" as a pump. As we all know, the blood circulation is one of the primary life-supporting processes not only in a chicken but also in all other animals including us human being. Note that human blood circulation was conceived as a rather peculiar idea first by William Harvey (1578–1657) and it took some time for other people to recognize its importance (Wright 2012); this is probably because the circulation is something invisible and it is exactly a kind of function (*Kata*) to read.

Figure 11.2 schematically demonstrates the flow and circulation of blood from the viewpoint of thermodynamics, that is, the embryotic body as "system" on focus,

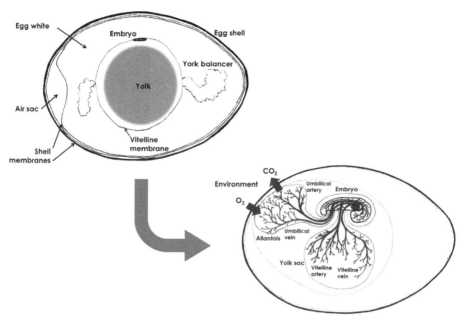

Fig. 11.1: Emergence of an active-transport system in the development of a chicken embryo (These drawings were made referring to Wolpert et al. (1998)).

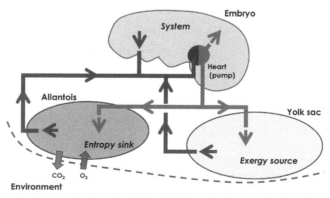

Fig. 11.2: An active-transport system functioning for the cycle of exergy intake and entropy disposal.

yolk sac as "exergy source" gifted by the mother chicken, and allantois as "entropy sink", which is connected with the environmental space of the eggshell. The whole of active-transport system develops so that it holds its own form (*Katachi*) and function (*Kata*) that follows the thermodynamic principle.

11.2 Friction and dispersion

As we have come to know in 8.2, Chapter 8, if two systems being in contact with each other are in the condition of thermodynamic equilibrium, the temperature, pressure, and chemical potential of one system are equal to those of the other system, respectively. Because of the same temperature, pressure and chemical potential, neither passive transport nor active transport can emerge. On the contrary, if there is a difference in temperature or a difference in chemical potential, there emerges passive transport. Likewise, if there is a difference in pressure, there emerges active transport.

Figure 11.3 shows the portions of a duct and a pipe, through which air and water are flowing, respectively. Here, the sectional areas of the duct and the pipe, which are normal to the direction of air and water flows, are assumed to be identical. In the duct wall, there are two manometers attached: one in the upstream side and the other in the downstream side, with which the pressure inside the duct space relative to the external pressure in the surrounding space can be measured. The measuring principle of a manometer is to compare the levels of liquid (water or mercury) filled in the U-shaped tube. The difference in the two heights becomes proportional to the difference between internal pressure and external pressure. This is based on what we came to know in the discussion made in Chapter 8.

In the case of water pipe as shown with the lower drawing in Fig. 11.3, making two tiny holes, one in the upstream side and the other in the downstream side, and thereby putting two vertical open-ended tubes filled with water, respectively, lets us know the height of water in the upstream side being higher than that in the downstream side. If there is no flow of water, the heights of water within the two vertical tubes become the same as each other as we confirmed the characteristic of a communicating vessel with Fig. 8.4 in Chapter 8. This in turn implies that the movement of water causes a decrease in the water pressure at the downstream side in comparison to the water pressure at the upstream side.

Why the air or water pressures at the downstream space are lower than those at the upstream space is due to the friction necessarily causing the generation of heat, which spontaneously dissipates into the surrounding space. As we learnt from the discussion in Chapters 7, 8, and 9, the total amount of energy is necessarily conserved in any natural phenomena and this of course applies to the phenomena involving friction. Therefore, the amount of heat as energy dispersion from the duct and the pipe has to be exactly equal to the decrease in energy held by the bulk of air or water in association with their pressure.

In order to deepen our qualitative understanding so far about the characteristic of friction, let us take a closer look at the flow patterns of fluid schematically demonstrated in Fig. 11.4. The fluid is flowing from the left-hand side to the right-hand side and there are two rigid bodies in the flows. The arrows heading from left to

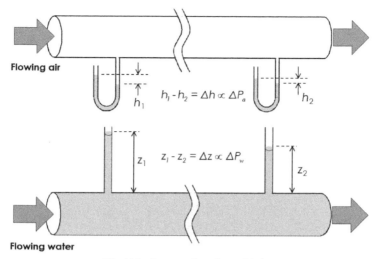

Flowing air

$$h_1 - h_2 = \Delta h \propto \Delta P_a$$

$$z_1 - z_2 = \Delta z \propto \Delta P_w$$

Flowing water

Fig. 11.3: Pressure drop due to friction.

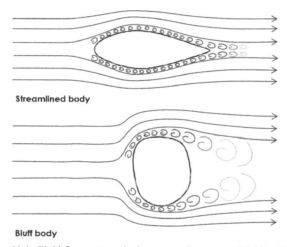

Streamlined body

Bluff body

Fig. 11.4: Fluid-flow patterns in the surrounding space of rigid bodies.

right are called streamlines representing the imaginary movement of fluid particles, though the fluid itself is a conceptually continuous body. The two rigid bodies disturb, more or less, the streamlines and there emerges a lot of vortices along their surfaces. The reason why such vortices emerge is due to the fluid characteristic called "viscosity". How vortices emerge is influenced by the fluid viscosity in addition to the shape of rigid bodies and the surrounding fluid velocity. In general, bluff bodies generate more friction than streamlined bodies as can be imagined by comparing the two drawings in Fig. 11.4.

The space between the surface of a rigid body and the closest streamline contains vortices. This space along the rigid body is called boundary layer (Imai 1970, Bejan 2004). In fact, the viscous nature of the fluid inside the boundary layer is exactly the cause of heat generated by friction. Outside the boundary layer, if a fluid is so viscous, for instance, as honey, chocolate syrup, or ketchup, there is also the generation of heat due to the internal viscous fluid motion. But, if a fluid is much less viscid as water or air, the fluid outside the boundary layer may well be approximated with an imaginary fluid called "perfect fluid", which is entirely inviscid.

Vortices do not stay in the location where they are born, but instead they move towards the downstream space and in due course they sooner or later disperse into nothing; that is, they disappear. Disappearance of vortices, that is exactly dissipation, results in the generation of heat. We all know through our daily experiences that the atmospheric convection occasionally bears tornados, typhoons, or cyclones, and they are never active forever, but necessarily disappear. The same applies to tiny vortices emerging and disappearing here and there in ducts, pipes, and also inside and outside buildings. Such nature with the examples shown in Fig. 11.4 also applies to what happens in the internal space within a duct and a pipe, in which the resulting generation of heat reduces the internal pressure as schematically shown in Fig. 11.3.

11.3 Energy, pressure and exergy balance

Based on the qualitative understanding so far reached in the last section, let us quantify the very basic principle, first on so-called perfect fluid flow, which is totally inviscid, and then on how the effect of friction can be included.

11.3.1 Flow without friction

Suppose that there is an imaginary stream tube as shown in Fig. 11.5. A mass of fluid is flowing into the tube at the velocity of v_1 [m/s] through the inlet, whose area is A_1 [m²], while at the same time, the corresponding equal mass of fluid is flowing out at the air velocity of v_2 [m/s] through the outlet, whose area is A_2 [m²]. This stream tube may be regarded as a portion of the streamlines outside the boundary layers as was shown in Fig. 11.4. The mass coming in for the period of time, Δt [s], through inlet 1 is $\rho(A_1 v_1)\Delta t$, where ρ is the density of fluid [kg/m³]. Similarly, the mass coming out from the stream tube at outlet 2 is $\rho(A_2 v_2)\Delta t$. These masses are at height h_1 [m] and h_2 [m], respectively, as shown in Fig. 11.5. Their potential energies held are $\{\rho(A_1 v_1)\Delta t\}gh_1$ and $\{\rho(A_2 v_2)\Delta t\}gh_2$, respectively, where g is gravitational acceleration rate [m/s²].

The mass of $\rho(A_1 v_1)\Delta t$ exerts work on the fluid inside the stream tube since the fluid is continuous. Assuming that pressure at inlet 1 is P_1 [Pa], this work amounts to $P_1 A_1(v_1 \Delta t)$. In the same way, the mass of $\rho(A_2 v_2)\Delta t$ exerts work outwards as it flows out from the stream tube. This work amounts to $P_2 A_2(v_2 \Delta t)$, where P_2 is the pressure at outlet 2 [Pa].

Since the masses of $\rho(A_1 v_1)\Delta t$ and $\rho(A_2 v_2)\Delta t$ are moving in and out at their respective velocities, v_1 and v_2, they carry kinetic energies, respectively, in addition to exerting work as just mentioned above. They are expressed as $\frac{1}{2}(\rho A_1 v_1 \Delta t)v_1^2$ and $\frac{1}{2}(\rho A_2 v_2 \Delta t)v_2^2$, the form of which is identical to kinetic energy held by a raindrop discussed in Chapter 9.

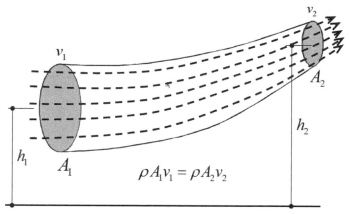

Fig. 11.5: Imaginary stream tube, through which an incompressible and inviscid fluid is flowing.

Assuming that the whole mass inside the stream tube is constant, that is, the fluid is incompressible, mass balance can be expressed as follows

$$\rho(A_1 v_1)\Delta t = \rho(A_2 v_2)\Delta t, \text{ that is, } A_1 v_1 = A_2 v_2 \tag{11.1}$$

Energy balance can be expressed as follows

$$P_1 A_1 (v_1 \Delta t) + \frac{1}{2}(\rho A_1 v_1 \Delta t) v_1^2 + \rho (A_1 v_1 \Delta t) g h_1$$

$$= P_2 A_2 (v_2 \Delta t) + \frac{1}{2}(\rho A_2 v_2 \Delta t) v_2^2 + \rho (A_2 v_2 \Delta t) g h_2 \tag{11.2}$$

Equation (11.2) can be reduced to the following equation by dividing all terms with $A_1 v_1 \Delta t$, which is the flowing volume of fluid for the period of Δt, and also by taking the relationship expressed by Eq. (11.1) into consideration

$$P_1 + \frac{1}{2}\rho v_1^2 + \rho g h_1 = P_2 + \frac{1}{2}\rho v_2^2 + \rho g h_2 \tag{11.3}$$

This relationship is called Bernoulli theorem, which was originally conceived by Daniel Bernoulli (1700–1782) and fully developed by Leonhard Euler (1707–1783) as it is. What we should note here is that the energy balance within the stream tube set up as Eq. (11.2) has been reduced to the equation on pressure balance expressed as Eq. (11.3). The unit of pressure is Pascal (Pa) as we discussed in Chapter 8; Pa is equal to the unit of N/m² and it may be converted to J/m³ by multiplying metre both to the denominator and the numerator (= (N·m)/(m²·m)). This implies that pressure is in fact equivalent to the density of energy held by the unit volume of a fluid.

Equation (11.3) is definitely valuable and useful, but we should keep its temporal position in relation to heat-transfer and thermodynamic sciences. Bernoulli theorem expressed as Eq. (11.3) was found several decades before heat transfer science was begun by Black (1728–1799), Fourier (1768–1830), and others, and also before thermodynamics was begun by Carnot (1792–1832), Joule (1818–1889), Thomson (1824–1907) and Clausius (1822–1888). If such succession of the birth of fluid dynamics, heat transfer science, and thermodynamics is still associated with the contemporary scant connectivity between them, then an adequate connectivity should be sought.

If the height at inlet 1 is the same as that at outlet 2, then Eq. (11.3) can further be reduced to the following equation

$$P_1 + \frac{1}{2}\rho v_1^2 = P_2 + \frac{1}{2}\rho v_2^2 \qquad (11.4)$$

Suppose that there is a fan injecting a volume of air at a certain rate into a duct as shown in Fig. 11.6. The continuous injection of air results in the constant mass flow of air inside the duct at the velocity of v_1 and the internal pressure exerted on the interior surface of the duct being at P_1, which is higher than the external pressure at P_o.

In order to introduce the effect of external pressure, P_o, into Eq. (11.4), let us add $(-P_o)$ to both sides of Eq. (11.4). Then, we can rewrite Eq. (11.4) as follows

$$(P_1 - P_o) + \frac{1}{2}\rho v_1^2 = (P_2 - P_o) + \frac{1}{2}\rho v_2^2 \qquad (11.5)$$

The whole of the left-hand side of Eq. (11.5) is what we can measure by the central manometer attached to the duct shown in Fig. 11.6. In this central manometer, the internal opening is perpendicular to the flow of air inside the duct; this allows the total pressure to be measured. Two other manometers positioned on the left and on the right of the central manometer are for the measurement of the first and second terms of the left-hand side of Eq. (11.5), respectively. The left one corresponding to the first term of the left-hand side of Eq. (11.5) is to measure "static pressure", which is isotropic, and the right one corresponding to the second term of the left-hand side of Eq. (11.5) is to measure "dynamic pressure", which is anisotropic. In other words, the former is associated with the stagnancy of the fluid, while the latter is associated with the motion of the fluid.

Equation (11.5) indicates that the sum of static and dynamic pressure is constant provided that two positions of the duct are close enough so that the effect of friction is negligible. This characteristic of Eq. (11.5) can be applied to the measurement of air velocity or the volumetric flow rate. The example is so-called Venturi tube, which originated from the idea conceived by Giovanni B. Venturi (1746–1822).

We may consider such a tube, whose shape is not bluff but streamlined as shown in Fig. 11.7. Substituting the relationship expressed by Eq. (11.1) into Eq. (11.5) and making a bit of algebraic operation yields the following formula for the calculation of the inlet air velocity, v_1

$$v_1 = \sqrt{\frac{2(P_1 - P_2)}{\rho\left\{\left(\dfrac{A_1}{A_2}\right)^2 - 1\right\}}} \qquad (11.6)$$

Since the two openings of manometers attached to the tube are parallel to the flow of air, the difference in the height of fluid (water or mercury) turns out to be the difference in static pressures at sections 1 and 2.

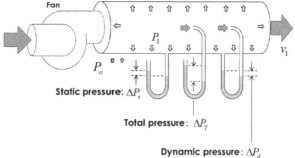

$$\Delta P_T = \Delta P_s + \Delta P_d = \left(P_1 - P_o\right) + \frac{1}{2}\rho v_1^2$$

Fig. 11.6: Total pressure comprising static and dynamic pressures.

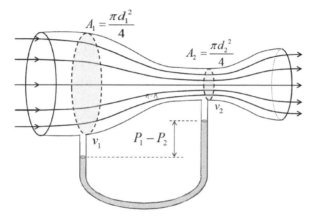

Fig. 11.7: The principle of a Venturi tube for the measurement of air velocity.

The characteristic that we should learn from Eq. (11.6), in addition to the possibility of air-velocity measurement, is that if a reduced outlet area, in comparison to the inlet area, is prepared, we can increase the air velocity. The opposite is also true, that is, setting up a rather small opening for inlet together with a larger opening enables us to have a rather high air velocity within the internal space near the inlet. Whether it is inlet or outlet, in order to realize such an increase of air velocity, there has to be a source of energy for an increase in kinetic energy of the flowing fluid. This, in fact, comes from the energy carried by the air associated with its static pressure. The increase in air velocity is compensated by the corresponding decrease in static pressure to be measured by the manometer set up as shown in Fig. 11.7. These features are worth keeping in mind for making natural ventilation attractive and effective as will be described later in the present chapter.

11.3.2 *Flow with friction*

As we discussed in 11.2, actual fluids are more or less viscous so that a long duct or a long pipe necessarily induces friction that results in the generation of heat.

Nevertheless, we discussed the characteristics of inviscid fluids first in the previous sub-section. Here in this subsection, let us come up with a way of pressure balance including the effect of pressure decrease caused by friction and also how it is related to exergy balance.

Referring again to Fig. 11.3, we may rewrite Eq. (11.3) so that the effect of pressure decrease is inclusive within the pressure balance. Assuming the height h_1 to be equal to the height h_2,

$$P_1 + \frac{1}{2}\rho v_1^2 - \Delta P_f = P_2 + \frac{1}{2}\rho v_2^2 \qquad (11.7)$$

where ΔP_f is the decrease in total pressure due to friction occurring between positions 1 and 2 [Pa = N/m² = J/m³] shown in Fig. 11.3. Multiplying the volumetric flow rate, V [m³/s], to both sides of Eq. (11.7) yields the energy balance equation, which includes the effect of heat generated by friction

$$P_1 V + \left(\frac{1}{2}\rho v_1^2\right)V - \Delta P_f V = P_2 V + \left(\frac{1}{2}\rho v_2^2\right)V \qquad (11.8)$$

Since the unit of pressure is equivalent to J/m³ as mentioned in the previous sub-section and the unit of volumetric flow rate is m³/s, each term of Eq. (11.8) is in the unit of J/s, that is, W. Therefore, the last term of the left-hand side of Eq. (11.8), $\Delta P_f V$, is confirmed to be exactly the rate of heat generation due to friction. Now, we may say that fluid dynamics is connected with thermodynamics.

Having the exergy-consumption theorem described in Chapters 7 and 9 in mind, $\Delta P_f V$ should correspond to the rate of exergy consumption. This implies that we should be able to come up with exergy balance equation, which is connected with the energy balance equation expressed as Eq. (11.8). For this purpose, we include the effect of the surrounding pressure, P_o, into Eq. (11.8) as we did for coming up with Eq. (11.5). That is,

$$(P_1 - P_o)V + \left(\frac{1}{2}\rho v_1^2\right)V - \Delta P_f V = (P_2 - P_o)V + \left(\frac{1}{2}\rho v_2^2\right)V \qquad (11.9)$$

11.3.3 High-pressure and low-pressure exergies

Equation (11.9) can be read as exergy balance equation. The last term on the left-hand side, $\Delta P_f V$, is nothing other than exergy consumption rate, X_C [W], to be expressed as follows

$$X_C = \Delta P_f V = S_g T_o \qquad (11.10)$$

where S_g is entropy generation rate [W/K] due to friction and T_o is the environmental temperature [K].

The two terms in front of $\Delta P_f V$ of Eq. (11.9) are exergy inputs and the two terms on the right-hand side are exergy outputs. The second term on each side of Eq. (11.9) is kinetic exergy. The first term on each side of Eq. (11.9) is work to be performed due to the difference in pressure between the air inside the duct and the environment. The manner in which they are expressed, in relation to the pressure difference, is similar to how thermal exergy and chemical exergy are determined, respectively, as described in Chapters 7 and 8. They are the exergy to be determined by the difference in pressure, which turns all the flow and circulation into action.

In order to clarify the nature of exergy in relation to the pressure difference between a system in question and its environment, let us take a look at a duct system comprizing two subsystems A and B as shown in Fig. 11.8. Within subsystem A, there is a wind turbine to extract work from the flow of air inside the duct at the rate of W_{out} [W], while on the other hand, within subsystem B, there is a fan to be operated with the input of work by electricity at the rate of W_{in} [W] to let the bulk of air move from left to right.

The fan pressurizes the air moving forward so that the air pressure at outlet 2, P_2 [Pa], is higher than the environmental pressure, P_o [Pa], while on the other hand, since the fan sucks the air from the rear side, the air pressure at inlet 1, P_1 [Pa], is lower than P_o. The duct surface is assumed to be thermally perfect-insulated, but a portion of the surface of respective subsystems is thermally conductive so that heat transfer takes place at the rates of Q_{in} [W] in subsystem A and Q_{out} [W] in subsystem B.

In order to make the present discussion simple, let us assume that the air temperature inside the duct and the environmental temperature are the same, that is, $T_1 = T_2 = T_o$, and also the sectional areas are the same throughout the duct, that is, $v_1 = v_2 = v_o$ according to the relationship expressed by Eq. (11.1). With these assumptions in mind, exergy balance equation for subsystem B can be set up as follows

$$W_{in} + (P_1 - P_o)V - X_{CB} = (P_2 - P_o)V \tag{11.11}$$

$$X_{CB} = \Delta P_{fB}V = S_{gB}T_o, \tag{11.12}$$

where X_{CB} is the exergy consumption rate within subsystem B [W], ΔP_{fB} is the pressure difference between inlet 1 and outlet 2 [Pa], and S_{gB} is the entropy generation rate within subsystem B [W/K]. The reason that there is no thermal exergy flow with Q_{in} and Q_{out} in Eq. (11.11) is due to the assumption of $T_1 = T_2 = T_o$.

Since $P_1 < P_o < P_2$ and $0 < V$, $(P_1 - P_o)V < 0$ and $0 < (P_2 - P_o)V$. Namely, the exergy in relation to pressure becomes either positive or negative. Such feature of exergy in relation to pressure, negativity, has already been pointed out by Ishigai (1977) and Bejan (2006), but a further explanation on its implication has not yet been given. Therefore, let us do it here. For this pupose, we first add $\{-(P_1 - P_o)\}V$, which is necessarily positive, to both sides of Eq. (11.11) and find the following equation

$$W_{in} - X_{CB} = \{-(P_1 - P_o)\}V + (P_2 - P_o)V. \tag{11.13}$$

Equation (11.13) is the final form of exergy balance equation for subsystem B under the condition of $T_1 = T_2 = T_o$ and $v_1 = v_2 = v_o$, since this form is exactly identical to other exergy balance equations described in Chapter 7.

What Eq. (11.13) implies is that there is an exergy input as W_{in} to subsystem B, its portion as X_{CB} $(= \Delta P_{fB}V)$ is consumed and thereby two exergy outputs as $\{-(P_1 - P_o)\}V$ and $(P_2 - P_o)V$ are produced. Here, let us call $(P_2 - P_o)V$ as "high-pressure" exergy, since $P_2 > P_o$, and $\{-(P_1 - P_o)\}V$ as "low-pressure" exergy, since

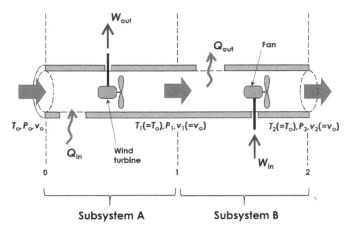

Fig. 11.8: A duct system comprising two subsystems.

$P_1 < P_o$. The characteristic of "high-pressure" exergy is similar either to "warm" exergy or "wet" exergy, while that of "low-pressure" exergy is either to "cool" exergy or "dry" exergy.

The exergy balance equation for subsystem A to be consistent with Eq. (11.13) for subsystem B can be expressed as follows

$$\{-(P_1 - P_o)\}V - X_{CA} = W_{out}, \tag{11.14}$$

$$X_{CA} = \Delta P_{fA}V = S_{gA}T_o, \tag{11.15}$$

where X_{CA} is the exergy consumption rate within subsystem A [W], ΔP_{fA} is the pressure difference between inlet 0 and outlet 1 [Pa], and S_{gA} is the entropy generation rate within subsystem A [W/K]. Equation (11.14) implies that there is the input of "low-pressure" exergy, $\{-(P_1 - P_o)\}V$, to subsystem A, its portion, X_{CA} $(= \Delta P_{fA}V)$, is consumed and thereby work is produced at the rate of W_{out} by the wind turbine.

Figure 11.9 schematically demonstrates the exergy balances in subsystems A and B. In summary, the fluid flow through subsystems A and B is realised by the exergy input into a fan in subsystem B, where both "high-pressure" and "low-pressure" exergies are produced with the exergy consumption rate at X_{CB}. The wind turbine in subsystem A functions with the supply of "low-pressure" exergy provided by subsystem B and it produces work at the rate of W_{out} as the result of exergy consumption rate at X_{CB}. Note that the flow of "high-pressure" exergy is in the same direction as the flow of air, while the flow of "low-pressure" exergy is in the opposite direction to the flow of air. The latter relationship between exergy flow and air flow is comparable to the relationship between "cool" exergy flow and heat flow or that between "dry" exergy flow and moisture flow.

The characteristics of "high-pressure" and "low-pressure" exergies given here for a duct system with a fan and a wind turbine can be applied to a pipe system with a pump and a waterwheel. They can also be applied to describing how natural ventilation systems function, either with wind effect or with buoyancy effect as will be discussed in the next section.

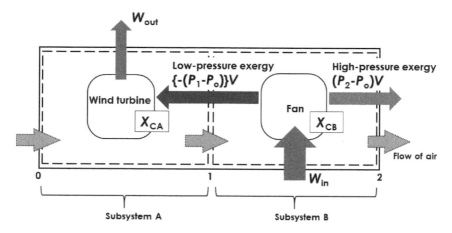

Fig. 11.9: Exergy balances in two subsystems.

Note
1. Flow of air is from left to right.
2. Friction occurring due to the flow of air and the rotation of wind turbine cause the exergy consumption in subsystem A, X_{CA}.
3. Rotation of the blades in the fan and friction occurring due to the flow of air cause the exergy consumption in subsystem B, X_{CB}.

11.3.4 Pressure decrease in ducts and pipes

In order to make it possible to calculate exergy consumption rate, X_C ($= \Delta P_f V$), we need to characterize how the pressure decreases in the course of air (or water) flowing inside a duct (or pipe). In general, the decrease in pressure expressed as ΔP_f becomes larger as the duct is made longer or its diameter smaller. The pressure decrease also becomes larger as the fluid velocity increases, that is, the dynamic pressure increases. The whole of such relationship may be expressed with one single formula as

$$\Delta P_f = f \frac{L}{D}\left(\frac{1}{2}\rho v^2\right), \tag{11.16}$$

where f is the proportional coefficient, which is called Darcy-Weisbach friction factor or Darcy's friction factor. The form of Eq. (11.16) was first conceived by J. Weisbach (1806–1871) and H. Darcy (1803–1859) identified the importance of surface roughness in determining the value of friction factor (Simmons 2008). L is the duct (or pipe) length [m], D is the duct (or pipe) diameter [m], ρ is the fluid density [kg/m³], and v is the fluid velocity [m/s].

The friction factor is characterized in relation to "shear stress", whose physical implication is positioned in parallel to pressure discussed in Chapters 7 and 8. The stress within fluids without motion is expressed by pressure alone, but that with motion is expressed not only by pressure but also by "shear stress". Shear stress, τ_s, is the force exerted parallelly on the unit area of a surface, while pressure, P, is the force exerted perpendicularly on the unit area of a surface. Therefore, both τ_s and P have the same unit as Pa ($= N/m^2$).

The shear stress, τ_s, can be assumed to be proportional to the space-wise gradient of velocity emerged within a viscous fluid, dv/dz, as shown in Fig. 11.10. At the solid surface,

$$\tau_s = \mu \left(\frac{dv}{dz} \right)_{z=0}, \tag{11.17}$$

where μ is the proportional coefficient called "dynamic viscosity" or simply "viscosity" [Pa·s], which is the physical characteristic unique to fluid dynamics positioning in parallel to thermal conductivity in heat-transfer science. The order of dynamic viscosity under the condition of temperature at 20°C is as follows: 18×10^{-6} Pa·s for air; 1×10^{-3} Pa·s for water; and 6×10^{0} Pa·s for honey. The human blood is in the order of 3.5×10^{-3} Pa·s at 37°C. The blood is more than three times viscous than water due to its containing various elements other than water and plasma in it.

There is one other intrinsic coefficient called "kinematic viscosity", κ_v, defined as $\kappa_v = \mu/\rho$ [m²/s]. Its order is 15×10^{-6} m²/s for air and 1×10^{-6} m²/s for water (both at 20°C) because of the density being 1.2 kg/m³ for air and 1000 kg/m³ for water. This indicates that air is fifteen-times more viscous than water; it is owing to the characteristic difference between gas and liquid.

In general, the viscosity of gaseous matters tends to increase as their temperature increases, while on the other hand, the viscosity of liquid matters tends to decrease as their temperature increases. Such difference comes from the following facts. Gaseous molecules bounce on and off each other and also the solid surface of their container. If their temperature increases and thereby molecular speed becomes higher, then gaseous viscosity necessarily becomes higher because molecular action of bouncing on and off becomes more vigorous. On the other hand, liquid matters, whose molecules are bonded each other more or less with intermolecular force, become loose if the effect of thermal motion becomes more significant as their temperature becomes higher. This makes the viscosity of liquid lower.

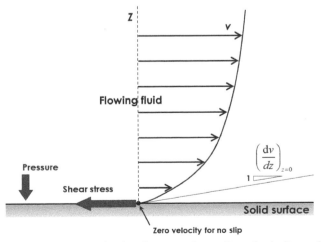

Fig. 11.10: Shear stress to be proportional to the space-wise gradient of velocity parallel to the solid surface.

Figure 11.11 schematically demonstrates how the shear stress relates to the infinitesimal decrease in pressure, dP, within the infinitesimally short length of a duct, dx. For the infinitesimally short period of time, dt, the incoming momenta, one due to upstream pressure and the other due to shear stress, has to be balanced with the outgoing momentum due to downstream pressure. Following what was described in Chapter 9, the momentum balance equation can be expressed as follows

$$\left\{\left(\frac{\pi D^2}{4}\right)P\right\}dt + \left\{-(\pi D dx)\tau_s\right\}dt = \left\{\left(\frac{\pi D^2}{4}\right)(P-dP)\right\}dt. \qquad (11.18)$$

Equation (11.18) can be reduced to

$$dP = \frac{4\tau_s}{D}dx. \qquad (11.19)$$

Although this equation has been obtained from the momentum balance equation, the same result can be obtained from the energy balance equation, which can be set up by replacing dt in Eq. (11.18) with dx. Integrating Eq. (11.19) for the whole decrease in pressure, ΔP_f, occurring in the duct length of L yields the following equation

$$\int_0^{\Delta P_f} dP = \frac{4\tau_s}{D}\int_0^L dx, \text{ that is, } \Delta P_f = 4\tau_s\frac{L}{D}. \qquad (11.20)$$

Comparing Eq. (11.16) with Eq. (11.20), we come to know that the friction factor is four times the ratio of shear stress to the dynamic pressure.

$$f = \frac{4\tau_s}{\frac{1}{2}\rho v^2} \qquad (11.21)$$

Experimental research work in relation to Darcy's friction factor was intensively made by quite a few scientists and engineers such as G. Hagen (1797–1884), J. Poiseuille (1797–1869), O. Reynolds (1842–1912), L. Prandtl (1875–1953), T.

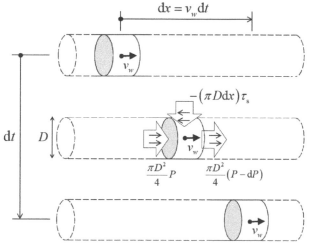

Fig. 11.11: Momentum balance in a volume of flowing matter while moving for a short distance in a short period of time.

von Karman (1881–1963), J. Nikuradse (1894–1979), and others in the first half of 20th century. L.F. Moody (1944) complied those experimental data in the form of so-called Moody's chart, which illustrates the relationship between friction factor and Reynolds number as shown in Fig. 11.12. This graph was produced by using a set of empirical equations for calculating the values of friction factor (Turns 2006) given in Appendix 11.A.

Reynolds number, *Re*, which was named after Reynolds, characterizes the relative magnitude of viscosity, velocity and the size of solid body as follows

$$Re = \frac{\rho v D}{\mu} = \frac{vD}{\kappa_v}. \tag{11.22}$$

Reynolds number is the non-dimensional number to be derived from exact mathematical expression of the momentum balance with respect to viscous fluid given as so-called Navier-Stokes equation, which was named after C. Navier (1785–1836) and G. Stokes (1819–1903) who developed the equation independently from each other.

Reynolds number, *Re*, can be regarded as a kind of index representing the relative roles of inertial momentum held by a fluid mass and the viscous momentum exerted on it. The fluid flow with smaller Reynolds number is relatively more viscous, while on the other hand, the fluid flow with larger Reynolds number is relatively less viscous. The fluid flow with Reynolds number being infinity corresponds to perfect fluid. The flow with *Re* < 2320 inside a duct (or pipe) is relatively viscous and laminar, while the flow with 5000 < *Re* is less viscous and turbulent. Critical zone shown in Fig. 11.12 roughly represents the range in which laminar flow disappears and instead turbulent flow starts emerging and growing.

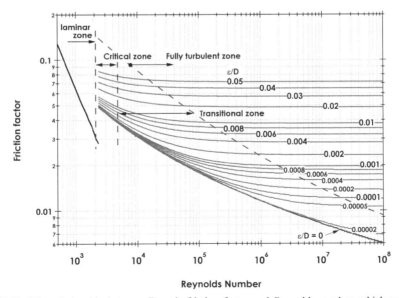

Fig. 11.12: The relationship between Darcy's friction factor and Reynolds number, which was first presented as shown by Moody (1944). This graph was made by calculating the values of friction factor from a set of empirical formulae given in Turns (2006).

The parameter, ε, represents the surface roughness of solid: for example, $\varepsilon = 0.0015$ mm for glass tube, $\varepsilon = 0.007$ mm for polyethylene pipe, $\varepsilon = 0.15$ mm for galvanized iron pipe, $\varepsilon = 2$ mm for eroded iron pipe or concrete-made duct, and $\varepsilon = 4$ mm for brick-made duct. The ratio of ε to D, ε/D, indicates the relative roughness of the internal surface of the duct or pipe; for example, a glass tube with its diameter of 5 mm is equivalent to a concrete-made duct with its diameter of 13.3 m (= 13300 mm).

Since the volumetric flow rate is expressed by

$$V = \frac{\pi D^2}{4} v, \tag{11.23}$$

the exergy consumption rate, $X_C \, (= \Delta P_f V)$, can then be expressed as

$$X_C = \Delta P_f V = \frac{\pi}{8} f D L \rho v^3. \tag{11.24}$$

Equation (11.24) indicates that making the velocity smaller should result in a significant reduction in exergy consumption rate due to friction, since the exergy consumption rate is proportional to the velocity cubed. But, a lower velocity could result in thermal exergy to be carried and transferred becoming smaller, since the volumetric rate to be calculated from Eq. (11.23) turns out to be smaller, though it should be made larger in order to increase the rate of thermal exergy to be transferred. This suggests that there is an optimization problem (the minimization of exergy consumption rate) in terms of a duct or a pipe system with heat exchangers. More will be discussed later in the present chapter.

11.4 Passive and active systems for ventilation

As described in Chapter 3, we humans live by inhaling and exhaling all the time. This is exactly the same as what happens in a chicken embryo shown in Fig. 11.1, though the chicken embryo does not have a lung. The eggshell, which is open to the surrounding air, corresponds to a building envelope system comprising walls, windows, doors, ceiling, and floor. This implies that the building envelope system has to be open to outdoor air so that indoor pollutants can be effectively removed, while at the same time fresh air can be taken in. Fresh air intake is also important to get rid of indoor thermal pollution by removing unnecessary "warm" exergy within the indoor environment. The eggshells are leaky so that ventilation is made relying on infiltration and exfiltration, but building walls should not be leaky. Instead, they should be sufficiently air-tight so that ventilation is properly made through appropriate window openings either by passive systems (natural ventilation) or by active systems (mechanical ventilation).

11.4.1 Indoor-air pollutants and air change rate

We humans are the primal source of indoor air pollution since we have to exhale carbon dioxide (CO_2) and moist air ceaselessly from our noses and mouths. The evaporated water is also being discharged from our skin surfaces. There are also other possible sources such as various volatile organic compounds (VOCs) coming out from electric appliances, furniture and building materials, moulds and mites that live with pets or foliage plants in pots with mud, and tobacco smoke if smoking is allowed and there are smokers. These sources should be minimized in ordinary indoor spaces so that the resulting pollution does not cause any health-related problems, that is, so-called sick-building syndrome (SBS) and also so that the required rate of fresh air to be taken in through mechanical heating and cooling systems can be minimized.

Once the sources of indoor pollution other than human beings are reduced to the minimum, then the required rate of ventilation may be determined based on the needs of building occupants. Figure 11.13 shows the relationship between the metabolic energy generation rate of a person and the rates of CO_2 emission and exhalated air. Both the rate of CO_2 emission and the volumetric rate of exhaled air increases linearly as the rate of metabolic energy generation rate increases.

With the activity level of 1 Met, one person discharges CO_2 approximately at the rate of 9.9 mg/s with the exhaled volume of air at the rate of 126.5 mL/s. This implies that the concentration of CO_2 in the exhaled air is 0.0783 mg/mL, which is equal to 78.3 g/m^3 and also to 43500 ppm, higher than the inhaled air. If the atmospheric CO_2 concentration is assumed to be 400 ppm, the exhaled air turns out to be 43900 ppm. Here we come to know that the concentration of CO_2 of exhaled air is very high, more than one-hundred times the outdoor concentration of CO_2. This confirms that ventilation is very much necessary to keep healthy and comfortable built environment.

The indoor air is also contaminated with moisture discharged from occupants. Figure 11.14 shows how much of moisture comes out from one adult person at the metabolic energy generation rate of 1.1 Met. Discharged moisture with the exhaled air amounts approximately to 36% and that from skin surface to 64% when the surrounding air temperature is 20°C. The total of moisture emission decreases as the surrounding air temperaure increases; this is because the difference in water-vapour concentration between the skin surface and the surrounding air becomes smaller as

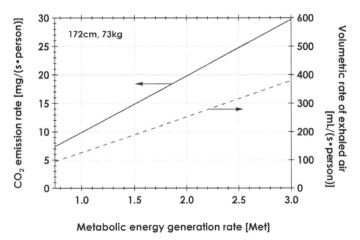

Fig. 11.13: Carbon-dioxide emission and volumetric exhalation rates as a function of metabolic energy generation rate.

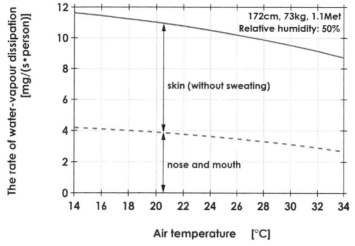

Fig. 11.14: Water-vapour discharged from a human being as a function of surrounding air temperature.

the surrounding air temperature increases with the condition of relative humidity being constant at 50%. This suggests that fresh air taken in should be decreased down to the required minimum rate in winter, while on the other hand, it may be increased up to a certain desired rate, especially in summer when natural ventilation is used, since it makes possible for the human body to rationalize its exergy consumption rate for restoring thermal comfort as discussed in Chapter 10.

The criterion of ventilation is usually expressed with air change rate in the unit per hour, which is the ratio of the air volume taken from outdoors for every one-hour period to the volume of room space. The air change rate of one time per hour (1 h^{-1}) implies that the whole volume of room air is replaced with outdoor air for the period of one hour.

Figure 11.15 demonstrates the variation of average concentration of CO_2 within a room having the volume of 93.6 m³ (floor area of 36 m² with the ceiling height of 2.6 m). The assumption is that five persons, 7.2 m²/person, occupy the room for the period of one hour and thereafter leave the room and the air change rate is kept constant throughout the two-hour period at 0.5, 1, 2, or 5 h^{-1}. The outdoor concentration of CO_2 is constant at 400 ppm. The air change rate at 0.5 h^{-1} corresponds to the minimum requirement in residential buildings with 15 to 30 m²/person and 5 h^{-1} corresponds to a case with natural ventilation as will be described later in this section.

The concentration of CO_2 shown in Fig. 11.15 was calculated from the following formulae. While the room space is occupied for the first one-hour period,

$$C_{cd}(t) = C_{cd\_out} + \frac{G}{n_{ach}V_r} + \left[\{C_{cd}(0) - C_{cd\_out}\} - \frac{G}{n_{ach}V_r} \right] e^{-n_{ach}t}. \qquad (11.25)$$

While nobody occupies for the second one-hour period,

$$C_{cd}(t) = C_{cd\_out} + \{C_{cd}(t_1) - C_{cd\_out}\} e^{-n_{ach}(t-t_1)}. \qquad (11.26)$$

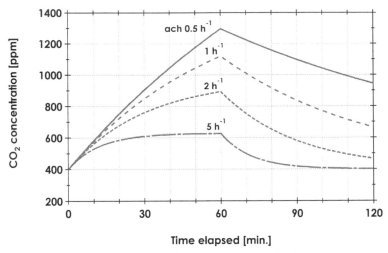

Fig. 11.15: Variation of average carbon-dioxide concentration in the room space.

$C_{cd}(t)$ is the concentration of CO_2 [g/m³] at time t [s], C_{cd_out} is the concentration of CO_2 outdoors, which is assumed to be constant [g/m³], G is the whole emission rate of CO_2 [g/s], n_{ach} is air change rate per second [s⁻¹], which is 1/3600 of air change rate per hour, V_r is room air volume [m³], and t_1 is the time when the occupants leave the room (= 3600 s). The results shown in Fig. 11.15 are the cases with $C_{cd}(0) = C_{cd\_out}$. These formulae are derived based on the mass balance of CO_2 by applying the method of mathematical operation described in 9.2.3, Chapter 9.

The concentration of CO_2 used in Eqs. (11.25) and (11.26) is gram per cubic meters, that is, g/m³. This can be converted into the unit of parts per million (ppm) by multiplying the conversion factor, 1.865 T_{ra} [ppm/(g/m³)], where T_{ra} is air temperature in Kelvin, to the value of the concentration in the unit of g/m³; for example, 1000 mg/m³ corresponds to 556 ppm assuming that T_{ra} = 298K (= 25°C).

As can be seen in Fig. 11.15, with the air change rate at 0.5 h⁻¹, the concentration of CO_2 can easily increase, while on the other hand, with the air change rate at 5 h⁻¹, the concentration of CO_2 does not increase more than 600 ppm. If the required ventilation rate of 20 L/(s·person) (Sepännen and Fisk 2004) is assumed, then the room space of 93.6 m³ with five persons needs 3.8 h⁻¹ of air change rate, with which the concentration of CO_2 remains lower than 700 ppm. If the required ventilation rate is reduced to 10 L/(s·person), then the air change rate turns out to be 1.9 h⁻¹. In general, the average indoor concentration rate of CO_2 should be kept lower than 1000 ppm.

Such a relationship between the concentration of CO_2 and the air change rate is basically the same as that between water-vapour concentration and air change rate, as far as desorptive and adsorptive nature of building materials is negligible.

11.4.2 Pressure decrease in window openings

Windows are important building elements for natural ventilation in addition to daylighting discussed in Chapter 5. As described in previous section, in order for a duct or pipe system to function properly, there has to be both inlet and outlet so that the air or water can flow through smoothly. The same applies to ventilation. There need to be two openings, through which air can flow through for making a portion of indoor air replaced smoothly with outdoor fresh air.

Since the form of a room space is usually not streamlined but rather blunt, how the room air is replaced with the outdoor air is quite complex. In order to know fully the horizontal and vertical distribution of pressure and air velocity, we need to subdivide the room space into discrete small cells, set up the Navier-Stokes momentum balance equation together with mass and energy balance equations and thereby solve them in terms of pressure and air velocity. Although it should be very useful, time-consuming calculation is required in general and also the basic knowledge of fluid-dynamic nature is necessary for a sufficient understanding of the results obtained from a large amount of calculation (Anderson 1995).

Although the detailed information in terms of the distribution of pressure and air velocity cannot be obtained, a kind of rough estimation of average indoor pressure and air velocity through the window openings based on the pressure balance equations described in 11.3.2 is still useful in knowing the overall air change rate to be realized in particular by natural ventilation. As we assumed that the pressure decrease by friction in a duct system is proportional to dynamic pressure, we also assume the pressure decrease within a window opening, ΔP_{fw} [Pa], by the following equation

$$\Delta P_{fw} = \xi \left(\frac{1}{2} \rho v_w^{\ 2} \right),$$ (11.27)

where ξ is the overall proportional factor characterized by the shape of a window opening, which is called "dynamic presurre-decrease (loss) coefficient", which is equivalent to the factor, $f\left(\frac{L}{D} \right)$, in Eq. (11.16) and v_w is the average velocity of air flowing through the window opening.

Using the volumetric air flow rate as V_w [m³/s] and the total window opening area as A_w [m²], Eq. (11.27) may be rewritten as

$$V_w = \alpha A_w \left(\frac{2}{\rho} \Delta P_{fw} \right)^{\frac{1}{2}}$$ (11.28)

where $\alpha \left(= \frac{1}{\sqrt{\xi}} \right)$ is called "discharge coefficient", which indicates the proportion of a window area effective for ventilation. Figure 11.16 shows some examples of the dynamic pressure-decrease coefficient and the discharge coefficient (Ishihara 1969, Aynsley et al. 1977). The more complex the window openings, the larger the dynamic pressure-decrease coefficient and the smaller the discharge coefficient.

11.4.3 *Wind-driven ventilation*

When a wind blows around a building, it exerts dynamic pressure, ΔP_W or ΔP_L, as shown in Fig. 11.17. The ratio of dynamic pressure, ΔP_W or ΔP_L, to the dynamic pressure given with the outdoor reference air velocity, v_o, at a certain height is called

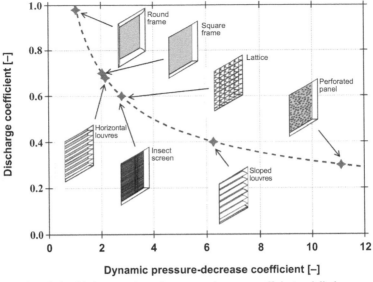

Fig. 11.16: The relationship between dynamic pressure-decrease coefficient and discharge coefficient.

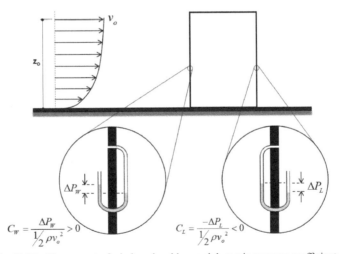

Fig. 11.17: The concept of windward and leeward dynamic pressure coefficients.

"dynamic pressure coefficient": that is, $C_W = \dfrac{\Delta P_W}{\left(\frac{1}{2}\rho v_o^2\right)}$ and $C_L = \dfrac{-\Delta P_L}{\left(\frac{1}{2}\rho v_o^2\right)}$.

The dynamic pressure on the windward side is in the direction to push the building wall inward, $C_W > 0$, while on the other hand, $C_L < 0$, the dynamic pressure on the leeward side is to pull the building wall outward. Their typical values are as shown in Fig. 11.18 (Ishihara 1969). The negativity of leeward dynamic-pressure coefficient implies that the wind blowing over a building acts so that the bulk of air near the leeward surface is peeled off.

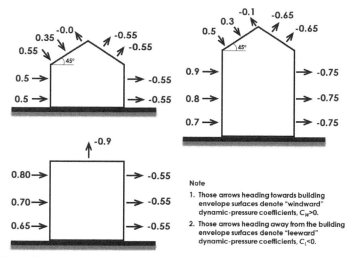

Fig. 11.18: Typical values of windward and leeward dynamic-pressure coefficients.

Assuming a room having two windows: one in windward and the other in leeward as shown in the drawing attached to Fig. 11.19, the pressure balance across the respective windows can be expressed as follows referring to Eq. (11.7)

$$P_o + C_W \left(\frac{1}{2} \rho_o v_o^2 \right) - \Delta P_{f\_w1} = P_r, \tag{11.29}$$

$$P_r - \Delta P_{f\_w2} = P_o + C_L \left(\frac{1}{2} \rho_o v_o^2 \right), \tag{11.30}$$

where $\Delta P_{f\_w1} = \xi_{w1} \left(\frac{1}{2} \rho_o v_{w1}^2 \right)$ and $\Delta P_{f\_w2} = \xi_{w2} \left(\frac{1}{2} \rho_r v_{w2}^2 \right)$, for which ξ_{w1} and v_{w1} are the dynamic pressure-decrease coefficient and air velocity at the windward window, respectively; ξ_{w2} and v_{w2} are those at the leeward window, respectively; and ρ_o and ρ_r are the density of outdoor air and indoor air, respectively. P_o and P_r are static pressure outdoors and indoors at the floor level [Pa], respectively.

If indoor air temperature is higher than outdoor air temperature, then $\rho_o > \rho_r$. It causes buoyancy effect provided that the heights of two windows are different. In such a case, P_o and P_r at windward and leeward windows should be taken at their respective heights. But, here we assume that the heights of windward and leeward windows are equal to each other so that the buoyancy effect does not matter.

The mass balance equation is,

$$\rho_o A_{w1} v_{w1} = \rho_r A_{w2} v_{w2} \tag{11.31}$$

where A_{w1} and A_{w2} are windward and leeward window areas [m²], respectively.

If the values of P_o, v_o, C_W, C_L, ξ_{w1}, ξ_{w2}, A_{w1}, A_{w2}, ρ_o and ρ_r are assumed, then the unknown variables are the average room air pressure, P_r, and two air velocities, v_{w1} and v_{w2}. Therefore, with three equations above, from Eqs. (11.29) to (11.31), the values of P_r, v_{w1} and v_{w2} can be determined.

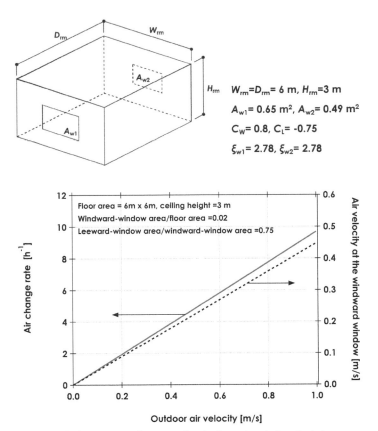

$W_{rm}=D_{rm}= 6$ m, $H_{rm}=3$ m

$A_{w1}= 0.65$ m^2, $A_{w2}= 0.49$ m^2

$C_W= 0.8$, $C_L= -0.75$

$\xi_{w1}= 2.78$, $\xi_{w2}= 2.78$

(a) air change rate and average air velocity at the windward window

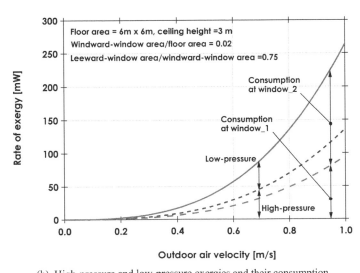

(b) High-pressure and low-pressure exergies and their consumption

Fig. 11.19: High-pressure and low-pressure exergies provided by wind effect and their consumption for ventilation.

Figure 11.19 demonstrates the results of example calculation for the outdoor reference air velocity from 0 to 1.0 m/s with the assumption of a room as follows: the floor area is 36 m²; the ceiling height is 3 m; the windward-window area is 0.7 m², which is 2% of the floor area; the leeward-window area is 75% of the windward-window area; the room is located in the middle of a multi-story building (C_W = 0.8 and C_L = –0.75); and both windows are covered by insect screen (ξ_{w1} = ξ_{w2} = 2.78).

The upper graph shows the relationship between the outdoor reference air velocity and air change rate together with the average air velocity at the windward window. As the outdoor reference air velocity increases, the air change rate linearly increases. The same tendency applies to the average air velocity at the windward window, which is a little lower than a half of the outdoor reference air velocity. Such a decrease in air velocity is due mainly to the use of insect screen on the windows. Since the air velocity from 0.3 to 0.8 m/s is sufficient for restoring thermal comfort in a naturally ventilated room as discussed in Chapter 10, the insect screen may be regarded as a kind of device to ease the incoming outdoor wind while preventing the insects.

Referring to what was described in 11.3.3, the whole exergy balance for wind-driven ventilation in the room assumed here can be expressed as follows

$$C_W\left(\frac{1}{2}\rho_o v_o^{\,2}\right)V_{w1} + \left\{-C_L\left(\frac{1}{2}\rho_o v_o^{\,2}\right)\right\}V_{w1} - X_C = 0. \tag{11.32}$$

where $V_{w1}(= A_{w1}v_{w1})$ is the volumetric air flow rate through the windward window [m³/s] and X_C is the exergy consumption rate [W], which is equal to $(\Delta P_{f\_w1} + \Delta P_{f\_w2})V_{w1}$. The first term of Eq. (11.32) represents the input of high-pressure exergy to the windward window and the second term the input of low-pressure exergy to the leeward window.

The lower graph of Fig. 11.19 shows how the rates of exergy inputs and consumption vary with the outdoor reference air velocity. They increase in proportion to the third power of velocity, since the dynamic pressure is proportional to the square of air velocity and the volumetric air rate is proportional to air velocity. The reason that high-pressure exergy input is larger than low-pressure exergy input here in this example is due to the difference in the absolute values of dynamic pressure coefficient between windward and leeward, which were assumed to be 0.8 and –0.75, respectively. The exergy consumption rate at the leeward window is larger than that at windward window; this is due to the leeward window being smaller than windward window area, with which the average air velocity turns out to be higher at the leeward window than at the windward window. The total of exergy input rates is in the order of 10 to 250 mW; this is 0.28 to 6.9 mW per floor area. The consumption of both high-pressure and low-pressure exergy may be regarded to bring about the air change rate shown in the upper graph of Fig. 11.19.

11.4.4 Buoyancy-driven ventilation

When there is no wind at all, ventilation may be made by buoyancy effect provided that the indoor air temperature is higher than outdoor air temperature. If two window openings have different heights as shown in the top plate of Fig. 11.20, then some ventilation effect emerges. A portion of indoor air can flow out from the upper window while the corresponding amount of outdoor air flows in from the lower window. This ascending movement of air is realised by the lower density of indoor air than that of outdoor air; this is buoyancy effect.

The pressure balance across the lower and upper windows can be expressed as follows

$$P_{o1} - \Delta P_{f\_w1} = P_{r1}, \tag{11.33}$$

$$P_{r2} - \Delta P_{f\_w2} = P_{o2}, \tag{11.34}$$

where P_{o1} and P_{o2} are outdoor static pressure at the heights of lower and upper windows measured from the floor surface level, h_1 and h_2, respectively; P_{r1} and P_{r2} are indoor static pressure at the heights, h_1 and h_2, respectively.

The indoor and outdoor static pressures at the floor surface level, P_o and P_r, are related to P_{o1}, P_{o2}, P_{r1} and P_{r2} as follows, respectively: $P_o = P_{o1} + \rho_o g h_1$, $P_o = P_{o2} + \rho_o g h_2$, $P_r = P_{r1} + \rho_r g h_1$ and $P_r = P_{r2} + \rho_r g h_2$. Substitution of these relationships into Eqs. (11.33) and (11.34) so that P_{o1}, P_{o2}, P_{r1} and P_{r2} are eliminated yields the following equations

$$(P_o - \rho_o g h_1) - \Delta P_{f\_w1} = P_r - \rho_r g h_1, \tag{11.35}$$

$$(P_r - \rho_r g h_2) - \Delta P_{f\_w2} = P_o - \rho_o g h_2. \tag{11.36}$$

Equations (11.35) and (11.36) together with Eq. (11.31) can be solved in terms of P_r, v_{w1} and v_{w2}, provided that the values of P_o, ρ_o, ρ_r, h_1, h_2, A_{w1}, A_{w2}, ξ_{w1} and ξ_{w2} are given.

When the buoyancy effect takes place, there is the height from the floor surface, where the indoor static pressure becomes exactly the same as outdoor air pressure. Denoting this height to be h_N,

$$P_r - \rho_r g h_N = P_o - \rho_o g h_N, \text{ that is, } h_N = \frac{P_r - P_o}{(\rho_r - \rho_o)g}. \tag{11.37}$$

Reflecting the neutral height, h_N, expressed by Eq. (11.37) to Eqs. (11.35) and (11.36), we find the following two relationships

$$-(\rho_o - \rho_r)g(h_1 - h_N) - \Delta P_{f\_w1} = 0, \tag{11.38}$$

$$(\rho_o - \rho_r)g(h_2 - h_N) - \Delta P_{f\_w2} = 0. \tag{11.39}$$

Since $\Delta P_{f\_w1} > 0$, $-(\rho_o - \rho_r)g(h_1 - h_N) > 0$, and also since $\Delta P_{f\_w2} > 0$, $(\rho_o - \rho_r)g(h_2 - h_N) > 0$.

Combining Eqs. (11.38) and (11.39) and then multiplying the volumetric air flow rate through the lower window, V_{w1}, we can come up with the following exergy balance equation for the case of buoyancy-driven ventilation

$$\{-(\rho_o - \rho_r)g(h_1 - h_N)\}V_{w1} + (\rho_o - \rho_r)g(h_2 - h_N)V_{w1} - X_C = 0, \tag{11.40}$$

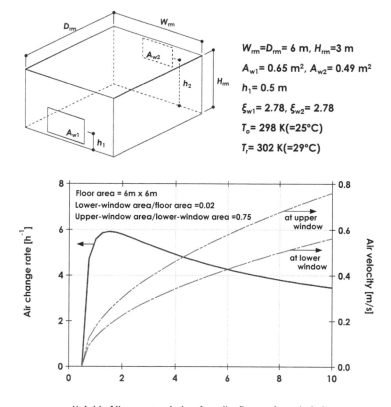

$W_{rm}=D_{rm}= 6$ m, $H_{rm}=3$ m

$A_{w1}= 0.65$ m^2, $A_{w2}= 0.49$ m^2

$h_1= 0.5$ m

$\xi_{w1}= 2.78$, $\xi_{w2}= 2.78$

$T_o= 298$ K$(=25°C)$

$T_r= 302$ K$(=29°C)$

(a) air change rate and average air velocities at the windows

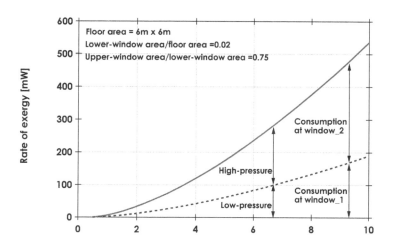

(b) High-pressure and low-pressure exergies and their consumption

Fig. 11.20: High-pressure and low-pressure exergies provided by buoyancy effect and their consumption for ventilation.

where $X_C = (\Delta P_{f\_w1} + \Delta P_{f\_w2})V_{w1}$. The first term of the left-hand side of Eq. (11.40) is low-pressure exergy and the second term high-pressure exergy. Buoyancy effect occurs when indoor air temperature is higher than outdoor air temperature, that is, when $\rho_r < \rho_o$, and thereby makes the low-pressure exergy outflow through the lower window and the high-pressure exergy outflow through the upper window. These outflowing low-pressure and high-pressure exergies realize a certain air-change rate in the room space. This is schematically demonstrated in Fig. 11.21 together with the flows of high-pressure and low-pressure exergies in the case of wind-driven ventilation.

Coming back to Fig. 11.20, the two graphs demonstrate the results of example calculation for the height of upper window from 0.5 to 10 m with the same assumption of a room as the case of wind-driven ventilation: the floor area is 36 m²; the ceiling height is 3 m; the lower-window area is 0.7 m², which is 2% of the floor area; the upper-window area is 75% of the lower-window area; the centre of the lower window is positioned at 0.5 m above the floor surface; both windows are covered by insect screen; and outdoor and indoor air temperatures are 25 and 29°C, respectively.

The upper graph shows the air change rate, air velocities at lower and upper windows as a function of the height of upper window. The air velocities increase gradually as the upper window is located at a higher height, though their increase tends to diminish. The reason for the air velocity at upper window being higher than that at lower window is that the size of upper window is smaller than that of lower window. The maximum air change rate of 6 h⁻¹ emerges as the height of upper window which is located about 1.5 m above the floor surface and the air change rate decreases as the height of upper window increases more than 1.5 m. This is due to the fact that the volume of room space becomes larger while the sizes of lower and upper windows are kept unchanged.

Figure 11.20b shows how the high-pressure and low-pressure exergies increase as the height of upper window increases. They increase in a quadratic manner as the height of upper window increases. High-pressure exergy emerging at upper window is larger than low-pressure exergy so that the air velocity at upper window becomes higher than that at lower window. Both high-pressure and low-pressure exergies are

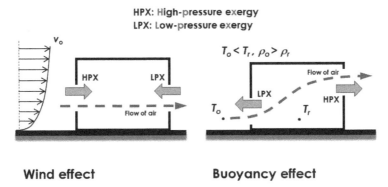

Fig. 11.21: Wind and buoyancy bringing about high-pressure and low-pressure exergies that actuate ventilation.

consumed totally by friction in the course of ventilated air flowing through the lower and upper windows as schematically shown in Fig. 11.21.

11.4.5 Mechanical ventilation

Instead of passive systems for ventilation relying either on wind effect or buoyancy effect, an active system for ventilation driven by electricity input, that is, mechanical ventilation, is also possible. While the former's performance is much dependent on local weather conditions, the latter's performance may be independent from ever-changing local weather conditions, if the electricity supply is guaranteed.

For mechanical ventilation, Eq. (11.13) is directly applied to its exergy balance. That is,

$$E_{fan} - X_C = \{-(P_1 - P_o)\}V_w + (P_2 - P_o)V_w, \tag{11.41}$$

where $E_{fan} = \dfrac{\Delta P_f V_w}{\eta_{fan}}$, $X_C = \Delta P_f V_w$, $\Delta P_f = (\xi_1 + \xi_2)\left(\frac{1}{2}\rho v_{fan}^2\right)$. E_{fan} is electric power input to a fan [W], V_w is the volumetric air flow rate [m³/s], η_{fan} is the fan efficiency, ξ_1 and ξ_2 are pressure-decrease coefficient of inlet and outlet elements, v_{fan} is outlet air velocity [m/s], and P_1 and P_2 are the static pressures at the inlet and outlet of the fan [Pa]. Table 11.A.2 shows the VBA code for the calculation of the electricity to be supplied to a fan.

Figure 11.22 shows the results of exergy-balance calculation of a mechanical ventilation system as a function of air change rate. The assumptions made for the calculation are $\eta_{fan} = 0.65$, $\xi_2 = 2.78$ for an opening with insect screen, $\xi_1 = 0.2\xi_2$, and $v_{fan} = 2$ m/s; other assumptions are the same as previous calculation of natural ventilation.

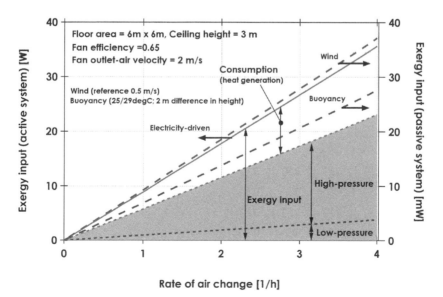

Fig. 11.22: Exergy input, consumption and output as a function of air change rate obtained by a fan.

The exergy input increases linearly as the air change rate increases. This implies that a larger exergy input is required in order to make air change rate larger. For the air change rate at 2 h^{-1}, the electric power input is about 17.8 W as shown on the left-side vertical axis, with which high-pressure and low-pressure exergies produced are 9.6 W and 1.4 W by the exergy consumption of 6.2 W. For comparison, the exergy inputs for wind-driven and buoyancy-driven ventilations are also presented in Fig. 11.22 with the right-side vertical axis. The relationship between exergy input and air change rate in the case of wind-driven ventilation is for the outdoor reference air velocity at 0.5 m/s and that in the case of buoyancy-driven ventilation is for the outdoor and indoor air temperature of 25 and 29°C, respectively, and the height difference between upper and lower windows of 2 m.

For 2 h^{-1} of air change rate, exergy inputs for wind-driven and buoyancy-driven ventilations are 18.6 and 13.8 mW, respectively. They are 0.10% and 0.08% of the electricity input for the same air change rate with mechanical ventilation; in other words, mechanical ventilation requires 960 to 1300 times larger exergy input than natural ventilation. Such a huge difference in the exergy requirement between mechanical and natural ventilation systems suggests that they have to be designed holistically and operated properly so that the respective system advantages are maximized while their disadvantages minimized. For wind-driven ventilation, the air change rate of 4 h^{-1} corresponds to the ratio of windward-window area to floor area at 1.65%, while on the other hand, for buoyancy-driven ventilation, 4 h^{-1} corresponds to 1.43%.

11.5 Heat pumps

Heat pumps are probably the best wide-spread example of mechanical device for active transport developed on the basis of thermodynamic principle. In general, they are regarded as a mechanical device that can pump up an amount of thermal energy from a source with a certain level of temperature to a sink whose temperature is higher than that of the source. What we are familiar with is a refrigerator pumping up thermal energy to the surrounding kitchen space or an air-conditioning unit pumping up thermal energy from a room space to be conditioned to the surrounding outdoor space in summer.

Figure 11.23 shows how air-conditioning units for home use have spread over the period of forty years from 1970 to 2010 in Japan (JRI 2009): in early 1970s, a fewer than ten out of one-hundred families, that is, one out of ten families, owned air-conditioning units, but in the year 2000, one family possesses more than two units on average. After 2000, the number of possession still grew, but it seemed almost saturated by 2010. On the contrary, the number of kerosene stoves gradually decreased after 1980 when it reached the maximum. The kerosene stoves must have decreased due to the replacement with air-conditioning units. Interestingly, the growing trend in the number of air-conditioning units resembles a logistic curve, which usually represents the growth of living creatures in a certain environmental space.

Here in this section, based on what was described in Chapter 7 with respect to "warm" and "cool" exergies, let us discuss how an air-conditioning unit functions for active transport.

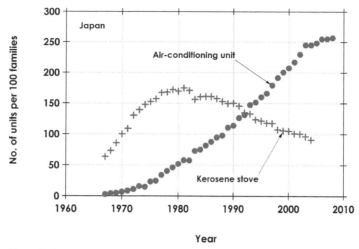

Fig. 11.23: The variation of the number of home-use air-conditioning units together with that of kerosene stoves in Japan (data quoted from JRI 2009).

11.5.1 Air-conditioning units coupled with outdoor air and with underground soil

A typical air-conditioning unit consists of four subcomponents as shown in Fig. 11.24a: an internal heat exchanger with a fan, a compressor, a throttling valve, and an external heat exchanger with another fan coupled with outdoor air. The external heat exchanger may be coupled with an underground heat exchanger as shown in Fig. 11.24b; in this case, the whole of the system is usually called ground-coupled heat pump.

A portion of room air at the rate of V_1 [m³/s], having the temperature at T_{ra} [K], sucked by the internal fan with the exergy supply rate at E_1 [W] is cooled by the heat exchanger, through which the refrigerant at low temperature, T_{Rf1} [K], flows and is then released with the air temperature at T_1 [K] from the outlet back into the room space. In order to make the present discussion simple, we here focus only on the effect of decrease in temperature, but not on the effect of decrease in humidity.

The refrigerant is circulated within the conduit and in due course it keeps on expanding at the throttling valve and contracting at the compressor. Among a variety of materials that can function as refrigerants, difluoromethane, CH_2F_2 in chemical symbols, is usually used for home-use air-conditioning units. In the mode of cooling, the refrigerant temperature is decreased down to T_{Rf1} by expansion at the throttling valve, while on the other hand, it is increased by contraction at the compressor up to T_{Rf2} [K], which has to be higher than the surrounding temperature of the external heat exchanger. In the case coupled with outdoor air, T_{Rf2} has to be higher than outdoor air temperature, T_o [K], and in the case coupled with underground, T_{Rf2} has to be higher than the temperature of water-antifreeze mixture coming into the external heat exchanger, T_{gout} [K]. The circulation of refrigerant is made by the input of exergy at the rate of E_3 [W] into the compressor.

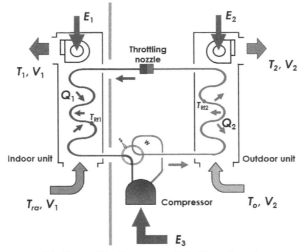

(a) Air-conditioning unit coupled with outdoor air

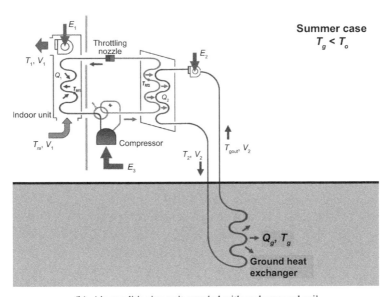

(b) Air-conditioning unit coupled with underground soil

Fig. 11.24: An air-conditioning unit coupled with outdoor-air heat exchanger and that coupled with underground heat exchanger.

At the external heat exchanger coupled with outdoor air, a tiny portion of atmospheric air is sucked by the external fan at the rate of V_2 [m³/s] with the exergy supply rate at E_2 [W]. In the case coupled with underground heat exchanger, an amount of water-antifreeze mixture is circulated at the rate of V_2 between the external heat exchanger of the heat pump and the underground heat exchanger so that heat can be discharged into the ground instead of outdoor air. The circulation of water-antifreeze mixture is made by a pump with the exergy supply rate at E_2. The values of

V_2 and E_2, in the case of water-antifreeze mixture, are smaller than those in the case of outdoor air because of a huge difference in the volumetric heat capacity between water and air: water has 3470 times larger volumetric heat capacity than air.

Exergy balance of the whole of air-conditioning unit coupled with outdoor air can be written as follows

$$\sum_{i=1}^{3} E_i - X_C = \left(x_1 - x_r \right) V_1 + x_2 V_2, \tag{11.42}$$

where the variables E_1, V_1, and V_2 are as already explained above, respectively, and X_C is the exergy consumption rate [W] within the whole of air-conditioning unit. The variables x_1 and x_r are thermal exergy contained by a unit volume of room air with the outdoor air temperature at T_o and with T_1 and T_{ra}, respectively. Since we assume $T_1 < T_{ra} < T_o$ in the present discussion, both x_1 and x_r are "cool" exergy and $0 < (x_1 - x_r) V_1$. Similarly, x_2 is thermal exergy contained by a unit volume of exhaust air at T_o and T_2; it is "warm" exergy, since $T_o < T_2$. The values of x_1, x_2 and x_r can be calculated from Eq. (7.33) for a unit volume of air; their unit is J/m³.

Equation (11.42) expresses that there are three inputs of exergy to two fans and to one compressor, $\sum_{i=1}^{3} E_i$, their portion as X_C is consumed in the whole process and thereby two outputs of exergy are generated: one is "cool" exergy, $(x_1 - x_r) V_1$, delivered to the room space and the other is "warm" exergy as $x_2 V_2$ discarded into the outdoor space. This implies that an air-conditioning unit coupled with outdoor air is exergy separator, rather than a device to be called as "heat pump". It should be worth noting that the whole form of Eq. (11.42) is congruent to that of Eq. (11.41); the positions of "cool" and "warm" exergies correspond to those of "low-pressure" and "high-pressure" exergies. In this sense, an air-conditioning unit coupled with outdoor air is definitely a kind of pump as one of the active-transport devices, but should be regarded as the exergy separator.

In the case of air-conditioning unit coupled with the underground heat exchanger, exergy balance equation is expressed as follows

$$\left(x_g - x_2 \right) V_2 + \sum_{i=1}^{3} E_i - X_C = \left(x_1 - x_r \right) V_1, \tag{11.43}$$

where x_g and x_2 are not "warm" but "cool" exergy contained by a unit volume of water-antifreeze mixture [J/m³], since $T_g < T_{gout} < T_2 < T_o$.

Let us compare Eq. (11.43) with Eq. (11.42). Equation (11.43) expresses that there is one more exergy input as $(x_g - x_2) V_2$ in addition to $\sum_{i=1}^{3} E_i$. A portion of these exergy inputs as X_C is consumed and thereby one single output of exergy as $(x_1 - x_r)$ V_1 is generated. Both $(x_g - x_2) V_2$ and $(x_1 - x_r) V_1$ are "cool" exergy. This implies that an air-conditioning unit coupled with underground soil is not an exergy separator, but a heat pump, since a portion of cool exergy stored underground as $(x_g - x_2) V_2$ is pumped up by the heat-pump process into the room space at the rate of $(x_1 - x_r) V_1$.

If the coefficient of performance of the whole of an air-conditioning unit is given as η_{cop_c}, then the rate of thermal energy extracted from the internal heat exchanger, Q [W], is expressed as follows

$$Q_1 = \eta_{cop\_c} \left(\sum_{i=1}^{3} E_i \right). \tag{11.44}$$

Cool exergy demand emerged in the room space, X_{demand} [W], can be written as

$$X_{demand} = -\left(1 - \frac{T_o}{T_{ra}}\right) Q_1. \tag{11.45}$$

When cooling is made by air conditioning unit, $T_{ra} < T_o$; therefore, the Carnot factor in Eq. (11.45) is negative. Since $0 < Q_1$ and there is negative sign in front of Carnot factor, the value of X_{demand} is necessarily greater than zero: that is, X_{demand} is "cool" exergy requirement of the room space. In order to keep the room air temperature at a desired value, it is necessary for the room to let a certain amount of exergy be consumed. Denoting the rate of this exergy consumption as X_{C_room} [W], the relationship between exergy supplied to the room space, $(x_1 - x_r)V_1$, and exergy demand, X_{demand}, can be expressed as follows

$$(x_1 - x_r)V_1 - X_{C\_room} = X_{demand}. \tag{11.46}$$

The performance of an air-conditioning unit, η_{cop_c}, may vary depending on the indoor and outdoor conditions. If the refrigerant temperature at internal heat exchanger, T_{Rf1}, has to be very low and that at external heat exchanger, T_{Rf2}, has to be very high, then the value of η_{cop_c} must become small. On the other hand, if the refrigerant temperature does not need to be very low at the internal heat exchanger and not very high at the external heat exchanger, then the value of η_{cop_c} must become larger. Such characteristics can be quantified as shown in Fig. 11.25; this chart is made for typical home-use air conditioning units on the basis of experimental data available (Ueno et al. 2010). We can see that η_{cop_c} becomes smaller as the room air temperature is lowered and also the ambient temperature of the external heat exchanger increases.

Let us suppose that there is a room having 36 m² of floor area, which is identical to the room assumed for the discussion on natural ventilation in previous section. The room is cooled by an air-conditioning unit, whose coefficient of performance, η_{cop_c}, is given as shown in Fig. 11.25 so that room air temperature is kept at a certain set-point value between 24 and 30°C under the condition of outdoor air temperature at 32°C or the ground temperature at 20°C. The rate of thermal energy to be extracted for keeping the indoor temperature constant, Q_1, is assumed to be the sum of solar heat gain from the window, internal heat gain from the occupants and miscellaneous sources of heat, and heat gain caused by thermal-energy conduction through the wall and the window; it is assumed to be from 29 to 41 W/m², depending on the set-point value of indoor air temperature.

Figure 11.26 shows the results of calculation made with the assumption just described above: (a) is the case coupled with outdoor air and (b) is the case coupled with underground soil. What we can see first in both cases is that total exergy input increases as the room air temperature is lowered. This is due to the increase in exergy demand and also to the decrease in the coefficient of performance, η_{cop_c}. The electricity input, in the case coupled with underground soil, is about 60% of the

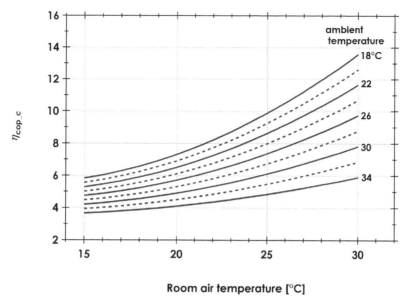

Fig. 11.25: Coefficient of performance (COP) as a function of room air temperature and ambient temperature of the external heat exchanger (data quoted from Ueno et al. 2010).

electric power input in the case coupled with outdoor air, regardless of any values of room air temperature.

In the case coupled with outdoor air, "warm" exergy has to be generated for the entropy disposal into the outdoor air. This causes the necessity of a larger exergy input and results in larger exergy consumption. The "warm" exergy produced is approximately a half of "cool" exergy, which is produced to keep the room air temperature at a desired level.

In the case coupled with underground soil, the exergy input originates not only from electric power but also "cool" exergy stored underground. Total exergy input in the case coupled with underground soil is about 15% less than that in the case coupled with outdoor air, but the difference in the exergy consumption rate between the two cases is only 7%, since the "cool" exergy to be delivered from the ground is partially consumed in addition to the smaller electric power supplied and consumed, that is, as mentioned above, 60% of the electric power input in the case coupled with outdoor air. As a whole, this implies that the exergy supplied to the air-conditioning unit coupled with underground soil is more rationally consumed than in the case coupled with outdoor air.

Keeping this quantitative tendency so far discussed in mind together with what we have come to know through the discussion made in Chapters 6 and 10, we come to recognize that an active-transport option can function properly once the corresponding passive-transport option is properly taken and designed so as to make the flow of light, heat and air through building envelope systems desirable to occupants' well being. Exactly the same discussion must apply to the cases for heating.

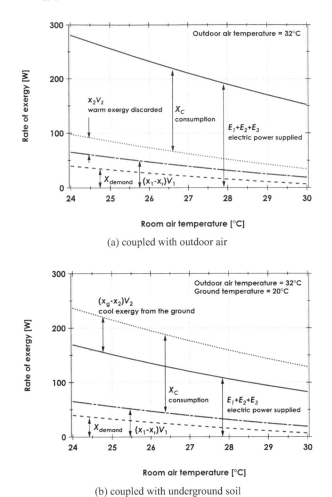

(a) coupled with outdoor air

(b) coupled with underground soil

Fig. 11.26: Exergy balance with the whole of air-conditioning units in the cases of (a) coupled with outdoor air and (b) coupled with underground soil.

11.5.2 Cool tubes

The use of "cool" exergy via a ground-coupled heat pump is attractive as we discussed in the previous subsection, but there is another simpler option called "cool tube", which should function as a kind of heat pump, as schematically shown with the top plate in Fig. 11.27. It may be equipped as one of the openings for ventilation discussed earlier in the present chapter.

Let us assume that outdoor air sucked by a fan at a certain rate, V_{tb} [m³/s], with the exergy supply rate at E_{fan} [W] is cooled while it flows through the tube which is buried underground. The outlet air temperature, T_{tb_out} [K], becomes lower than outdoor air temperature T_o [K], since the ground temperature, T_g [K], is lower than T_o during summer season. It means that the amount of air taken in turns out to have

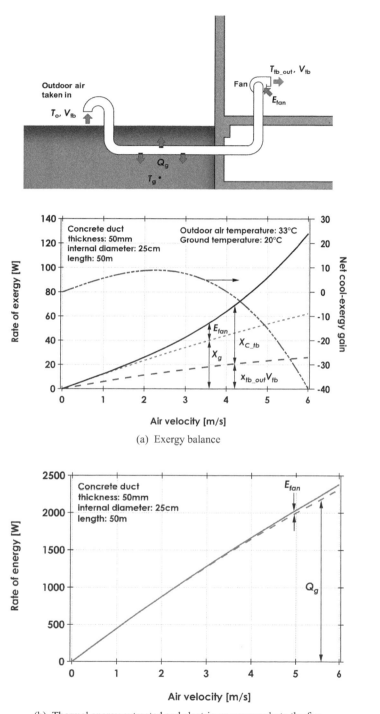

(a) Exergy balance

(b) Thermal energy extracted and electric-power supply to the fan

Fig. 11.27: Exergy balance of the cool-tube system working with different flow rates and the associated rate of thermal energy extracted from the tube.

"cool" exergy due to the extractions of thermal energy originally contained by the outdoor air at the rate of Q_g [W]. This is exactly the same as what is expected in the case of an air-conditioning unit coupled with underground soil.

If the flow rate is small, then the outlet temperature becomes low. This results in "cool" exergy contained by one unit volume of air becoming large, but the whole rate of "cool" exergy available must become small, since the volume of air coming into the room space must be very small for making the outlet temperature low. In order to increase "cool" exergy available, we need to increase the volumetric rate of air up to a certain required value so that outlet temperature is sufficiently low, while at the same time making the rate of exergy supply to the fan as small as possible. If the availability of "cool" exergy is sufficiently large in comparison to the supply of exergy to the fan, then the whole system may be regarded as properly functioning, but otherwise not. Therefore, it is necessary to find the optimal design solution among various possible combination of tube materials, length, diameter, and air velocity. This can be done by taking a look at the exergy balance of the whole cool-tube system.

The rate of "cool" exergy to be supplied from the ground to the air flowing through the cool tube is expressed as

$$X_g = -\left(1 - \frac{T_o}{T_g}\right)Q_g, \tag{11.47}$$

$$Q_g = c_{pa}\rho_a V_{tb}\left(T_o - T_{tb\_out}\right)$$
$$V_{tb} = \frac{\pi}{4}D_{tb}^2 v_{tb} \tag{11.48}$$

where $c_{pa}\rho_a$ is the volumetric heat capacity of air (≈ 1205) [J/(m^{3}K)], D_{tb} is the cool-tube diameter [m], and v_{tb} is air velocity inside the cool tube [m/s]. In the case of a cool tube system, $T_g < T_{tb\_out} < T_o$; therefore, the Carnot factor in Eq. (11.47) is negative. Since $Q_g > 0$ and there is negative sign in front of Carnot factor, $X_g > 0$.

The exergy balance can be written as follows

$$X_g + E_{fan} - X_{C\_tb} = x_{tb\_out}V_{tb}, \tag{11.49}$$

where X_{C_tb} is the exergy consumption rate within the whole of cool-tube system [W] and X_{tb_out} is "cool" exergy contained by a unit volume of air to be calculated from Eq. (7.33).

Figure 11.27a shows the result of example calculation demonstrating the exergy balance as a function of air velocity. In this example, it was assumed that the cool tube is made of concrete with the thickness of 50 mm, where the internal diameter and the length are 25 cm and 50 m, respectively. Pressure decrease in the cool tube was given by Eq. (11.16) referring to the relationship between friction factor and Reynolds number. The electric power to be supplied to the fan was calculated by the formula given for Eq. (11.41).

As the air velocity increases, total exergy input rises sharply; this is due to the sharp increase in the electric-power input to the fan, which is proportional to the third power of air velocity. The available "cool" exergy, $x_{tb\_out}V_{tb}$, also increases, but it looks attenuating as the air velocity becomes higher.

The right-side axis shows the net "cool" exergy gain, which is the difference between the "cool" exergy obtained and the exergy input to the fan. As can be seen, the net "cool" exergy gain gradually increases, but it reaches the maximum at about 2.4 m/s of air velocity. With the air velocity higher than 4 m/s, the net "cool" exergy gain becomes negative; this implies that the exergy supplied to the fan is larger than the "cool" exergy obtained by the whole cool-tube system. Figure 11.27b demonstrates the rate of thermal energy extraction together with the electric power supply to the fan. With the concept of energy alone, it looks that there is no optimum air velocity; in other words, it looks all right to make air velocity as high as possible, since the rate of thermal energy extraction is much larger than the electric power supplied. But this could result in a misleading solution.

Optimum air velocity, which can be determined from the viewpoint of exergy, may vary depending on the tube size, tube material, local climatic conditions, building types, and also types of ventilation system. As we have come to know through the discussion made in 11.4.5, a passive system for ventilation, whether it is wind-driven or buoyancy-driven, works with much smaller exergy input than mechanical ventilation system with the electric-power input. Therefore, an optimal passive-system configuration for ventilation coupled with an underground cool-tube is worth seeking. Optimization problems to be solved similar to the one described above are here and there within a variety of built-environmental conditioning systems.

References

Anderson J. D. 1995. Computational Fluid Dynamics—The basics with applications. McGraw-Hill.

Ansley R. M., Melbourne E. and Vickery B. J. 1977. Architectural Aerodynamics. Applied Science Publishers. London.

Bejan A. 2004. Convective Heat Transfer. Third Ed. John Wiley & Sons, Inc.

Bejan A. 2006. Advanced Engineering Thermodynamics. Third Ed. John Wiley & Sons, Inc.

Imai I. 1970. Fluid Dynamics. Iwanami Publishers (in Japanese).

Ishigai S. 1977. Energy Evaluation and Management with the Concept of Exergy. Kyoritsu Publishers (in Japanese).

Ishihara M. 1969. Design Method for Architectural Ventilation (in Japanese).

Jyukankyo Research Institute (JRI). 2009. Handbook on Home Energy Use. The Energy Conservation Center Japan (in Japanese).

Moody L.F. 1944. Friction factors for pipe flow. Transactions of ASME 66(8): 671–684.

Seppänen O. A. and Fisk W. J. 2004. Summary of human responses to ventilation. Indoor Air 14(7): 102–118.

Simmons C. T. 2008. Henry Darcy (1803–1858): Immortalised by his scientific legacy. Hydrology Journal 16: 1023–1038. DOI: 10.1007/s10040-008-0303-3.

Turns S. R. 2006. Thermal Fluid Sciences—An Integrated Approach. Cambridge University Press.

Ueno T., Miyanaga T., Urabe W. and Kitahara H. 2010. Development of heat source characteristic model of home user air conditioner—part 1. Characteristic model for cooling period. Central Research Institute of Electric Power Industry (CRIEPI). Research report: No. R09017 (in Japanese).

Wolpert L., Beddington R., Brockes J., Jessell T., Lawrence P. and Meyerowitz E. 1998. Principles of Development. Current Biology Ltd.

Wright T. 2012. Circulation: William Harvey's Revolutionary Idea. London: Chatto & Windus.

Appendix

11.A VBA codes for the calculation of friction factor and fan power

Table 11.A.1 shows the VBA code for the calculation of Darcy's friction factor using a set of empirical formulae (Turns 2006) together with the VBA code for the calculation of kinematic viscosity of air, which is necessary for the calculation of Reynolds number.

Table 11.A.2 shows the VBA code for the calculation of fan power making use of the VBA codes shown in Table 11.A.1.

Table 11.A.1: VBA codes for the calculation of Darcy's friction factor and kinematic viscosity of air.

```
Function fD(eps_D, DD, Re)
'
' Calculation of Darcy-Weisbach (Darcy's) friction factor
'
' eps_D is surface roughness [m];
' DD is pipe diameter [m];
' Re is Reynold's number.
'
'                    27th March, 2013; 27th February, 2014 MS
'
If Re < 2300 Then
     fD = 64 / Re
ElseIf eps_D = 0 Then
     fD = 1# / ((0.79 * Log(Re) - 1.64) ^ 2)
Else
     Log10_value = Log(eps_D / (3.7 * DD) + 5.74 / (Re ^ 0.9)) / Log(10#)
     fD = 0.25 / ((Log10_value) ^ 2)
End If
End Function
```

```
Function Nyu_air(t_air)
'
'Kinematic viscosity of air
'
't_air is air temperature in degree Celsius.
'
'                    27th March, 2013; 27th February, 2014 MS
'
Nyu_air = (0.0991 * t_air + 13.764) / 1000000
End Function
```

Table 11.A.2: VBA code for the calculation of Fan power.

```
Function Fan_power(vv, temp, D_char, eps_D, LL, eff)
'
'Calculation of power required by a fan for sucking air
'
'vv is air velocity [m/s]; temp is air temperature in the unit of degree Celsius;
'D_char is duct diameter [m]; eps_D is roughness of duct surface;
'LL is duct length [m]; eff is fan efficiency [-].
'
'                    27th March, 2013; 27th February, 2014 MS
'
pai = 3.1415926: roh_air = 1.2
Re = vv * D_char / Nyu_air(temp)
fDD = fD(eps_D, D_char, Re)
Fan_power = pai / 8# * fDD * D_char * LL * roh_air * vv ^ 3 / eff
End Function
```

Chapter 12

Global Environmental System Enfolding Built-Environmental Systems

12.1 The Earth in between Venus and Mars

In Chapter 1, it was described that average atmospheric temperature near the ground surface of the global environmental system, from the Arctic via the equator to the Antarctic and also throughout one year, is approximately 15°C. This value is in fact a kind of nominal value. The actual value is considered to be within the range between 12 and 18°C referring to various proxy data based on isotopic atoms such as Carbon-14, Berillium-10 or Oxygen-18 available in relation to the global climate over the last 900 thousand years; furthermore, if including a period in the far past over the last 500 million years, it is considered to be within the range between 12 and 20°C (Hoffman and Simmons 2008). According to the updated meteorological information available on the near-ground air temperature, the present average global temperature is 13.9 ~ 14.2°C (Jones and Harpham 2013). Either in the far past or near past, the average global temperature is realized due to the ceaseless inflow of short-wavelength radiation from the Sun together with the ceaseless outflow of long-wavelength radiation towards the Universe.

The corresponding temperatures on Venus and on Mars are considered to be about 500°C and –60°C, respectively, both of which are far different from 15°C as the nominal value on the Earth. This is mainly due to the following two facts: one is that the Venus is closer to the Sun than the Earth is and Mars is farther away from the Sun than the Earth is; the other is that their respective conditions of gravitation and near-surface constituents are largely different from those of the Earth. Note that the water temperature at 500°C can only be realized with forty-times higher pressure than our atmospheric pressure, 1013 hPa; in this condition, there is no distinction between liquid and vapour. Note also that the water temperature at –60°C is realised only with the state of solid despite of any pressure conditions. We can easily imagine that living creatures including us humans containing a lot of liquid water, which weighs more than 65% of the body mass, cannot live with such condition of temperature either at

500°C or at –60°C. In other words, the Earth is really in a good position relative to the Sun in the Universe and also in an appropriate mass that allows the atmospheric air and water not to escape, not to be prisoned too much, but to circulate.

Biological systems live in general by feeding on foods including liquid water, consuming them and disposing of the wastes into their environmental space. The same applies to built-environmental systems. In this process, essential is the flow and circulation of water and air as described in some of the previous chapters. It is, therefore, important for us to know how the average global environmental temperature is realized so as to be at 15°C as the nominal value or to be within a rather narrow range from 12 to 18°C in order for the flow and circulation of water either as liquid or as vapour to be secured.

12.2 Extraterrestrial nested structure

The first discovery of "vacuum" made by Pascal and others in the mid 17th century, as discussed in Chapter 8, had gradually led to the recognition of Universal space, either interstellar or interplanetary, being vacuum. But almost 300 years later, in early 20th century, it was re-discovered that the Universal space was not perfect vacuum, but there are cosmic-ray particles flying around, some of which are coming deeply into the Earth's atmosphere; the existence of such cosmic rays was first found in 1911 by Hess (1884–1964) and his colleagues with their ascending balloons that detected the phenomena of ionization high above within the atmosphere, from 1 to 5 km above the ground surface (Hoffman and Simmons 2008). In the very beginning, as its name suggests, cosmic rays were thought to be a kind of electro-magnetic radiation, but they were later confirmed to be the flow of subatomic particles consisting mainly of protons, the nuclei of hydrogen atoms, and some others. Most of cosmic rays, whose energy level is extremely high from 1 to 10 GeV per particle, are considered to originate from supernovae explosion happening occasionally within the galactic space (Eddy 2009). Note that the unit of eV is equivalent to 1.6×10^{-19} J. For comparison, if the kinetic energy held by each of nitrogen or oxygen molecules in the atmospheric air is expressed by the unit of eV instead of J, then its order becomes the level of 0.03 to 0.04 eV; we come to know how intense the cosmic-ray particles are.

Almost 50 years later since the rediscovery of the Universal space being not perfect vacuum, it was also found by Parker (1927~) in 1958 that there is a ceaseless flow of subatomic particles called "solar wind" blown by the Sun, which forms a kind of atmospheric space called "heliosphere" surrounding the Sun. The energy level of solar wind is considered to be at the level of 0.1 to 2 MeV per particle (SWPC/NOAA 2018), lower than the level of galactic cosmic rays, but still very high in comparison to nitrogen and oxygen molecules bouncing around in our closest environmental space. Aurora Borealis (northern light), seen over the sky vault in the regions at higher latitudes such as northern parts of Norway, Sweden, Finland and Canada, which still looks mysterious even today, is caused by solar wind (Eddy 2009). The Sun also emits more intense subatomic particles occasionally at the events called coronal mass ejection in addition to the solar wind; this is not so intense as galactic cosmic rays, but more intense than solar wind blown rather regularly.

The kinetic behaviour of galactic cosmic rays and solar wind is very complex because of the electro-magnetic nature of interstellar and interplanetary space together with cosmic-ray particles and solar wind, both of which are electrically charged. This complexity is beyond the subject of this treatise, but before proceeding the discussion in the present chapter, since we came to know the existence of galactic cosmic rays and solar wind, let us redefine the nested structure described in Chapter 1 so as to include the magnetosphere and the heliosphere as shown in Fig. 12.1.

The reach of heliosphere formed by solar wind ceaselessly blown from the Sun is considered to be approximately 1.5×10^{13} m, one hundred times larger than the distance between the Earth and the Sun. Thanks to the heliosphere, the Earth is protected from the galactic cosmic rays; the intensity of those rays penetrating through the barrier of heliosphere is decreased to some extent by solar wind. The whole of the Earth is of course within the heliosphere so that it is exposed to solar wind in addition to the galactic cosmic rays penetrating into the heliosphere. But thanks to the terrestrial magnetosphere, which is a huge magnet whose reach is in the order of 5×10^7 m, the solar wind coming onto the edge of magnetosphere shields its entry to the global atmosphere, while it causes the emergence of Aurorae Borealis and Australis near the polar regions.

Under the heliosphere and also the magnetosphere within, our immediate shallow atmosphere, whose thickness is in the order from 1 to 5×10^4 m, exists. All the terrestrial living creatures including us human beings live at the bottom of atmosphere, which protects them from the bombardment of solar wind via the magnetosphere and also from the extremely intense galactic cosmic rays via the heliosphere. Thus we come to recognize now that there are two environmental spaces, that is, the magnetosphere and the heliosphere, extended as parts of terrestrial nested structure shown in Fig. 1.4 of Chapter 1.

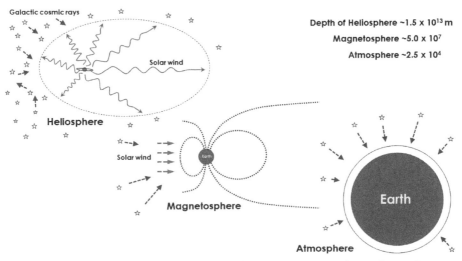

Fig. 12.1: Nested structural relationships between heliosphere, magnetosphere and atmosphere.

12.3 Relative sizes of the Earth and the Sun

As described in Chapter 4, the diameter of the Earth is 1.27×10^7 m, the distance between the Sun and the Earth is 1.49×10^{11} m, and the visible diameter of the Sun is 1.39×10^9 m. These large figures, together with large radii of the heliosphere and the magnetosphere shown in Fig. 12.1, hardly enable us to imagine the relative position and sizes of the heliosphere, the magnetosphere, the Sun and the Earth. In order to grasp them with a kind of imaginary scaled-down modelling, let us assume that the Earth is the size of a soccer ball, whose diameter is about 22.3 cm. Then, the edge of magnetosphere on the solar side turns out to be about 88 cm. The diameter of the Sun relative to the scaled-down-model Earth becomes 24.4 m, equivalent to the height of a six-story building, and the distance between the Sun and the Earth becomes 2.6 km, a forty-minute walking distance. The edge of heliosphere is about 260 km, equivalent to the distance between Tokyo and Nagoya or between London and Manchester.

Imagine such a situation watching the Earth from the position of the Sun. A soccer ball located at a point 2.6 km away must look nothing. This implies that most of the solar radiation and also the solar wind scatters away into the interplanetary space. Imagine next the other situation watching the Sun from the Earth; this is what we experience in reality. The Sun, almost equivalent to the appearance of a six-story building that we see from a point 2.6 km away, looks tiny. Nevertheless, the intensity of solar radiation that we sense on a day with fine weather is very strong. This lets us recognize how intense the short-wavelength radiation emitted by the Sun is.

All the radiation emitted by the Sun is originated from the thermo-nuclear fusion reaction taking place under the condition of very high temperature and very high pressure, both of which are realized by the huge gravitational force. This results in the surface temperature of the Sun to be about 5700°C, which is the estimated blackbody-radiation value based on the spectral characteristics of the measured solar radiation as was shown in Fig. 6.11 of Chapter 6.

As mentioned above, the distance between the Sun and the Earth is large and the Earth looks almost a faintest circle so that the radiation reaching just outside the upper atmosphere of the Earth is very directional, even though the radiation emitted by the Sun is not directional. This is the reason why solar entropy is very small and thereby the exergy-to-energy ratio is relatively large; it is from 0.6 to 0.92 depending on the weather conditions (Shukuya 2013).

The thickness of the atmosphere surrounding the ground surface is about 45 km up to the upper edge of stratosphere, that is the upper edge of ozone layer, with which most of harmful ionizing ultra-violet radiation is absorbed and thereby ozone molecules are formed from oxygen. The circulation of air and water taking place is within about 10 km up to the upper edge of the troposphere from the ground surface. In the case of a soccer-ball-sized model Earth having the diameter of 223 mm, the corresponding thickness of the atmosphere is 0.79 mm inclusive of ozone layer and 0.18 mm inclusive of troposphere alone. Let us imagine drawing a circle, whose diameter is 223 mm, on an A3-size sheet of paper using a mechanical pencil with 0.2 or 0.8 mm lead. The circular trace having the width of 0.2 mm is larger than the thickness of troposphere, 0.18 mm, and that of 0.8 mm is larger than the thickness of the atmosphere including ozone layer, 0.79 mm, in the model Earth.

We come to recognize that the terrestrial atmosphere is extremely thin, though what we sense by our own sensory organs and then perceive is rather opposite.

Let us add a couple of figures obtained from similar calculation to the above. The depth of crust, the hard shell above the liquid mantle, is equivalent to 0.1 to 0.7 mm, and the depth of sea water, whose surface covers 70% of the whole of terrestrial surface is equivalent to 0.07 mm in the model Earth of 223 mm diameter. The core of the Earth, from the centre to one half of the radius, which corresponds to 55 mm of the model Earth, is believed to consist mostly of nickel and iron. Its internal part, which corresponds to 21 mm from the centre of the model Earth, is considered to be in the state of solid and the rest, up to 55 mm, in liquid state. The temperature of the liquid core is believed to be in the order of 4000°C at the outermost boundary and from 6000 to 7000°C at the centre due to nuclear fission reactions (Maruyama and Isozaki 1998). Above the core is the layer called "mantle", most of which is in the state of liquid, and further above it is the solid shell corresponding to the thickness of 0.1 to 0.7 mm in the model Earth.

Here we can have an image of the Earth having the overall thin moist-air layer, under which there is very thin hard shell, two thirds of which is slightly concaved and covered by a thin film of liquid water. Such a globe is the Earth, on which we live.

12.4 Selective shielding by heliosphere, magnetosphere and atmosphere

Although the Sun appears steadily bright to our naked eyes with naïve minds, it was known that there were some dark-looking spots on the Sun already in ancient time; it seems that the ancient people saw the spots through hazy sky near the sunrise or sunset with their naked eyes. The systematic observation in terms of the sunspot number using telescopes was started in the era of Galilei and has been continued till present at astronomical observatories throughout the world (Eddy 2009).

Figure 12.2 shows the variation of monthly average sunspot number for a thirty-year period from 1975 to 2004 (WDC-SILSO 2018). We can see that the sunspot number increases and decreases in almost eleven-year cycle. Shown together is the extraterrestrial solar irradiance measured by spacecrafts (Fröhlich 2006, RMOD/WRC 2018). The extraterrestrial solar irradiance also increases and decreases in a consistent manner with the increase and decrease in sunspot number. This suggests that the electro-magnetically and fluid-dynamically complex and vigorous behaviour originated from the thermo-nuclear fusion reaction at the Sun which varies from time to time and its degree of vigorousness is indicated by the sunspot numbers. But the change in the extraterrestrial solar irradiance is less than 2 W/m^2, which is only 0.1% of its absolute value. Therefore, the extraterrestrial short-wavelength radiation received by a surface normal to the direction of the Sun is usually taken to be constant with the name of solar constant at 1366 W/m^2. This is one of the reasons that annual average of near-ground atmospheric temperature all over the Earth stays around 15°C every year.

Even though the variation of extraterrestrial solar irradiance is small, the fact that the way it changes looks very consistent with the variation of sunspot number suggests that the solar wind must also vary accordingly. Figure 12.3 demonstrates

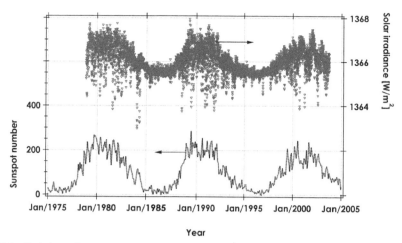

Fig. 12.2: Variations of monthly average sunspot numbers (WDC-SISLO 2018) and extraterrestrial solar irradiance (RMOD/WRC 2018).

that it is so by comparing the variation of sunspot number (WDC-SILSO 2018) and that of cosmic-ray count (CRSUO/SGO 2014) in the period from 1965 to 2015. As the sunspot number increases, the cosmic-ray count decreases and vice versa. This implies that an increase in the sunspot number indicates the corresponding increase in solar-wind intensity, which leads to decreasing the penetration of galactic cosmic rays into the heliosphere and thereby also into the atmosphere.

Figure 12.3 also shows that the change in cloud-cover fraction observed from satellites is consistent with the change in the cosmic-ray count within the atmosphere (Svensmark and Friis-Christensen 1997, Calder 1997). The cloud-forming mechanism in relation to the availability of galactic cosmic rays penetrating into the terrestrial atmosphere has gradually been clarified by Svensmark et al. (Svensmark and Calder 2007, Svensmark et al. 2013). The key phenomena of cloud formation is what can be seen in a cloud chamber originally invented by C. T. R. Wilson (1869–1959) (Longair 2014), that is, the generation of cloud condensation nuclei, on which water vapour condenses and grows into droplets of water, is basically due to the ionization of supersaturated moist air, for which intense subatomic particles have to pierce through and thereby generate the shower of secondary particles such as pions, muons, electrons, and photons.

Figure 12.4 schematically demonstrates the role of galactic-cosmic-ray shower in the cloud formation found by a series of research work led by Svensmark (Svensmark and Calder 2007). About 60% of the galactic cosmic rays, whose intensity is grouped as strong, are not influenced by the variation of solar-wind intensity, while on the other hand, the rest of them, 40%, whose intensity is grouped as mostly intermediate, are more or less influenced by the variation of solar-wind intensity.

Whether the penetrated galactic cosmic rays are strong or intermediate does not differ in the basic role of cloud-forming phenomena, but it does differ in when and how long the cloud formation is likely to occur (Svensmark 2012, 2015). The long-term climate change with the cycle of 100 million years or so is considered to have

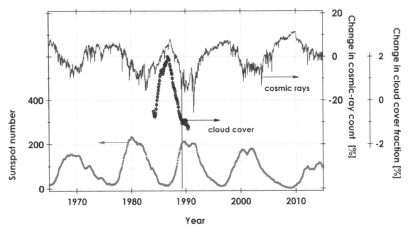

Fig. 12.3: Variation of sunspot numbers (WDC-SILSO 2018) together with the changes in cosmic-ray counts (CRSUO/SGO 2014) and in cloud-cover fraction (Svensmark and Friis-Christensen 1997).

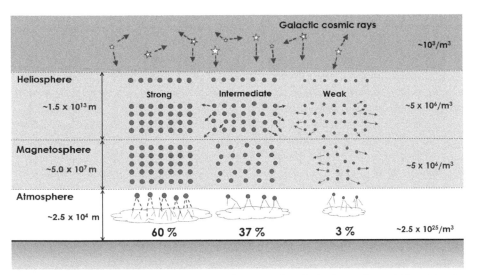

Fig. 12.4: Selective shading systems in relation to the penetration of galactic cosmic rays influential to the cloud formation; this diagram was made referring to Svensmark and Calder (2007).

been influenced quite a lot by the amount of clouds formed in relation to the location of the whole solar system including the Earth within the Milky-Way galaxy since the density of galactic cosmic ray particles varies depending on where the solar system locates inside the whole of Milky-Way galaxy.

The shorter-term climate change with the cycle of one hundred years or one thousand years is considered to have been influenced by the activity of the Sun since the penetration of galactic cosmic rays with the intermediate intensity is more or less regulated by the solar wind. While the sunspot number is large, the Earth tends to be in the warming mode because the intense solar wind tends to cause a decrease in the penetration of galactic cosmic rays at intermediate strength into the atmosphere and thereby to result in less cloudy sky. The opposite becomes true, while the sunspot number is small.

In Fig. 12.4, the densities of particles in four spaces are also shown. In the interstellar space, it is about $10^3/m^3$, while on the other hand in the interplanetary space, it is about $5 \times 10^6/m^3$. The order becomes one thousand times more in the interplanetary space than in the interstellar space. Near the bottom of terrestrial atmosphere, where we live, the density of air molecules is about $2.5 \times 10^{25}/m^3$. This implies that we are immersed in the very shallow but very dense atmospheric sea in comparison to the scarce but very intense particles existing outside the atmosphere.

We thus come to recognize that the denseness of atmosphere is important to protect the living creatures including us humans from the galactic cosmic-ray shower and solar wind, while at the same time its shallowness is important for the water vapour together with the galactic cosmic rays to play a role in regulating the average near-ground atmospheric temperature within the range from 12 to 18°C by modulating the amount of clouds as a kind of shading devise to be opened or closed for the global environmental system.

12.5 Average near-ground atmospheric temperature

According to the updated global energy balance archive (GEBA) (Wild 2017), the quasi-steady state averaged energy balance per one-squared-meter ground surface is as shown in Fig. 12.5. The overall average solar irradiance incident on the extraterrestrial surface near the edge of troposphere is 340 Wm², which is one fourth of solar constant, 1366 W/m². This is because the ratio of the surface area of a circle to that of a sphere is 1/4. The diameter of the Earth measured between the North and South poles is slightly shorter than the diameter measured along the equator because of the rotation around the North-South axis, but we can assume that the section is a true circle for the discussion here. For the energy balance to be at steady state, the output has to be exactly 340 W/m², which is the sum of reflected solar radiation as 100 W/m², and re-radiated by long-wavelength radiation as 240 W/m², which is of course the sum as the total rate of solar energy absorbed.

Taking a look at the energy transfer by radiation, convection and evaporation, we may regard the global environmental system as the composite of two subsystems: one is the near-ground air subsystem and the other the mid-tropospheric subsystem. Since the total of incoming rates of energy onto the near-ground air subsystem has to be equal to the outgoing rates of energy, energy balance equation can be written as follows, referring to Fig. 12.5

$$160 + 341 = 103 + 398. \tag{12.1}$$

The first and second terms on the left-hand side of Eq. (12.1) are the rates of solar radiant energy absorbed and the long-wavelength radiant energy emitted downward by the mid troposphere, which is absorbed by the ground surface. Those as output on the right-hand side are the rates of energy transferred by convection and evaporation and the outgoing long-wavelength radiation emitted by the ground surface.

Strictly speaking, there is an input of super long-wavelength radiation filled in the Universal space shown in Fig. 6.12 in Chapter 6, but it is ignored in Fig. 12.5

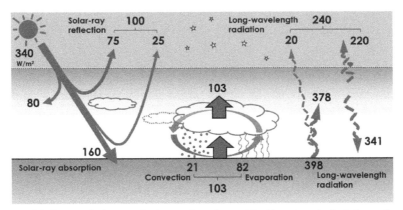

Fig. 12.5: Global energy balance so far known from the world-wide meteorological observation (Wild et al. 2017).

and in Eq. (12.1) because of its negligible smallness in comparison to other inputs and outputs. As mentioned in 12.3, the Earth's hard shell is as thin as troposphere so that geothermal energy must be flowing up to the ground surface from the core of the Earth at a certain rate. In reality, the internal temperature under the ground surface increases as we go deeper at the gradient of about 1°C each 25 m according to the underground-temperature measurement (Toda 1995). Nevertheless, it is known that the temperature at 3 m deep below the ground surface is almost the same as the annual average of local near-ground air temperature. Therefore, we may assume that a layer below the ground surface with the thickness of 3 m, which is 50 nm in the soccer-ball-sized model Earth, is the super thin and adiabatic boundary layer for the present discussion.

The energy balance equation for the mid-tropospheric subsystem can be written as follows in the same manner as Eq. (12.1)

$$80 + 103 + 378 = 341 + 220. \qquad (12.2)$$

The first, second, and third terms on the left-hand side of Eq. (12.2) are the rates of solar radiant energy absorbed, energy transferred by convection and condensation from the near-ground air subsystem, and long wavelength radiant energy absorbed, which is emitted by the overall ground surface including the sea-water surface. The right-hand side of Eq. (12.2) represents two radiant energy outputs: one is downward to the near-ground air subsystem and the other is upward to the Universe.

Taking a look at Fig. 12.5, what we can also see in addition to setting up Eqs. (12.1) and (12.2) is that the rate of energy delivered by the circulation of water by changing its phase from liquid to vapour in the course of evaporation and vice versa in the course of condensation is almost four times larger than the rate of energy delivered by the circulation of air alone. This fact lets us recognize the importance of water on the Earth surface to make its temperature remain unchanged around 15°C, with which the phase change of water from liquid to vapour, or vice versa, can take place.

Let us remake these two energy balance equations with numerals into the corresponding general energy balance equations expressed by mathematical symbols representing the respective rates of energy. Referring to Fig. 12.6,

$$a_{NG} I_{ET\_sr} + \varepsilon_{NG} R_{MT\_dn} = (Q_{cv} + Q_{ev}) + R_{NG}, \tag{12.3}$$

$$a_{MT} I_{ET\_sr} + (Q_{cv} + Q_{ev}) + \varepsilon_{MT} R_{NG} = R_{MT\_dn} + R_{UT\_up}, \tag{12.4}$$

where a_{NG} is overall solar absorptance of the near-ground air subsystem; I_{ET_sr} is extra-terrestrial solar irradiance [W/m²]; ε_{NG} is overall absorptance of long-wavelength radiation at the ground surface including sea water; R_{MT_dn} is long-wavelength radiation emitted downward by the mid troposphere [W/m²]; Q_{cv} is the rate of energy transferred by convection [W/m²]; Q_{ev} is the rate of energy transferred by evaporation at the ground surface, which also represents the rate of energy transferred by condensation within the mid troposphere [W/m²]; R_{NG} is long-wavelength radiation emitted by the ground surface including the sea-water surface; a_{MT} is overall solar absorptance of the mid troposphere; ε_{MT} is mid-tropospheric absorptance against long-wavelength radiation emitted by the ground surface; R_{MT_dn} is long-wavelength radiation emitted downward by the mid troposphere; and R_{UT_up} is long-wavelength radiation emitted upward by the upper troposphere [W/m²].

In order to estimate the average near-ground air temperature, t_{NG} [°C], and upper tropospheric temperature, t_{UT} [°C], let us assume that both ground surface and upper troposphere are quasi blackbody: that is, $\varepsilon_{NG} = \varepsilon_{MT} = 1.0$, $R_{NG} = \sigma T_{NG}^4$, and $R_{UT\_up} = \sigma T_{UT}^4$, where σ is Stephan-Boltzmann constant (= 5.67×10^{-8} W/(m²K⁴)), and T_{NG} and T_{UT} are the average near-ground and upper tropospheric absolute temperatures [K], respectively, with which $t_{NG} = T_{NG} + 273.15$ and $t_{UT} = T_{UT} + 273.15$. With these assumptions in mind, Eqs. (12.3) and (12.4) can be rewritten as follows

$$a_{NG} I_{ET\_sr} + R_{MT\_dn} = (Q_{cv} + Q_{ev}) + \sigma T_{NG}^4, \tag{12.5}$$

$$a_{MT} I_{ET\_sr} + (Q_{cv} + Q_{ev}) + \sigma T_{NG}^4 = R_{MT\_dn} + \sigma T_{UT}^4. \tag{12.6}$$

According to the numerical data shown in Fig. 12.5, overall solar absorptance of near-ground air and mid-tropospheric subsystems are 0.47 and 0.235, respectively, and the sum of convective and evaporative heat transfer rates is 103 W/m². They may vary depending on how much of clouds are formed due to the extra-terrestrial conditions described above. As can also be seen in Fig. 12.5, the rate of downward long-wavelength radiation from the mid troposphere, 341 W/m², is smaller than that of long-wavelength radiation emitted by the ground surface, 398 W/m², that is, $R_{MT\_dn} < R_{NG} = \sigma T_{NG}^4$. Their intensities must be subject to the amount of clouds formed in the sky.

The long-wavelength radiation coming from all over the sky, R_{MT_dn}, consists of two portions: one is from the cloudless sky and the other from the clouds. Let us denote them as $R_{MT\_dn}^{clear}$ and $R_{MT\_dn}^{cloud}$, respectively. Their unit is of course the same as R_{MT_dn} in W/m². $R_{MT\_dn}^{cloud}$ is dependent on the position of clouds in the troposphere;

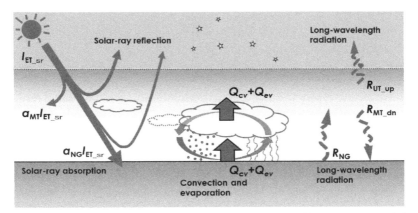

Fig. 12.6: Two-subsystem modelling of global environmental system with respect to energy balance.

it decreases as the cloud height increases as schematically shown with the spectral distribution curves in the left plate of Fig. 12.7. This characteristic can be quantified in relation to the long-wavelength radiation emitted by the clear sky, $R_{\text{MT\_dn}}^{\text{clear}}$, and the blackbody radiation at near-ground air temperature, R_{NG}, as follows (Phillips 1940)

$$\kappa = \frac{R_{\text{MT\_dn}}^{\text{cloud}} - R_{\text{MT\_dn}}^{\text{clear}}}{R_{\text{NG}} - R_{\text{MT\_dn}}^{\text{clear}}}, \tag{12.7}$$

where κ is the ratio of two values with respect to the net radiation downward: the numerator is the difference between the long-wavelength radiation emitted by the clouds and that by the clear sky; and the denominator is the difference between the long-wavelength radiation emitted by the blackbody at the near-ground air temperature and that by the clear sky. According to Phillips, as the height of clouds increases, the value of κ decreases almost linearly as shown in the right graph of Fig. 12.7.

The mass-wise major constituents of tropospheric air, nitrogen and oxygen, are almost perfectly transparent so that they themselves have nothing to do with the emission of long-wavelength radiation. But, water vapour and carbon dioxide, though they are mass-wise minor constituents, are the major players in the emission of long-wavelength radiation from the sky vault; they are partly opaque and partly translucent within the whole range of long-wavelength blackbody radiation from 3 to 60 μm at the temperature between –20 and 20°C as demonstrated in Fig. 6.10 in Chapter 6. The molecules of water vapour and carbon dioxide are not circular-shaped, while nitrogen and oxygen are circular: such difference in their molecular structure (*Katachi*) is the origin of their difference in radiative characteristics as their function (*Kata*).

The spectral emissivity of water vapour and carbon dioxide is low in the range from 8 to 13 μm as schematically shown in the left plate in Fig. 12.7. Because of such spectral characteristic, this range of wavelength is called "infra-red atmospheric window". For this atmospheric-window effect as the major reason, the value of κ decreases for the clouds at high altitude. It must be worth noting that a foggy

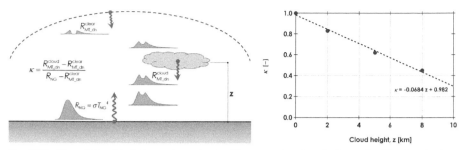

Fig. 12.7: The relationship between the net rate of long-wavelength radiation emitted by the clouds at some altitude and that at the ground level (Phillips 1940).

condition is just like the height of clouds being 0 km, which corresponds to κ being unity; in such a case, the water vapour behaves just like blackbody at near-ground air temperature.

The long-wavelength radiation available from the clear sky can be expressed as

$$R_{MT\_dn}^{clear} = \bar{\varepsilon}_{sky}^{clear}\, \sigma T_{NG}^{4} \tag{12.8}$$

where $\bar{\varepsilon}_{sky}^{clear}$ is effective clear-sky emissivity. According to the measurements of atmospheric radiation made under clear sky conditions (Brunt 1932, Berdahl and Martin 1984, Iziomon et al. 2003, Wang and Liang 2009, Li et al. 2017), $\varepsilon_{sky}^{clear}$ is correlated well with near-ground air temperature and humidity. Let us take here an empirical formula given by Berdahl and Martin (1984)

$$\varepsilon_{sky}^{clear} = 0.711 + (0.56 \times 10^{-2})t_{dp} + (0.73 \times 10^{-4})t_{dp}^{2} \tag{12.9}$$

where t_{dp} is the near-ground air dewpoint temperature [°C].

The whole of mid-tropospheric downward radiation, R_{MT_dn}, can be expressed as the weighted average of two components, $R_{MT\_dn}^{clear}$ and $R_{MT\_dn}^{cloud}$. Using the cloud-cover fraction, c_f, where $0 \le c_f \le 1.0$, as the weighting factor,

$$R_{MT\_dn} = c_f R_{MT\_dn}^{cloud} + (1 - c_f) R_{MT\_dn}^{clear}. \tag{12.10}$$

Eliminating $R_{MT\_dn}^{cloud}$ in Eq. (12.10) by substituting the relationship expressed by Eq. (12.7) and also reflecting the relationship expressed by Eq. (12.8) and $R_{NG} = \sigma T_{NG}^{4}$, R_{MT_dn} can finally be expressed as follows

$$\left. \begin{aligned} R_{MT\_dn} &= \bar{\varepsilon}_{sky}\, \sigma T_{NG}^{4} \\ \bar{\varepsilon}_{sky} &= c_f \kappa + \left(1 - c_f \kappa\right) \bar{\varepsilon}_{sky}^{clear} \end{aligned} \right\}, \tag{12.11}$$

where $\bar{\varepsilon}_{sky}$ is the overall effective sky emissivity.

Figure 12.8 demonstrates the values of $\bar{\varepsilon}_{sky}$ in the cases of clear sky ($c_f = 0$) and partly cloudy sky ($c_f = 0.5$ and the cloud height of 3 km with $\kappa = 0.76$). As the near-ground air temperature is lower, overall effective sky emissivity, $\bar{\varepsilon}_{sky}$, tends to be lower accordingly. As the relative humidity becomes higher and also if there are clouds, it tends to be higher. Comparing the two graphs in Fig. 12.8, whether there

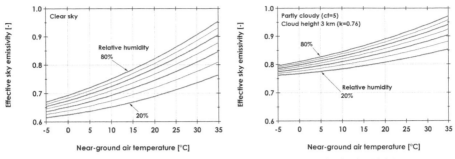

Fig. 12.8: Effective emissivity of clear sky (left) and partly cloudy sky (right).

are clouds or not affects the values of $\bar{\varepsilon}_{sky}$ quite a lot. At 15°C with 50%rh, $\bar{\varepsilon}_{sky} = 0.74$ for clear sky, while $\bar{\varepsilon}_{sky} = 0.84$ for partly cloudy sky.

The emission of long-wavelength radiation from mid-tropospheric air is, as mentioned above, owing mostly to the existence of water vapour and carbon dioxide gas within the tropospheric air. The spectra-averaged emissivities of water vapour as a function of relative humidity (Kotari 2009) and that of carbon dioxide gas as a function of its concentration (Farag and Allan 1982), both for the thickness of 3 km and its average temperature at 10°C, are as shown in Fig. 12.9. The emissivities of water vapour and carbon dioxide are in the order of 0.61 and 0.16, respectively. It is worth noting that the emissivity of water vapour is nearly four-times higher than that of carbon dioxide gas. Taking a look at the emissivity of carbon dioxide as a function of its concentration shown in the right graph in Fig. 12.9, it seems not very likely that a large change in the concentration of carbon dioxide gas alone can cause a significant change in the effective sky emissivity demonstrated in Fig. 12.8. Although the values of emissivity shown in Fig. 12.9 is not exactly the same as the effective sky emissivity, the relative magnitude of water vapour within the whole of effective sky emissivity shown in Fig. 12.8 must be similar to what can be seen in Fig. 12.9.

The value of average sky temperature, T_{sky} [K], assuming the sky vault being blackbody, can be calculated from the following formula

$$\left.\begin{array}{l} T_{sky} = \bar{\varepsilon}_{sky}^{1/4} T_{NG} \\ t_{sky} = T_{sky} - 273.15 \end{array}\right\} \tag{12.12}$$

Figure 12.10 shows the values of t_{sky} in relation to near-ground air temperature and relative humidity in the same cases as in Fig. 12.8. Average sky temperature, t_{sky}, is necessarily lower than the near-ground air temperature; this implies that the sky is the source of coolness, which is well quantified by "cool" radiant exergy as will be discussed later in the present chapter. On wintry days with low air temperature and clear sky condition, the sky is really cold; the sky temperature may decrease to as low as –25°C or so. On hot and humid summer days with partly cloudy sky conditions, the sky temperature is not so low, but may still decrease to as low as 20°C or so, which is 5 to 10°C lower than near-ground air temperature.

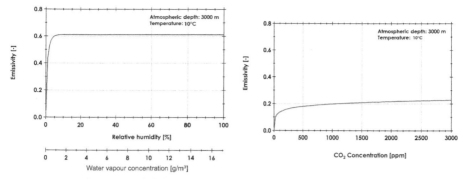

Fig. 12.9: Emissivity of water vapour (left) and carbon-dioxide (right).

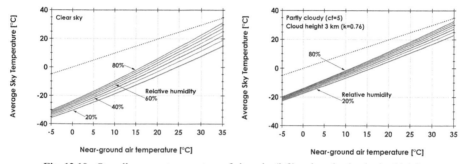

Fig. 12.10: Overall average temperature of clear sky (left) and partly cloudy sky (right).

Substituting the relationship expressed by Eq. (12.11) into Eq. (12.5) yields the following formula for the average near-ground air temperature, t_{NG} [°C],

$$t_{NG} = \left\{ \frac{a_{NG} I_{ET\_sr} - \left(Q_{cv} + Q_{ev}\right)}{\left(1 - \bar{\varepsilon}_{sky}\right)\sigma} \right\}^{1/4} - 273.15. \tag{12.13}$$

As can be seen in Eq. (12.13), the value of t_{NG} is dependent on the solar absorptance of near-ground troposphere, a_{NG}, the extraterrestrial solar irradiance, I_{ET_sr}, being almost constant at 340 W/m², effective sky emissivity, $\bar{\varepsilon}_{sky}$, and the rate of tropospheric convection, Q_{cv}, and evaporation, Q_{ev}. The variables, a_{NG}, $\bar{\varepsilon}_{sky}$, Q_{cv}, and Q_{ev} are dependent on each other; if cloud cover increases, a_{NG} must decrease in general, but in such a condition $\bar{\varepsilon}_{sky}$ may increase. Whether their effects are either likely to cancel each other and bring about little change in the value of t_{NG} or likely to magnify each other and bring about rather a large change in t_{NG} is quite complex, since the changes in a_{NG} and $\bar{\varepsilon}_{sky}$ are considered to be further associated with the changes in the rate of tropospheric convection, Q_{cv}, and evaporation, Q_{ev}, which must definitely have much influence on the values of t_{NG}. In short, the determination of t_{NG} must not be so simple, though the form of Eq. (12.13) appears rather simple. It should be worth noting that the global environmental system must have its own self-organizing mechanism as one of the typical nonlinear phenomena, by which the near-ground air temperature remains within the narrow range, from 12 to 18°C, as the state of dynamic equilibrium.

Figure 12.11 shows the near-ground air temperature calculated from Eq. (12.13) by assuming the values of solar absorptance, overall effective sky emissivity, and also the rate of convective and evaporative heat transfer. Solar absorptance was assumed in the range from 0.45 to 0.49 as shown in the left graph, while on the other hand, effective sky emissivity in the range from 0.84 to 0.88 as shown in the right graph. Three lines in the two graphs represent the cases of convective and evaporative heat transfer between the near-ground air subsystem and mid tropospheric subsystem at 99.7, 103, and 106.1 W/m², respectively. The value of 103 W/m² is the one taken from Fig. 12.5 and other two values are 3% less and 3% more, respectively. We can see that in order for the near-ground air temperature to remain within the range of 12 and 18°C, an increase or a decrease either in solar absorptance or in overall effective sky emissivity has to be accompanied with a change in the rate of energy transfer by convection and evaporation.

On the contrary to the calculation of t_{NG}, it is rather simple to calculate the average upper-tropospheric temperature, t_{UT} [°C]. Its formula can be derived by combining Eqs. (12.5) and (12.6) as follows

$$t_{UT} = \left\{ \frac{\left(1 - \rho_{glb}\right) I_{ET\_sr}}{\sigma} \right\}^{1/4} - 273.15, \qquad (12.14)$$

where ρ_{glb} is overall global reflectance, which is also called "albedo"; $\rho_{glb} = 1 - \left(a_{NG} + a_{MT}\right)$. The value of t_{UT} is dependent on ρ_{glb} and I_{ET_sr}. The value of ρ_{glb} is estimated to be 0.294 referring to the data given in Fig. 12.5; this value is the one representing the present state of the Earth. It could be higher or lower depending on how much of clouds tend to cover the ground surface due to the cosmic-ray effects in the long run.

The value of I_{ET_sr} is rather constant as described in 12.4 with Fig. 12.2. The substitution of 340 W/m² for I_{ET_sr} and 0.294 for ρ_{glb} into Eq. (12.14) results in the value of t_{UT} to be approximately –18°C. This is lower than the average sky temperature in the cases of near-ground air temperature, t_{NG}, from 12 to 18°C, as can be seen in Fig. 12.10. According to a balloon measurement, the air temperature near the upper

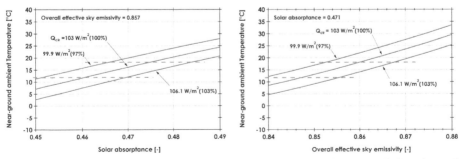

Fig. 12.11: Estimated near-ground air temperature as a function of solar absorptance (left) and overall effective sky emissivity (right).

edge of troposphere at the height of 10 km is about $-45°C$ as was shown in Fig. 9.10 in Chapter 9. Therefore, t_{UT} being at $-18°C$ is just in the middle of two temperatures: one at the edge of troposphere, $-45°C$, and the nominal value of near-ground air temperature, $15°C$.

12.6 Exergy balance of global environmental system

As we have discussed through some of the previous chapters, the thermodynamic behaviours of any systems, whether they are natural or artificial, are well described with the concept of exergy. Such description necessarily starts with setting up entropy balance equation together with energy balance equation and ends up with yielding the exergy balance equation with the chosen appropriate environmental temperature for a system on focus.

What was discussed in the previous section is based on energy, but not on exergy. Therefore, let us set up here the entropy balance equations that should stand in parallel to Eqs. (12.5) and (12.6) and then come up with the set of exergy balance equations for the two subsystems. Referring to Fig. 12.12, which is the entropy version in parallel to Fig. 12.6 for energy, we may express the entropy balance equations of two subsystems: the near-ground subsystem and the mid-tropospheric subsystem as follows

$$a_{NG}s_{ET\_sr} + s_{MT\_dn} + s_{g\_NG} = \left(\frac{1}{T_{NG}}\right)(Q_{cv} + Q_{ev}) + \frac{4}{3}\sigma T_{NG}^{3}, \qquad (12.15)$$

$$a_{MT}s_{ET\_sr} + \left(\frac{1}{T_{MT}}\right)(Q_{cv} + Q_{ev}) + \frac{4}{3}\sigma T_{NG}^{3} + s_{g\_MT} = s_{MT\_dn} + \frac{4}{3}\sigma T_{UT}^{3}. \qquad (12.16)$$

where s_{ET_sr} is the rate of extra-terrestrial solar entropy; s_{MT_dn} is long-wavelength radiant entropy emitted downward by the mid troposphere; s_{g_NG} is the rate of entropy generation within the near-ground air subsystem; and s_{g_MT} is the rate of entropy generation within the mid tropospheric subsystem. All of these variables are in the unit of $W/(m^2K)$.

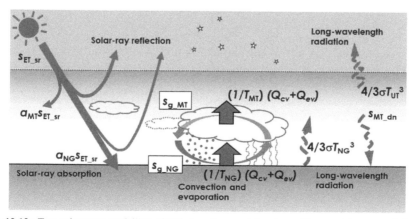

Fig. 12.12: Two-subsystem modelling of global environmental system with respect to entropy balance.

The rate of extra-terrestrial solar energy, I_{ET_sr}, and entropy, s_{ET_sr}, may be given as follows based on the fact that the whole of solar radiation is very well approximated by the blackbody radiation with the surface temperature, T_{sun}, at 5773 K (= 5500°C) or 5973 K (= 5700°C) as was shown in Fig. 6.11 in Chapter 6

$$I_{ET\_sr} = \frac{1}{4} f_{se}\left(\sigma T_{sun}^{4}\right), \qquad s_{ET\_sr} = \frac{1}{4} f_{se}\left(\frac{4}{3}\sigma T_{sun}^{3}\right), \qquad (12.17)$$

where f_{se} is the factor for converting the values of emission expressed by the respective formulae of energy and entropy in the brackets to those of incidence. $f_{se} = d_{sun}^{2} \big/ \left(4D_{se}^{2}\right)$, where d_{sun} is the diameter of the Sun, 1.39×10^{9} m, and D_{se} is the distance between the Sun and the Earth, 1.5×10^{11} m. The factor of 1/4 is to convert the values of extra-terrestrial solar energy and entropy into the overall average for 1 m^2 of ground surface. Assuming T_{sun} to be 5788.15 K (= 5515°C), I_{ET_sr} and s_{ET_sr} become 341.6 W/m^2 and 0.079 W/(m^{2}K), respectively. With these values, the extra-terrestrial solar radiant temperature, T_{ET_sr} [K], may be estimated as follows

$$T_{ET\_sr} = \frac{I_{ET\_sr}}{s_{ET\_sr}} = \frac{3}{4}T_{sun} \approx 4341. \qquad (12.18)$$

Equation (12.18) indicates that solar radiation emitted by the Sun approximately at 5790 K scatters in the course of its travel through the interplanetary space and its temperature decreases by 1449 K (= 5790–4341) until it reaches the extra-terrestrial surface right above the Earth's atmosphere.

The rate of radiant entropy emitted downward by the sky vault, s_{MT_dn}, can be expressed as follows using the sky temperature, T_{sky}, assuming the sky vault as blackbody

$$s_{MT\_dn} = \frac{4}{3}\sigma T_{sky}^{3}. \qquad (12.19)$$

Substitution of the relationship expressed by Eq. (12.12) into Eq. (12.19) yields the following equation

$$s_{MT\_dn} = \frac{4}{3}\sigma \bar{\varepsilon}_{sky}^{3/4} T_{NG}^{3}. \qquad (12.20)$$

Using this relationship, Eq. (12.15) can be rewritten as Eq. (12.22), whose corresponding energy balance equation that should stand in parallel is expressed by Eq. (12.21). Note that for the derivation of Eq. (12.21) the relationship expressed by Eq. (12.11) is used.

$$a_{NG}I_{ET\_sr} + \bar{\varepsilon}_{sky}\,\sigma T_{NG}^{4} = (Q_{cv} + Q_{ev}) + \sigma T_{NG}^{4}, \qquad (12.21)$$

$$a_{NG}s_{ET\_sr} + \frac{4}{3}\sigma \bar{\varepsilon}_{sky}^{3/4} T_{NG}^{3} + s_{g\_NG} = \left(\frac{1}{T_{NG}}\right)(Q_{cv} + Q_{ev}) + \frac{4}{3}\sigma T_{NG}^{3}. \qquad (12.22)$$

Combining Eqs. (12.21) and (12.22) together with the near-ground air temperature, T_{NG}, as the thermodynamic environmental temperature yields the following exergy balance equation for the near-ground air subsystem

$$a_{NG}x_{ET\_sr} + x^{crad}_{MT\_dn} - x_{C\_NG} = 0 \qquad (12.23)$$

where
$$\begin{cases} x_{ET\_sr} = I_{ET\_sr} - s_{ET\_sr}T_{NG} \\ x^{crad}_{MT\_dn} = \left(\frac{1}{3} + \bar{\varepsilon}_{sky} - \frac{4}{3}\bar{\varepsilon}_{sky}^{3/4} \right)\sigma T_{NG}^{4} \end{cases} \qquad (12.24)$$

x_{ET_sr} is the rate of extra-terrestrial solar exergy [W/m²], which is the primary source of light and heat at the near-ground air subsystem. $x^{crad}_{MT\_dn}$ is the rate of "cool" radiant exergy [W/m²], which is the cold source for a variety of cooling purposes necessary within the near-ground air subsystem. x_{C_NG} is the exergy consumption rate within the near-ground air subsystem, which is equal to $s_{g\_NG}T_{NG}$. Equation (12.23) indicates that solar exergy and "cool" radiant exergy absorbed by the near-ground air subsystem are totally consumed and in this course of exergy consumption, the average near-ground air temperature is kept within the range from 12 to 18°C.

Assuming T_{NG} to be 288.15 K (= 15°C), x_{ET_sr} defined by the first formula in Eq. (12.24) turns out to be 318.8 W/m² using the aforementioned values of $I_{ET\_sr} = 341.6$ W/m² and $s_{ET\_sr} = 0.079$ W/(m²K). The exergy to energy ratio is 0.93, which is higher than the ratio available at the ground surface, 0.6 to 0.92 (Shukuya 2013). This is considered reasonable since 0.93 is the extra-terrestrial value.

Figure 12.13 demonstrates the rate of "cool" radiant exergy, $x^{crad}_{MT\_dn}$, is expressed by the second formula in Eq. (12.24) as the function of near-ground air temperature and relative humidity. The cases assumed for its calculation are the same as in Figs. 12.8 and 12.10. $x^{crad}_{MT\_dn}$ is large when the sky is clear and also near-ground air relative humidity is low. The existence of clouds and higher relative humidity let $x^{crad}_{MT\_dn}$ be smaller, but $x^{crad}_{MT\_dn}$ still remains in the order from 0.5 to 1 W/m², which is much larger than the rate of "cool" radiant exergy to be available for providing the coolness within the built environment under hot and humid summer conditions, that is, from 20 to 100 mW/m² as discussed in Chapter 10.

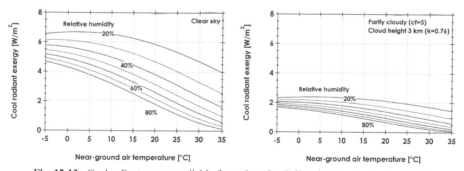

Fig. 12.13: Cool radiant exergy available from clear sky (left) and partly cloudy sky (right).

Let us next set up the exergy balance equation for the mid-tropospheric subsystem. In the same manner as done for the near-ground air subsystem, energy balance equation (12.6) and entropy balance equation (12.16) are rewritten as follows taking the relationships expressed by Eqs. (12.11) and (12.20) into consideration

$$a_{MT}I_{ET\_sr} + (Q_{cv} + Q_{ev}) + \sigma T_{NG}{}^4 = \bar{\varepsilon}_{sky}\sigma T_{NG}{}^4 + \sigma T_{UT}{}^4, \qquad (12.25)$$

$$a_{MT}S_{ET-sr} + \left(\frac{1}{T_{MT}}\right)(Q_{cv} + Q_{ev}) + \frac{4}{3}\sigma T_{NG}{}^3 + s_{g\_MT} = \frac{4}{3}\sigma\bar{\varepsilon}_{sky}{}^{3/4}T_{NG}{}^3 + \frac{4}{3}\sigma T_{UT}{}^3, \quad (12.26)$$

where T_{MT} is the mid-tropospheric temperature [K], which may be assumed to be equal to T_{sky}, but different from the upper-tropospheric temperature, T_{UT}, defined by Eq. (12.14). Combining these two equations together with the near-ground air temperature, T_{NG}, the final form of the exergy balance equation for the mid-tropospheric subsystem is expressed as follows

$$a_{MT}x_{ET\_sr} + x_{ET}^{crad} - x_{C\_MT} = x_{MT\_dn}^{crad} + x_{cv+ev} \qquad (12.27)$$

where

$$\begin{cases} x_{ET}^{crad} = -\left(1 - \frac{4}{3}\frac{T_{NG}}{T_{UT}}\right)\sigma T_{UT}{}^4 \\ \\ x_{cv+ev} = -\left(1 - \frac{T_{NG}}{T_{MT}}\right)(Q_{cv} + Q_{ev}) \end{cases} \qquad (12.28)$$

x_{ET_sr} is the rate of extra-terrestrial solar exergy as explained above. x_{ET}^{crad} is the rate of "cool" radiant exergy available from extra-terrestrial space, the Universe, as an input to the mid-tropospheric subsystem. The reason that x_{ET}^{crad} is "cool" exergy is that $0 < T_{UT} < T_{NG}$ and $0 < x_{ET}^{crad}$. x_{C_MT} is the exergy consumption rate within the mid-tropospheric subsystem and equal to $s_{g\_MT}T_{NG}$. $x_{MT\_dn}^{crad}$ is the outgoing rate of "cool" radiant exergy toward the near-ground air subsystem, which is defined by Eq. (12.24). x_{cv+ev} is the rate of "cool" exergy to be transferred downward. The reason that x_{cv+ev} is "cool" exergy is that $0 < T_{MT} < T_{NG}$ and $0 < x_{cv+ev}$. It is transferred by convection and precipitation to the near-ground subsystem together with $x_{MT\_dn}^{crad}$.

Assuming T_{NG} and T_{UT} to be 288.15 K (= 15°C) and 255.15 K (= –18°C), respectively, x_{ET}^{crad} becomes 120.8 W/m². Referring to Fig. 12.5, a_{MT} is found to be 0.235. With this value of a_{MT} and x_{ET_sr} = 318.8 W/m² mentioned above, the rate of solar exergy absorbed by the mid-tropospheric subsystem becomes 74.9 W/m². For the mid-tropospheric subsystem, the "cool" radiant exergy input is more than one-and-a-half times larger than the solar exergy absorbed. Equation (12.27) indicates that the inputs of solar exergy and "cool" exergy from the Universe are partly consumed and thereby the "cool" radiant exergy to be available at the ground, $x_{MT\_dn}^{crad}$, and also "cool" exergy to be transferred downward by convection and condensation, x_{cv+ev}, are generated.

As already described in association with Fig. 12.13, $x_{MT\_dn}^{crad}$ is the source of radiant coolness to be available within the near-ground air subsystem. On the other hand, the role of x_{cv+ev} is the source of the following three types of exergy other than "cool" radiant exergy: (1) "cool" exergy delivered downward by tropospheric air circulation, x_{air}; (2) "wet" exergy carried by liquid water after condensation of moisture in the clouds, x_{wet}; (3) "potential" exergy held by this liquid water, x_{pt}. The production of these three exergies is expressed by the following exergy balance equation

$$x_{cv + ev} - x_{C\_cirl} = x_{air} + x_{wet} + x_{pt},$$ (12.29)

where x_{C_cirl} is exergy consumption rate for the production of three types of exergy. Referring to what was described in Chapters 7 and 8, x_{air}, x_{wet}, and x_{pt} can be expressed as follows

$$\left.\begin{array}{l} x_{air} = c_a \rho_a V_a \left\{ (T_{MT} - T_{NG}) - T_{NG} \ln \dfrac{T_{MT}}{T_{NG}} \right\} \\[2mm] x_{wet} = \dfrac{R}{\mathfrak{M}_w} m_{lw} T_{NG} \ln \dfrac{100}{\varphi_{NG}} \\[2mm] x_{pt} = m_{lw} gz \end{array}\right\},$$ (12.30)

where c_a is specific heat capacity of air (= 1005) [J/(kg·K)]; ρ_a is the density of air (= 1.2) [kg/m³]; V_a is the average volumetric rate of circulated air [m³/(m²s)]; R is gas constant (= 8.314) [J/(mol·K); $\mathfrak{M}_w$ is molar mass of water (= 18.015 × 10⁻³) [kg/mol]; m_{lw} is the rate of precipitation [kg/(m²s)]; g is the gravitational acceleration rate (= 9.8) [m/s²], and z is the height of liquid water at the upstream of rivers or at reservoirs [m].

Let us estimate the order of x_{air}, x_{wet}, and x_{pt}. Referring to the numbers given in Fig. 12.5, the average rate of heat transferred by convection is 21 W/m², which has to be equal to $c_a \rho_a V_a (T_{NG} - T_{MT})$. Assuming the near-ground air temperature to be 14 or 15°C, the sky temperature is approximately 1.2 or 2.5°C under the partly cloudy condition as can be seen in Fig. 12.10. Then, the average volumetric rate of air circulated, V_a, is estimated to be approximately 0.0014 m³/(m²s) (= 5 m³/(m²h)) for the rate of thermal energy transfer at 21 W/m². With this volumetric rate of air circulation, x_{air} becomes approximately 0.4 W/m².

Since the water circulates within the troposphere by changing its state from liquid to vapour and again to liquid in the course of evaporating, ascending, condensing, and falling down as rain and no portion of water escapes away from the atmosphere, the rate of liquid water falling annually, that is, the annual precipitation rate, can be estimated from the rate of thermal energy transferred by evaporation at 82 W/m² quoted from Fig. 12.5. Assuming the specific latent heat of evaporation to be 2465 kJ/kg for 15°C, the rate of evaporated liquid water turns out to be 33.3 mg/(m²s), which is 1041 kg/(m²year); this fits very well with the nominal value of annual precipitation, 1000 mm/(m²year). With this value, m_{lw} = 0.033 × 10⁻³ kg/(m²s), we can estimate the values both of x_{wet}, and x_{pt}. x_{wet} becomes 3.0 W/m² assuming the near-ground air relative humidity to be 50%, and x_{pt} at 0.5 W/m² assuming the average height of liquid water to be 1500 m.

With the assumption of T_{MT} (= T_{sky}) and T_{NG} being equal to 274.35 K (= 1.2°C) and 288.15 K (= 15°C), respectively, $x_{cv + ev}$ becomes 4.2 W/m² using Eq. (12.28). Substituting all values of x_{air}, x_{wet}, x_{pt} and $x_{cv + ev}$ so far obtained into Eq. (12.29), we come to know that the rate of exergy consumption, x_{C_cirl}, is 0.3 W/m².

Although $x_{cv + ev}$ certainly plays a key role in driving the circulation of air and water, the ascending air flow from the near-ground air subsystem to the mid tropospheric subsystem can hardly occur with x_{cv+ev} alone. One other driving agent for the circulation of air and water is "warm" exergy generated as the result of solar exergy absorption within the near-ground air subsystem. The exergy balance equation for this "warm" exergy production is

$$a_{NG} x_{ET\_sr} - x_{C\_ab} = x_{warm},\tag{12.31}$$

$$\left.\begin{array}{l} x_{C\_ab} = s_{g\_ab} T_{NG} \\[2ex] x_{warm} = \left(1 - \dfrac{T_{NG}}{T_{sur}}\right) a_{NG} I_{ET\_sr} \\[2ex] T_{sur} = T_{NG} + \dfrac{a_{NG} I_{ET\_sr}}{h} \end{array}\right\},\tag{12.32}$$

where x_{C_ab} is the exergy consumption rate due to solar-exergy absorption [W/m²], x_{warm} is the rate of "warm" exergy produced [W/m²], s_{g_ab} is the rate of entropy generation due to solar-exergy absorption [W/(m²K)], T_{sur} is the ground-surface temperature which is higher than the near-ground air temperature because of solar absorption [K]; and h is overall heat-transfer coefficient along the ground surface [W/(m²K)].

Provided that $a_{NG} I_{ET\_sr} = 0.47 \times 340$ W/m² and $\bar{h} = 10$ W/(m²K), then T_{sur} turns out to be 304.15 K (= 31°C) and thereby "warm" exergy, x_{warm}, becomes 8.4 W/m². At this rate of "warm" exergy and at that of "cool" exergy at 4.2 W/m² described above, the circulation of air and water is driven. Using $a_{NG} = 0.47$ and $x_{ET\_sr} = 318.8$ W/m² given above, x_{C_ab} becomes 141 W/m². This rate of exergy consumption is about 94% of solar exergy absorbed. Solar water heating, space heating and others are to make a portion of this exergy consumption for such specific purposes. Note that "warm" exergy produced at the ground surface, 8.4 W/m², is of the same order as "cool" exergy produced within the mid troposphere, 4.2 W/m².

Figure 12.14 summarises the whole exergy balance of global environmental system. In the mid-tropospheric subsystem, the inputs of exergy are 74.6 W/m² of solar exergy and 120.8 W/m² of "cool" radiant exergy available from the Universe. The portion of their sum, 190 W/m², is consumed and thereby 1.2 W/m² of "cool" radiant exergy and 4.2 W/m² of "cool" exergy to be transferred by air and water circulation are generated. The former output, 1.2 W/m², becomes one of the two inputs to the near-ground air subsystem and the latter becomes the input for the production of "cool" exergy delivered downward by air circulation at 0.4 W/m², "wet" exergy contained by liquid water as rain at 3.0 W/m², and "potential" exergy held by liquid water at 0.5 W/m², a portion of which is being utilized for hydro-electric power plants. Similarly, a portion of tropospheric air in circulation is also being utilized for wind-power generation.

All of the three types of exergy are sooner or later consumed altogether at the ground surface. This is expressed as follows

$$x_{air} + x_{wet} + x_{pt} - x_{C\_cir2} = 0,\tag{12.33}$$

where x_{C_cir2} is the exergy consumption rate due to "cool" exergy transfer by air circulation, the evaporation of liquid water, and the fall of liquid water down to the sea. It amounts to 3.9 W/m² as can be seen in Fig. 12.14. Among the exergies available within the near-ground air subsystem, it is worth noting that "cool" radiant exergy, 1.2 W/m², and "wet" exergy, 3.0 W/m², are not small in comparison to "cool" exergy transferred by air circulation, 0.4 W/m², and "potential" exergy, 0.5 W/m².

In the near-ground air subsystem, the inputs are 149.5 W/m² of solar exergy and 1.2 W/m² of "cool" radiant exergy. Their sum, 150.7 W/m², is totally consumed, but beforehand, a portion of the former is consumed for the photovoltaic power generation and for a variety of heating and a portion of the latter for radiative cooling, outdoors and indoors. Aforementioned "warm" exergy of 8.4 W/m², which is produced by solar exergy absorption, is a portion of total exergy consumption rate at 150.7 W/m².

Combining Eqs. (12.23), (12.27), (12.29) and (12.33) altogether yields the exergy balance equation for the whole of two subsystems as follows

$$\left(a_{NG}+a_{MT}\right)x_{ET\_sr}+x_{ET}^{crad}-\left(x_{C\_NG}+x_{C\_MT}+x_{C\_cir1}+x_{C\_cir2}\right)=0. \quad (12.34)$$

Similarly, combining the two entropy balance equations (12.22) and (12.26) yields

$$\left(a_{NG}+a_{MT}\right)s_{ET\_sr}+\frac{T_{NG}-T_{MT}}{T_{NG}T_{MT}}\left(Q_{cv}+Q_{ev}\right)+\left(s_{g\_NG}+s_{g\_MT}\right)=\frac{4}{3}\sigma T_{UT}^{3}. \quad (12.35)$$

Equation (12.34) together with Eq. (12.35) indicates that the global environmental system sustains its thermodynamic state in the range of global average temperature from 12 to 18°C by consuming both solar exergy and "cool" radiant exergy available from the Universe, while at the same time by discarding the generated entropies as the result of exergy consumption altogether accompanied by the outgoing long-wavelength radiant entropy expressed as $\frac{4}{3}\sigma T_{UT}^{3}$ on the right-hand side of Eq. (12.35). Namely, the global environmental system functions as an "exergy-entropy process", which is a series of process from exergy input, its consumption,

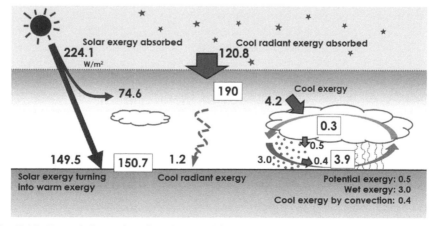

Fig. 12.14: Exergy balance of quasi steady-state global environmental system. The numbers given in the rectangles denote the exergy consumption rate.

the resulting entropy generation, and finally the disposal of the generated entropy as one cycle (Shukuya 2013).

Note that what we have confirmed so far with two subsystem modelling of the global environmental system here is equivalent to what we discussed with a small actual model experiment using a drinking bird described in 2.5.3, Chapter 2.

12.7 The nature to be emulated by built-environmental systems within

The Earth is heavy enough to keep all molecules of air and water within the height of atmospheric layer over the Earth surface. If the Earth were much lighter, it must have given up the molecules of air and water long before as the Moon did before. But, it does not imply that heavier is better. More than enough is too much. The Earth is in fact sufficiently light so that the water on the Earth can exist either in the state of solid (ice), liquid, or gas (vapour). The density of liquid water being approximately 1000 kg/m^3 is much larger than that of tropospheric air being 1.2 kg/m^3 so that liquid water as rain can fall down to the bottom of troposphere. While on the other hand, the tropospheric moist air is dry enough so that liquid water can rather easily diffuse mutually with the dry air and be delivered up to the mid troposphere until it condenses into liquid water to form clouds and thereafter falls as rain. The concentration (density) of water vapour ranging from 5 to 20 g/m^3 as demonstrated with Fig. 8.14 in Chapter 8 indicates that the tropospheric air is capable of letting liquid water evaporate.

The melting point at 0°C and boiling point at 100°C of water are in fact peculiarly higher than the substances having similar molecular structure such as hydrogen sulfide (H$_2$S), hydrogen selenide (H$_2$Se), or hydrogen telluride (H$_2$Te); their melting and boiling points are lower than 0°C (Pauling 1970, Katsuki 1999, Shukuya 1999). If the difference in the melting and boiling points of those molecules had to be mostly mass dependent, then the water, hydrogen oxide (H$_2$O), should have had the melting point at about –85°C and the boiling point at about –65°C, respectively, because of the molar mass of H$_2$O being smaller than those of H$_2$S, H$_2$Se and H$_2$Te. But the actual values of melting and boiling points of water (H$_2$O) are, as we all know, 0°C and 100°C, that is, much higher than those values of H$_2$S, H$_2$Se and H$_2$Te.

This implies that each of water molecules is sticky due to the characteristic called "hydrogen bonding". Water molecules have to be made more vigorous and become freer to be liquidized from solid state or to be vapourized from liquid state. Sulfur (S), selenium (Se), and tellurium (Te) are too large for them to have hydrogen bonding as they form a molecule with two hydrogen atoms, but oxygen (O) is small enough to have the hydrogen bonding. We all know that ice floats in liquid water. The reason for this phenomenon is also due to the effect of hydrogen-bonding. The molecules of water form hollow spaces in the state of solid at the temperature being equal to 0°C or below, but no such space in the state of liquid; this makes the density of solid water be smaller than that of liquid water.

Thanks to the hydrogen bonding, the water molecules have to carry collectively a lot of latent heat as they turn themselves from the state of liquid to that of vapour. This is the reason that Q_{ev} is almost four times larger than Q_{cv} as shown in

Fig. 12.5. The fact that liquid water is approximately 10^5 times denser than water vapour makes "wet" exergy very large and its consumption makes it possible to discard very effectively the generated entropy emerged all over the near-ground air subsystem to the mid-tropospheric subsystem. Such effectiveness also originates from the characteristic of large latent heat for vapourization due to hydrogen bonding.

The molecular nature of water that makes the effective disposal of entropy generated in the course of solar-exergy absorption at the rate of 224 W/m² and "cool" radiant exergy available from the Universe at the rate of 121 W/m² realizes the average global environmental temperature to be within the narrow range from 12 to 18°C. Within this whole process, water plays one other key role, that is of the cloud formation in relation to the solar-wind and the galactic-cosmic-ray effects as described in 12.4.

The ceaseless repetition of evaporation and condensation of water as the working fluid within the troposphere is basically the same as what is happening inside the artificial heat pumps in operation. Therefore, the global environmental system may also be regarded as a huge heat pump realized by the Nature. Heat pumps, whether they are refrigerators or air-conditioning units, function by feeding on exergy supplied as electric power, consuming its portion to circulate the working fluid inside the closed space formed by a compressor, a heat exchanger for condensation, a throttling valve, and the other heat exchanger for evaporation. "Warm" exergy and "cool" exergy are produced in the circulation of working fluid being contracted at the compressor and being expanded at the throttling valve as described in Chapter 11.

The expansion of the working fluid and its resulting absorption of thermal energy, which necessarily accompanies entropy, in artificial heat pumps, corresponds to the evaporation of liquid water and its resulting absorption of thermal energy, which also necessarily accompanies entropy at the near-ground air subsystem. Similarly, the compression of the working fluid and its resulting disposal of thermal energy together with entropy in the heat pumps correspond to the condensation of water vapour and its resulting disposal of thermal energy together with entropy into the mid-tropospheric subsystem.

The troposphere, in which the water and air as the working fluid circulates, corresponds to the whole of internal space within artificial heat pumps comprising the condenser, evaporator, compressor, and throttling valve as a closed space with high pressure in one part and with low pressure in the other. Equivalent to the firmness required for the closed pipe space within the artificial heat pumps is the gravitational field on the Earth, which keeps air and water within.

Having in mind the existence of such a huge heat pump realised by the global environmental system, one other system to be recalled is, in a similar implication, the existence of mitochondria within the biological cells including human-body cells. Mitochondria are nothing other than the tiny micro-cogeneration systems equipped in the living cells as described in 10.2.1 of Chapter 10. We thus come to recognize that we humans are living in between two active systems made by the Nature: a huge heat pump and micro-cogeneration systems.

In parallel to these active systems made by the Nature, the global atmosphere, which is a manifold layer of selective transparency and insulation as described in 12.4, can be regarded as the passive system made by the Nature. This is owing to the following features: firstly, the global atmosphere lets the galactic cosmic rays transmit selectively down to the height of several kilometres above the ground surface so that the ionization of tropospheric air emerges now and then and thereby the cloud formation takes place; secondly, the global atmosphere lets the short-wavelength radiation transmit down to the ground surface for light and heat to be available for all of the living creatures from micro organisms to plants and animals including human beings; thirdly, the global atmosphere lets the upward emission of long-wavelength radiation from the ground surface emerge at an appropriate rate so that the global environmental temperature becomes neither too high nor too low. It is worth noting that the requirement for better building envelope systems to form the built-environmental space for human well-being is to follow such selective features embedded within the global atmosphere.

The key to better solutions for built-environmental systems to be equipped with passive and active subsystem components in rational and hence harmonious manner is, therefore, to emulate how the global environmental system has been developed on its own for sustaining its dynamic state.

Acknowledgment

In the production of Figs. 12.2 and 12.3, the data available from the following three open sources are referred and used: World data centre for sunspot index and long-term solar observation (WDC-SILSO), Royal Observatory of Belgium, Brussels; Cosmic Ray Station of the University of Oulu/Sodankyla Geophysical Observatory (CRSUO/SGO); Physikalisch-Meteorologisches Observatorium Davos (PMOD)/ World Radiation Center (WRC).

References

Berdahl P. and Martin M. 1984. Emissivity of clear skies. Solar Energy 32: 663–664.
Brunt D. 1932. Notes on radiation in the atmosphere. Quarterly Journal of the Royal Meteorological Society 58: 389–418.
Calder N. 1997. The Manic Sun—Weather Theories Confounded. Pilkington Press.
Cosmic Ray Station of the University of Oulu/Sodankyla Geophysical Observatory (CRSUO/SGO). http://cosmicrays.oulu.fi/ accessed on 4th/December/2014.
Eddy J. 2009. The Sun, the Earth, and Near-earth Space—A Guide to the Sun-Earth System. NASA NP-2009-1-066 GSFC. ISBN: 978-0-16-08308-8.
Farag I. H. and Allam T. A. 1982. Carbon dioxide standard emissivity by mixed gray-gases model. Chemical Engineering Communications 14(3-6): 123–131.
Frölich C. 2006. Solar irradiance variability since 1978—Revision of the PMOD Composite during Solar Cycle 21. Space Science Review 125: 53–65.
Hoffman V. L. and Simmons A. 2008. The resilient earth—Science, global warming and the future of humanity. 1st Ed. Booksurge Publishing (http://www.theresilientearth.com).
Iziomon M. G., Mayer H. and Matzarakis A. 2003. Downward atmospheric longwave iiradiance under clear and cloudy skies: Measurement and parameterization. Journal of Atmospheric and Solar-terrestrial Physics 65: 1107–1116.
Jones P. D. and Harpham C. 2013. Estimation of the absolute surface air temperature of the Earth. J. Geophys. Res. Atmos. 118: 3213–3217. Doi: 10.1002/jgrd.50359.
Katsuki A. 1999. Fundamental theory on environment based upon theoretical physics. Kaimei-sha Publishers (in Japanese).

Kotari F. 2009. What engineering and scientific calculation to be made possible with excel spread sheet. Maruzen Publishers Ltd. (in Japanese).

Li M., Jiang Y. and Coimbra C. F. M. 2017. On the determination of atmospheric longwave irradiance under all-sky conditions. Solar Energy 144: 40–48.

Longair M. 2014. C.T.R. Wilson and the cloud chamber. Astroparticle Physics 53: 55–60.

Maruyama S. and Isozaki Y. 1998. The history of biological systems and the Earth. Iwanami Publisher (in Japanese).

Pauling L. 1970. General Chemistry. 3rd Ed. W. H. Freeman and Company.

Phillips H. 1940. Die Theorie der Wärmestrahlung in Bodennähe. Gerlands Beiträge zur Geophysik. Bd.56 H.3.S.229–319.

Physikalisch-Meteorologisches Observatorium Davos (PMOD)/World Radiation Center (WRC). The composite dataset: version d25_07_0310a. https://www.pmodwrc.ch/ accessed on 10th/May/2018.

Shukuya M. 1995. Exergy-entropy process of the global environmental system. Proc. of the Annual Meet. of Architectural Inst. of Jpn. 545–546 (in Japanese).

Shukuya M. and Komuro D. 1996. Exergy-entropy process of passive solar heating and global environmental systems. Solar Energy 58(1): 25–32.

Shukuya M. 1999. Seeking what the environment-conscious architecture is. Kajima Publishers (in Japanese).

Shukuya M. 2013. Exergy—Theory and its application in the built environment. Springer-Verlag. London.

Space Weather Prediction Center/National Oceanic and Atmospheric Administration (SWPC/NOAA). https://www.swpc.noaa.gov/products/ace-real-time-solar-wind accessed on 30th/May/2018.

Svensmark H. and Friis-Christensen E. 1997. Variation of cosmic ray flux and global cloud coverage-a missing link in solar-climate relationships. Journal of Atmosphere and Solar-Terrestrial Physics 59(11): 1225–1232.

Svensmark H. and Calder N. 2007. The chilling stars—a cosmic view of climate change. Icon Books Ltd. UK.

Svensmark H. 2012. Evidence of nearby supernovae affecting life on Earth. Monthly Notices of the Royal Astronomical Society 423: 1234–1253.

Svensmark H., Enghoff M. B. and Pedersen J. O. P. 2013. Response of cloud condensation nuclei (> 50 nm) to changes in ion-nucleation. Physics Letters A 377: 2343–2347.

Svensmark H. 2015. Cosmic rays, clouds and climate. Europhysics News 46(2): 26–29.

Toda M (1995) Thirty lectures on thermal physics. Asakura Publisher (in Japanese).

Wang K. and Liang S. 2009. Global atmospheric downward longwave radiation over land surface under all-sky conditions from 1973 to 2008. Journal of Geophysical Research 114: D19101. DOI: 10.1029/2009JD011800.

Wild M. 2017. The Global Energy Balance Archive (GEBA) version 2017: a database for worldwide measured surface energy fluxes. Earth Syst. Sci. Data 9: 601–613. https://doi.org/10.5194/essd-9-601-2017.

World Data Center—Sunspot Index and Long-term Solar Observations (WDC-SILSO): Sunspot data. the Royal Observatory of Belgium. Brussels. http://www.sidc.be/silso/ accessed on 5th/May/2018.

Index